Geranium and Pelargonium

Medicinal and Aromatic Plants – Industrial Profiles

Individual volumes in this series provide both industry and academia with in-depth coverage of one major medicinal or aromatic plant of industrial importance.

Edited by Dr Roland Hardman

Volume 1
Valerian, edited by Peter J. Houghton

Volume 2
Perilla, edited by He-ci Yu, Kenichi Kosuna and Megumi Haga

Volume 3
Poppy, edited by Jenö Bernáth

Volume 4
Cannabis, edited by David T. Brown

Volume 5
Neem, edited by H.S. Puri

Volume 6
Ergot, edited by Vladimír Křen and Ladislav Cvak

Volume 7
Caraway, edited by Éva Németh

Volume 8
Saffron, edited by Moshe Negbi

Volume 9
Tea Tree, edited by Ian Southwell and Robert Lowe

Volume 10
Basil, edited by Raimo Hiltunen and Yvonne Holm

Volume 11
Fenugreek, edited by Georgios Petropoulos

Volume 12
Gingko biloba, edited by Teris A. Van Beek

Volume 13
Black Pepper, edited by P.N. Ravindran

Volume 14
Sage, edited by Spiridon E. Kintzios

Volume 15
Ginseng, edited by W.E. Court

Volume 16
Mistletoe, edited by Arndt Büssing

Volume 17
Tea, edited by Yong-su Zhen

Volume 18
Artemisia, edited by Colin W. Wright

Volume 19
Stevia, edited by A. Douglas Kinghorn

Volume 20
Vetiveria, edited by Massimo Maffei

Volume 21
Narcissus and Daffodil, edited by Gordon R. Hanks

Volume 22
Eucalyptus, edited by John J.W. Coppen

Volume 23
Pueraria, edited by Wing Ming Keung

Volume 24
Thyme, edited by E. Stahl-Biskup and F. Sáez

Volume 25
Oregano, edited by Spiridon E. Kintzios

Volume 26
Citrus, edited by Giovanni Dugo and Angelo Di Giacomo

Volume 27
Geranium and Pelargonium, edited by Maria Lis-Balchin

Volume 28
Magnolia, edited by Satyajit D. Sarker and Yuji Maruyama

Geranium and Pelargonium

The genera *Geranium* and *Pelargonium*

Edited by

Maria Lis-Balchin

South Bank University, London, UK

Routledge
Taylor & Francis Group
LONDON AND NEW YORK

First published 2002 by Taylor & Francis

2 Park Square, Milton Park, Abingdon, Oxfordshire OX14 4RN
52 Vanderbilt Avenue, New York, NY 10017

Routledge is an imprint of the Taylor & Francis Group, an informa business

First issued in paperback 2019

Typeset in Garamond by
Integra Software Services Pvt. Ltd, Pondicherry, India

British Library Cataloguing in Publication Data
A catalogue record for this book is available from the British Library

Library of Congress Cataloging in Publication Data
A catlog record for this book has been requested

ISBN: 978-0-415-28487-5 (hbk)
ISBN: 978-0-367-39573-5 (pbk)

Contents

Contributors

Maria Lis-Balchin
School of Applied Science
South Bank University
Borough Road
London
SE1 OAA
UK

Adam Bogdan
Department of Biology and
Pharmaceutical Botany
Medical University of Gdansk
ul. Hallera 107
60-416 Gdansk
Poland

Barry V. Charlwood
Division of Life Sciences
Kings College London
150 Stamford Street
London
SE1 8WA
UK

Stanley G. Deans
Aromatic and Medicinal Plant Group
Food Systems Division
SAC Auchincruive
Ayr, KA6 5HW
UK

Frédéric-Emmanuel Demarne
ADRIAN S.A.
15, Rue de Cassis
F-13008 Marseille
France

Jeffrey B. Harborne
Department of Botany
School of Plant Sciences
The University of Reading
Whiteknights
Reading, RG6 6AS

Stephen Hart
Messenger & Signalling Research Group
School of Biomedical Sciences
King's College London
Guy's Campus
London
SE1 9RT
UK

Peter Houghton
Pharmacy Department
Kings College London
The Franklin-Watkins Building
150 Stamford Street
London
SE1 8WA
UK

Janet James
The Vernon Geranium Nursery
Cuddington Way
Cheam
Sutton SM2 7JB

Herbert Kolodziej
Institut für Pharmazie
Pharmazeutische Biologie
Freie Universität Berlin
Königin-Luise-Str. 2 + 4, D-14195
Berlin, Germany

Diana M. Miller
Royal Horticultural Society
Wisley
Woking Surrey GU23 6QB
UK

J. Renata Ochocka
Department of Biology and
Pharmaceutical Botany
Medical University of Gdansk
ul. Hallera 107
60-416 Gdansk, Poland

Arkadiusz Piotrowski
Department of Biology and
Pharmaceutical Botany
Medical University of Gdansk
ul. Hallera 107
60-416 Gdansk, Poland

B.R. Rajeswara Rao
Central Institute of Medicinal and
Aromatic Plants
Field Station
Boduppal
Uppal (PO)
Hyderabad 500 039, India

Monique S.J. Simmonds
Jodrell Laboratories
Royal Botanic Gardens
Kew, Surrey, TW9 3AB

Tatyana Stoeva
Laboratory of Biology and Chemistry of
Medicinal and Aromatic Plants
Institute of Botany
Bulgarian Academy of Sciences
23, Acad. G. Bonchev Str.
1113, Sofia
Bulgaria

Rhona Wells
Charabot et Cie
Mill Hill
London

Christine A. Williams
Department of Botany
School of Plant Sciences
The University of Reading
Whiteknights
Reading
RG6 6AS

Elizabeth M. Williamson
Centre for Pharmacognosy and
Phytotherapy
The School of Pharmacy
University of London
Brunswick Square
London
WC1N 1AX

Preface to the series

There is increasing interest in industry, academia and the health sciences in medicinal and aromatic plants. In passing from plant production to the eventual product used by the public, many sciences are involved. This series brings together information which is currently scattered through an ever increasing number of journals. Each volume gives an in-depth look at one plant genus, about which an area specialist has assembled information ranging from the production of the plant to market trends and quality control.

Many industries are involved such as forestry, agriculture, chemical, food, flavour, beverage, pharmaceutical, cosmetic and fragrance. The plant raw materials are roots, rhizomes, bulbs, leaves, stems, barks, wood, flowers, fruits and seeds. These yield gums, resins, essential (volatile) oils, fixed oils, waxes, juices, extracts and spices for medicinal and aromatic purposes. All these commodities are traded worldwide. A dealer's market report for an item may say 'Drought in the country of origin has forced up prices'.

Natural products do not mean safe products and account of this has to be taken by the above industries, which are subject to regulation. For example, a number of plants which are approved for use in medicine must not be used in cosmetic products.

The assessment of safe to use starts with the harvested plant material which has to comply with an official monograph. This may require absence of, or prescribed limits of, radioactive material, heavy metals, aflatoxin, pesticide residue, as well as the required level of active principle. This analytical control is costly and tends to exclude small batches of plant material. Large scale contracted mechanised cultivation with designated seed or plantlets is now preferable.

Today, plant selection is not only for the yield of active principle, but for the plant's ability to overcome disease, climatic stress and the hazards caused by mankind. Such methods as *in vitro* fertilization, meristem cultures and somatic embryogenesis are used. The transfer of sections of DNA is giving rise to controversy in the case of some end-uses of the plant material.

Some suppliers of plant raw material are now able to certify that they are supplying organically-farmed medicinal plants, herbs and spices. The Economic Union directive (CVO/EU No. 2092/91) details the specifications for the *obligatory* quality controls to be carried out at all stages of production and processing of organic products.

Fascinating plant folklore and ethnopharmacology leads to medicinal potential. Examples are the muscle relaxants based on the arrow poison, curare, from species of *Chondrodendron*, and the anti-malarials derived from species of *Cinchona* and *Artemisia*. The methods of detection of pharmacological activity have become increasingly reliable and specific, frequently involving enzymes in bioassays and avoiding the use of laboratory animals. By using bioassay linked fractionation of crude plant juices or extracts, compounds can be specifically targeted which, for example, inhibit blood platelet aggregation, or have anti-tumour, or anti-viral, or any other required activity. With the assistance of robotic devices, all the members of a genus may be readily screened. However, the plant material must be *fully* authenticated by a specialist.

The medicinal traditions of ancient civilisations such as those of China and India have a large armamentaria of plants in their pharmacopoeias which are used throughout South-East Asia. A similar situation exists in Africa and South America. Thus, a very high percentage of the World's population relies on medicinal and aromatic plants for their medicine. Western medicine is also responding. Already in Germany all medical practitioners have to pass an examination in phytotherapy before being allowed to practise. It is noticeable that throughout Europe and the USA, medical, pharmacy and health related schools are increasingly offering training in phytotherapy.

Multinational pharmaceutical companies have become less enamoured of the single compound magic bullet cure. The high costs of such ventures and the endless competition from 'me too' compounds from rival companies often discourage the attempt. Independent phytomedicine companies have been very strong in Germany. However, by the end of 1995, eleven (almost all) had been acquired by the multinational pharmaceutical firms, acknowledging the lay public's growing demand for phytomedicines in the Western World.

The business of dietary supplements in the Western World has expanded from the health store to the pharmacy. Alternative medicine includes plant-based products. Appropriate measures to ensure the quality, safety and efficacy of these either already exist or are being answered by greater legislative control by such bodies as the Food and Drug Administration of the USA and the recently created European Agency for the Evaluation of Medicinal Products, based in London.

In the USA, the Dietary Supplement and Health Education Act of 1994 recognised the class of phytotherapeutic agents derived from medicinal and aromatic plants. Furthermore, under public pressure, the US Congress set up an Office of Alternative Medicine and this office in 1994 assisted the filing of several Investigational New Drug (IND) applications, required for clinical trials of some Chinese herbal preparations. The significance of these applications was that each Chinese preparation involved several plants and yet was handled as a *single* IND. A demonstration of the contribution to efficacy, of *each* ingredient of *each* plant, was not required. This was a major step forward towards more sensible regulations in regard to phytomedicines.

My thanks are due to the staffs of Harwood Academic Publishers and Taylor & Francis who have made this series possible and especially to the volume editors and their chapter contributors for the authoritative information.

Roland Hardman

1 General introduction

Maria Lis-Balchin

The genera *Geranium* and *Pelargonium* are invariably confused by the general public and also plant sales personnel, health food shop workers and alternative medicine practitioners, especially aromatherapists. This confusion has existed before Linnaeus (1753) and his binomial system of classification, where both genera were put under the genus *Geranium*, and although Sweet (1820) and other botanists reclassified them under two genera, acceptance by the majority of laymen as well as nurserymen is still low. The flowers of typical *Geranium* and *Pelargonium* species are shown in Figure 1.1.

Geranium oil is extracted from the leaves of some *Pelargonium* species and cultivars, but its paramedical effects are often equated with those of the genus *Geranium* e.g. *G. robertianum* and *G. maculatum*. The latter are native to Europe and were used as herbal medicines for hundreds of years; they were written up by Gerard (1633), Culpeper (1652) and even Grieve (1937). The *Pelargonium* species, native to southern Africa, although introduced to European Botanic Gardens for example in Leiden as early as 1600, was used only as ornamental plants; their medicinal properties were known only to Hottentots, Zulus and the local Boers in South Africa till the early 1900s when there was some mention in the literature. The medicinal properties of the fat-soluble Geranium oil (from *Pelargonium*) are therefore largely unsubstantiated as

(a) (b)

Figure 1.1 Flowers of typical *Geranium* (a) and *Pelargonium* (b) species.

they are based solely on the properties attributed to the mainly water-soluble extracts (teas) of *Geranium* species. A further confusion arises with the Geranium oil derived from *G. macrorrhizum* in Bulgaria, which is entirely different to the commercial Geranium oil from *Pelargonium* species in both chemical composition and also medicinal properties etc.

The main usage of *Geranium* species is in herbal medicine, whilst that of the *Pelargonium*-derived Geranium oil is in perfumery, cosmetics and aromatherapy products. The production of commercial Geranium oil, from several *Pelargonium* cultivars, is now mainly in Reunion, Egypt and China; however the true sales of Geranium oil are greatly in excess of that derived from plants, due to the ever-increasing production of synthetic and 'nature-identical' Geranium oil.

Geranium oil contains mainly citronellol and geraniol and their esters, therefore can be easily concocted from cheaper essential oils (EOs) and adjusted to the recommended ISO standards. The antimicrobial activity of such EOs is much greater than that of some authentic oils but has a similar pharmacological effect on smooth muscle (spasmolytic) and the actual odour can be even more appreciated by perfumers than the real EO. It remains to be seen whether aromatherapy has any actual medicinal benefits, other than stress-alleviating, and whether these are attributable only to the true EOs, especially as there is a wide difference in the actual percentage chemical composition of EOs obtained from different geographical sources and also different samples from plants grown in various countries where differences in hybridization has occurred and even the same plants grown under different climatic conditions etc.

The pharmacological activity of the water-soluble extracts of the two genera are not very different: they both have a high proportion of tannins and have an antidiarrhoeal function. The lipophylic EOs of *Pelargonium* species have mainly a spasmolytic effect on smooth muscle, except for *P. grossularioides*, which was used as an abortifacient in Southern African folk medicine and has been shown to have a spasmogenic action on smooth and uterine muscle.

Pelargonium EOs from leaves of the many different species and cultivars (other than those grown to produce commercial Geranium oil) have very different odours and chemical compositions, but most of the floral-smelling ones act through cyclic AMP (cAMP) as the secondary messenger; others with odours which are more pine or menthol-like have a different mode of action. There is therefore some correlation between their mode of action and their odour and chemical composition. The numerous aromatherapeutic uses for Geranium oil are yet to be scientifically validated, although there is every reason to accept the scientific evidence that inhalation of the aroma and its action through the limbic system has a relaxing effect; theoretically, this could lead to the acceptance that many stress-related conditions like dermatitis, asthma, intestinal problems and headaches could be alleviated.

Part 1

Geranium

2 History of nomenclature, usage and cultivation of *Geranium* and *Pelargonium* species

Maria Lis-Balchin

The genera *Geranium* and *Pelargonium* have remained confused for over 200 years, even after Linnaeus (1753) and his binomial system of classification. Both genera were then placed under the genus *Geranium* and although Sweet (1820) re-classified them under two genera, acceptance by the general public as well as nurserymen is still minimal.

The majority of plants sold as 'geraniums' in garden centres, shops and supermarkets are *Pelargonium* species and cultivars. There are however some *Pelargonium* cultivars sold as Pelargoniums at garden centres: these are the large-flowered Regal Pelargoniums. All the scented-leaf Pelargoniums, on the other hand, even those with large flowers, are almost invariably sold as 'geraniums'.

Most of the books about *Pelargonium* are mis-headed e.g. 'Geraniums' (Delamain and Kendall, 1987); 'Miniature and Dwarf Geraniums' (Bagust, 1988); 'Geraniums for home and garden' (Shellard, 1981). The only difference between genera is seen when books are entitled 'Hardy Geraniums' (Bath and Jones, 1994; Yeo, 1985), which signifies that the true *Geranium* genus alone will be involved. This is because *Pelargonium* species are not hardy plants and are not able to survive the weather in Europe, unlike the hardy *Geranium* species found everywhere in hedgerows, wastelands and rockeries.

Pelargonium species originate from South Africa and different species are found in distinct habitats. The *Pelargonium* species related to the Geranium oil-producing cultivars are mainly located in the Cape area. The first *Pelargonium* species, *Pelargonium triste* was brought over from the Cape to Leiden before 1600, then John Tradescant obtained the species in 1631 (Miller, 1996), following this, other species were brought over by various botanists for the next 300 years and hybridization became very rampant. This occurred especially during Victorian times, where almost every rich landowner had conservatories and glasshouses dedicated to the tender and rare plants, which included the novelty *Pelargonium* species.

There are numerous misnomers given for the origin of 'Geranium oil' in the dozens of 'Aromatherapy' books appearing during the last few decades, as well as some scientific reference books (Fenaroli, 1997). The worst misnomer shows the total misconception of the genus as in many aromatherapy books and journals, *Geranium maculatum*, *Geranium robertianum* and other *Geranium* species are implicated either directly or indirectly. This arose due to the unfortunate original mistake by aromatherapy book authors, who read up the medicinal properties of true 'geranium' from the many Herbals (Culpeper, 1653, 1835; Gezard, 1597; Grieve, 1937) and thought that those were attributable to the *Pelargonium* species.

For example, we have an amazing botanical concoction in this quote: 'The oil is extracted not from the familiar brightly coloured geranium but from the species Pelargonium Geranium Robert or "lemon plant" – which is very often displayed in abundance in Greek restaurants' (Worwood, 1991).

Tisserand (1985) informs us that: '*Pelargonium odorantissimum graveolens* grows about 2 ft. in height, has serrated, pointed leaves, and small, pink, flowers. The whole plant is aromatic. It is found on wastelands, in hedgerows, and on the outskirts of woods. It was used by the ancients as a remedy for wounds and tumours.' He is undoubtedly referring to *G. robertianum* (Culpeper, 1835), as the quote is partly extracted from this source, and it continues: 'all geraniums are vulneraries but this herb more particularly so, only rather more detersive and diuretic, which quality is discovered by its strong, soapy smell'.

Another quote taken from Lawless (1992) again shows a lack of understanding of the origins of Geranium oil: 'The British plant herb Robert (*G. robertianum*) and the American cranesbill, (*G. maculatum*) are the most widely-used types in herbal medicine today, having been used since antiquity', as this has nothing to do with *Pelargonium*. In fact this statement is partly true, as the main usage of *Geranium* species is in herbal medicine, whilst that of the *Pelargonium*-derived 'geranium' oil is in perfumery, cosmetics and aromatherapy products. Furthermore, *Geranium* species are usually used as a tea or alcoholic extract which is taken orally; comparatively few external applications are mentioned and these again use the water-soluble or alcoholic extracts and not essential oils (EOs) (Culpeper, 1835; Grieve, 1937).

One of the misnomers most frequently used is '*Pelargonium odoratissimum*' or '*P. odorantissimum*' (Lawless, 1992; Valnet, 1984) and '*Pelargonium odorantissium*' (Westwood, 1991). *Pelargonium odoratissimum* is an actual species, with very small, white, apple-scented leaves (van der Walt, 1977), and not used for 'Geranium oil' production. The name '*Pelargonium odoratissimum*' probably came from a particular *P. graveolens* variety as '*odoratissimum*' i.e. it was not a true species but a very odoriferous variety. The name *P. odoratissimum* was then misquoted by Knuth (1912). The early writers about (EOs) (Guenther, 1950) as well as the trade distributors used the wrong name and it has been perpetuated by aromatherapists, who after all, are not botanists.

Another misnomer used is *P. asperum*: which is described as a cross between unknown parents (Knuth, 1912) and by Harvey (1860) as a garden variety of *P. quercifolium* (which has a camphoraceus smell). On the other hand, almost identical drawings are shown by Mastalerz (1982) for *P. graveolens* (L'Heritier, 1792) and *P.* × *asperum* adapted from *G. radula* Roth, but to confuse the issue further, *P. asperum* Ehrhart ex Willd is in fact a hybrid between *P. graveolens* × *P. radens*.

Yet another name often used is *P. roseum* Willd. The problem is that Willdenov's Herbarium (1800) shows two identical plants: one is labelled as *P. radula* var. *roseum* and the other as *P. graveolens* var. *roseum*, but the description matches that of the hybrid *P. radula* × *P. graveolens* which shows once more that it is not a species.

A fourth misnomer is *P. graveolens*. This may be partly true for some Geranium oil originating in Africa (Ducellier, 1933), but this is doubtful as the species has a more distinctive peppermint aroma (Demarne and van der Walt, 1989; Lis-Balchin, 1991). However, the main source of the oil is from a cultivar known as P. cv. 'Rosé' which gives rise to the commercial 'Geranium oil, Bourbon' and originated from hybridization in England, probably in the eighteenth century, and the cultivar was then exported to the

South of France and Reunion and also lately to China. The 'Rosé' cultivar has been found to be, most probably, a hybrid between *P. capitatum* × *P. radens* (Demarne and van der Walt, 1989). The cultivars used for the production of Geranium oil in many parts of the world remain confused; in some papers originating in India, the cultivar is stated to be that obtained as a cutting from the cultivar 'Rosé' from Reunion, however, many papers state that their Geranium oil source is from *P. graveolens*.

The main usage of *Geranium* species is in herbal medicine, whilst that of the *Pelargonium*-derived Geranium oil is in perfumery, cosmetics and aromatherapy products. The production of commercial Geranium oil, from several *Pelargonium* cultivars, is now mainly in Reunion, Egypt and China; however the true sales of Geranium oil are greatly in excess of that derived from plants, due to the ever-increasing production of synthetic and 'nature-identical' Geranium oil. Geranium oil contains mainly citronellol and geraniol and their esters, therefore can be easily concocted from cheaper EOs and adjusted to the recommended ISO standards. The actual odour can be even more appreciated by perfumers than the real essential oil. Synthetic 'Geranium oil' also has a more potent antimicrobial activity (Lis-Balchin *et al.*, 1996).

The pharmacological activity of the water-soluble extracts of the two genera are not very different: they both have a high proportion of tannins and have an antidiarrhoeal function (Watt and Breyer-Brandwijk, 1962). The lipophylic EOs of *Pelargonium* species have mainly a spasmolytic effect on smooth muscle, except for *P. grossularioides*, which was used as an abortifacient in southern African folk medicine (Watt and Breyer-Brandwijk, 1962) and has been shown to have a spasmogenic action on smooth and uterine muscle *in vitro* (Lis-Balchin and Hart, 1994).

Many *Pelargonium* species were used as folk medicines: herbal teas and extracts of leaves, tubers and roots were used (Watt and Breyer-Brandwijk, 1962) and one of these, Umckaloabo, from *P. sidoides* or *P. reniforme*, is now commercially produced for use in respiratory ailments (Kolodziej and Kaiser, 1997; Kolodziej *et al.*, 1995).

Pelargonium EOs from leaves of the many different species and cultivars (other than those grown to produce commercial Geranium oil) have very different odours and chemical compositions, but most of the floral-smelling ones act through cyclicAMP as the secondary messenger; others with odours which are more pine or menthol-like have a different mode of action (Lis-Balchin and Hart, 1998). There is therefore some correlation between their mode of action and their chemical composition. The antimicrobial effects have also been studied *in vitro* on a number of bacteria and *in vivo* as food preservatives, with promising results (Lis-Balchin *et al.*, 1998a). The antimicrobial effect could not be directly correlated with the chemical composition so far (Lis-Balchin and Roth, 2000), although an inverse relationship was found between the amount of 1,8-cineole and antifungal activity (Lis-Balchin *et al.*, 1998b). The DNA relationship with chemical composition has also been studied in a number of scented *Pelargonium* species and some correlation has become apparent (Piotrowski *et al.*, 1999). This has alsoprovided another aspect of chemotaxonomy to add to the alkaloid chemotaxomy (Lis-Balchin, 1996, 1997; Lis-Balchin *et al.*, 1996a).

The numerous aromatherapeutic uses for Geranium oil, both through massage and inhalation, are yet to be scientifically validated (Lis-Balchin, 1997a), although there is every reason to accept the scientific evidence that massage in itself can relax and that inhalation of a pleasant aroma and its action through the limbic system also has a relaxing effect (Buchbauer *et al.*, 1991; Jellinek, 1954, 1956; Kubota *et al.*, 1992; Manley, 1993; Stoddart, 1990; Torii *et al.*, 1988; Vickers, 1996). Theoretically, this could lead to the acceptance that many stress-related conditions like dermatitis, asthma, intestinal problems and headaches could be alleviated by the use of Geranium oil.

True *Geranium* species were used in the past all over the world. *G. maculatum* or American cranesbill root was used as 'styptic, astringent, tonic, for piles and internal bleeding; excellent as an injection for flooding and leucorrhoea; taken internally for diarrhoea, childhood cholera, chronic dysentery and for gargling (Grieve, 1937)'. It contains tannins and is said to be effective against stomach ulcers and inflammation of the uterus and has possibilities in treating cancer (Chevallier, 1996).

Geranium species are used nowadays mainly in Japan, USA and also in eastern Europe. *Geranium thunbergii*, which is one of the most important medicinal plants in Japan (gen-noshouko) is used as an antidiarrhoeal folk medicine (Ishimaru and Shimomura, 1995). The use is the same therefore as that for many *Pelargonium* species in South African folk medicine (Watt and Breyer-Brandwijk, 1962). It contains a large concentration of phenolic constituents in the form of hydrolyzable tannins such as corilagin, geraniin and elaeocarpsin. Although elaeocarpsin has a high potential for tannin activity, as measured in combination with proteins e.g. collagen, gelatin, casein, haemoglobin, it shows little astringency. Due to this characteristic, which only slightly stimulates the alimentary canal, it is used for treating digestive diseases. It is also used for various other diseases (Nishioka, 1983) including psychotropic diseases (Ueki *et al.*, 1985). *Geranium* has also shown antiviral action (Courtout *et al.*, 1991).

Both *Geranium* and *Pelargonium* species have been reproduced using micropropagation and some success has been made in producing active components by tissue culture (Charlwood and Charlwood, 1991; Ishimaru and Shimomura, 1995). The means are there therefore for producing plants as well as new active components when they are discovered.

REFERENCES

Bagust, H. (1988) *Miniature and Dwarf Geraniums*, Christopher Helm, London.

Bath, T. and Jones, J. (1994) *Hardy geraniums, A gardeners guide to growing*, David & Charles, Devon.

Buchbauer, G., Jirovetz, L., Jager, W., Dietrich, H., Plank, C. and Karamat, E. (1991). Aromatherapy: evidence for the sedative effects of the essential oils of lavender after inhalation. *Z. Naturforsch.*, 46, 1067–1072.

Charlwood, B.V. and Charlwood, K.A. (1991) *Pelargonium* spp. (Geranium): *In vitro* culture and the production of aromatic compounds. In Y.P.S. Bajaj (ed.), *Biotechnology in Agriculture and Forestry vol. 15. Medicinal and Aromatic Plants III*, Springer-Verlag, Berlin, pp. 339–352.

Chevallier, A. (1996) *Encyclopedia of Medicinal Plants*, Dorling Kindersley, London.

Courtout, J., Pieters, L.A., Claeys, M., Berghe, D.A.V. and Vlietinck, A.J. (1991) Antiviral ellagitanins from *Spondias mombin. Phytochemistry*, 30, 1129–1130.

Culpeper, N. (1653) *The English physitian Enlarged*, George Sawbridge, London.

Culpeper, N. (1835) *The Complete Herbal*. Reprinted, Th. Kelly, London.

Delamain, B. and Kendall, D. (1987) *Geraniums.*, Christopher Helm, London.

Demarne, F. and van der Walt, J.J.A. (1989) Origin of the rose-scented *Pelargonium* grown on Reunion Island. *S. Afr. J. Bot.*, 55, 184–191.

Ducellier, L. (1933) *Bull. Soc. Hist. Nat. Afrique du Nord*, 24, 142–148.

Fenaroli, G. (1997) *Handbook of flavour ingredients*, 3rd ed. CRC Press, London. Vol. 1.

Gezard, J. (1597) *The Herball or General History of Plants*, John North, London.

Grieve, M. (1937) *A Modern Herbal*. Reprinted 1992, Tiger Books International, London.

Guenther, E. (1950) *The Essential Oils*, Vol. 4. Van Nostrand Co., New York.

Harvey, W.H. (1860) *Flora capensis*, Hedges, Smith & Co., Dublin.

Ishimaru, K. and Shimomura, K. (1995) *Geranium thunbergii*: *In vitro* culture and the production of geraniin and other tannins. In: *Biotechnology in Agriculture and Forestry, Vol. 33. Medicinal and Aromatic Plants VIII* ed. Bajaj, Y.P.S. Springer-Verlag, Berlin, pp. 232–247.

Jellinek, P. (1954) *The Practice of Modern Perfumery*, Leonard Hill, London.

Jellinek, P. (1956) *Die Psychologischen Grundlagen der Parfumerie.* Alfred Hutig Verlag, Heidelberg.

Kolodziej, H. and Kaiser, O. (1997) *Pelargonium sidoides* D.C. – Neuste Erkenntnisse zum Verständnis des Phytotherapeutikums Umckaloabo. *Z. Phytother.*, 19, 141–151.

Kolodziej, H., Kaiser, O. and Gutman, M. (1995) Arzneilich verwendete Pelargonien aus Sudäfrika. *Dtsch. Apotheker Ztg.*, 135, 853–864.

Knuth, R. (1912) *Das Pflanzenreich*, 4, 129, Berlin.

Kubota, M., Ikemoto, T., Komaki, R. and Inui, M. (1992) Odor and emotion-effects of essential oils on contingent negative variation. *Proc. 12th Int. Congress on Flavours, Fragrances and Essential oils*, Vienna, Austria, Oct. 4–8. pp. 456–461.

L' Heritier, C.L. (1792) *Geraniologia*. Paris.

Lawless, J. (1992) *The Encyclopaedia of Essential Oils.*, Element, Dorset.

Linnaeus, C. (1753) *Species Plantarum*, Stockholm.

Lis-Balchin, M. (1991) Essential oil profiles and their possible use in hybridization of some common scented Geraniums. *J. Essent. Oil Res.*, 3, 99–105.

Lis-Balchin, M. and Hart, S. (1994) A pharmacological appraisal of the folk medicinal usage of *Pelargonium grossularioides* and *Erodium cicutarium (Geraniaceae). Herbs, Spices Med. Plants*, 2, 41–48.

Lis-Balchin, M. (1996) A chemotaxonomic reappraisal of the Section *Ciconium, Pelargonium* (Geraniaceae). *S. Afr. J. Bot.*, 62, 277–279.

Lis-Balchin, M., Deans, S.G. and Hart, S. (1996) Bioactivity of commercial Geranium oil from different sources. *J. Essent. Oil Res.*, 8, 281–290.

Lis-Balchin, M., Houghton, P. and Woldermariam, T. (1996a) Eleocarpidine alkaloids from *Pelargonium* species (*Geraniaceae*). *Nat. Prod. Letts.*, 8, 105–112.

Lis-Balchin, M. (1997) A chemotaxonomic study of *Pelargonium* (*Geraniaceae*) species and their modern cultivars. *J. Hort. Sci.*, 72, 791–795.

Lis-Balchin, M. (1997a) Essential oils and 'Aromatherapy': their modern role in healing. *J. Roy. Soc. Health*, 117, 324–329.

Lis-Balchin, M. and Hart, S. (1998) Studies on the mode of action of scented-leaf *Pelargonium* (*Geraniaceae*). *Phytother. Res.*, 12, 215–217.

Lis-Balchin, M., Buchbauer, G., Hirtenlehner, T. and Resch, M. (1998a) Antimicrobial activity of novel *Pelargonium* essential oils added to a quiche filling as a model food system. *Lett. Microbiol.*, 27, 207–210.

Lis-Balchin, M., Deans, S.G. and Eaglesham, E. (1998b) Relationship between the bioactivity and chemical composition of commercial plant essential oils. *Flav. Fragr. J.*, 13, 98–104.

Lis-Balchin, M. and Roth, G. (2000) Composition of the essential oils of *P. odoratissimum*, *P. exstipulatum* and *P.× fragrans* (*Geraniaceae*). *Flavor & Fragr. J.*, 15(6), 391–394.

Manley, C.H. (1993) Psychophysiological effect of odor. *Crit. Rev. Food Sci. Nutr.*, 33, 57–62.

Mastalerz, J.W. (1982) *Geranium – A Penn State Manual*, 3rd ed. Pennsylvania Flower Growers.

Miller, D. (1996) *Pelargoniums* B.T. Batsford Ltd, London.

Nishioka, I. (1983) Chemistry and biological activities of tannins, *Yakugaku Zasshi*, 103, 125–142.

Piotrowski, A., Bogdan, A., Lis-Balchin, M. and Ochocka, J.R. (1999) Genetic relationships among *Pelargonium* genus assessed by RAPD-PCR method. 2000 years of Natural Products Research: Past, Present and Future. Amsterdam, 26–30 July 1999, Poster.

Shellard, A. (1981) *Geraniums for Home and Garden*, David & Charles, Devon.

Stoddart, M. (1990) *The Scented Ape*, University Press, Cambridge.

Sweet, R. (1820–1830) *Geraniaceae*, London.

Tisserand, R. (1985) *The Art of Aromatherapy*, Revised ed. C.W. Daniel Co. Ltd, Saffron Walden, p. 231.

Torii, S., Fukuda, H., Kanemoto, H., Miyanchio, R., Hamauzu, Y. and Kawasaki, M. (1988) Contingent negative variation and the psychological effects of odor. In: *Perfumery: The psychology and Biology of Fragrance*. S. Toller and G.H. Dodds (eds), Chapman & Hall, New York.

Ueki, S., Nonaka, G., Nishioka, I. and Fujiwara, M. (1985) Psychotropic activity of GAO and its active substances. *J. Med. Pharm. Soc., Wakan-Yaku*, 2, 502–503.

Valnet, J. (1984) *Aromatherapie*, Maloine, S.A., Paris.

van der Walt, J.J.A. (1977) *Pelargoniums of Southern Africa*, Vol. 1. Purnell and Sons, Cape Town.

Vickers, A. (1996) *Massage and Aromatherapy. A Guide for Health Professionals*, Chapman & Hall, London.

Watt, J.M. and Breyer-Brandwijk (1962) *The Medicinal Plants of Southern Africa*, Livingstone Ltd, Edinburgh.

Westwood, C. (1991) *Aromatherapy. A Guide for Home Use*, Amberwood Publishing Ltd, Dorset.

Willdenov, C.L. (1800) *Species Plantarum*, Berlin.

Worwood, V.A. (1991) *The Fragrant Pharmacy*, Bantam Books, London.

Yeo, P.F. (1985) *Hardy Geraniums*, Croom Helm, London.

3 The taxonomy of *Geranium* species and cultivars, their origins and growth in the wild

Diana M. Miller

HISTORY OF DISCOVERY

A number of species are native to Europe and there are records of geraniums in old herbals since the late sixteenth and early seventeenth centuries. A few have been used in herbal remedies in the past but may not have been cultivated. Species such as *Geranium robertianum* are common wild plants in many parts of the country. More recently, species have been introduced into cultivation such as *G. maculatum* from North America which was used as a medicinal plant in the eighteenth century. Other species were introduced from further afield mainly for their ornamental features from countries around the Mediterranean, the Himalayas and the Far East and even towards the end of the twentieth century, new species were still being introduced.

GENERAL CHARACTERISTICS OF THE GENUS *GERANIUM*

In habit, the species are herbaceous, annual, biennial or perennial and sometimes woody at the base; a few have tubers. All have petiolate, usually glandular or aglandular hairy, palmate to deeply palmately divided leaves; the divisions toothed or lobed, with stipules. The stem leaves are normally opposite but the lower leaves may be alternate. The flowers of *Geranium* often pink or purplish or bluish-pink, are arranged in pairs or solitary or in small umbels subtended by bracteoles. Each radially symmetrical flower has five, usually hairy sepals with mucronate apex, usually increasing in size as the fruit develops. The five equal, free petals with a nectary at their base, may be clawed and are sometimes notched at the apex. The ten free stamens, about as long as the sepals, are in two whorls, the anthers of the outer dehiscing before those of the inner whorl. The style divides into five stigmas, which open after the anthers have dehisced, thus avoiding self pollination. Once the flower is fertilised, the five mericarps each containing a single seed develop. The method by which the awned mericarps is separated is important in division of the genus into three subgenera.

NATIVE HABITATS

The genus *Geranium* contains about 300 distinct species, the majority of which are to be found in the Northern Hemisphere, mainly in temperate climates. Some may be found in the tropics but at high altitudes in montane conditions. Many of those in cultivation are from Eurasia and North America but species also grow wild in North and South Africa, Australia, New Guinea and islands in the Atlantic and Pacific Oceans. Most of those found in cultivation are hardy but some are slightly less such as *G. maderense* from Madeira. In the warmer areas, many of the species are annual, passing the hottest part of the year as seed or, if perennial, may have a tuber and die down in the summer. Species native to alpine regions are frequently low growing. The perennial species, some reaching well over 1 m in height, tend to be found in grasslands or wood margins.

CLASSIFICATION HISTORY

The earliest references to the genus, included so few species that no attempts were made to classify them. However by 1912, Knuth produced his monograph, which included over 250 species and a means of separating groups became important. He used the characteristics of the rootstock to make basic separations between the perennial plants and kept the annuals in a separate group. More recently, Yeo (1984) recognised different methods of seed discharge as an important characteristic. The fruit can break in three different ways and this is used to create three subgenera within the genus, each subdivided into a number of smaller sections. However, this method results in the very large section *Geranium* within the subgenus, *Geranium*, which contained the majority of the known species. This has therefore been subdivided informally using other morphological features into groups (Yeo, 1985, 1992).

Subgenus *Geranium*

This subgenus includes over 250 annuals and perennials. The fruit breaks apart using a method described as *seed ejection* because as the fruit splits, the seed is thrown upwards out of the mericarps as the awn curves backwards. The awn may or may not remain attached to the remaining part of the fruit. The mericarps are smooth.

Section Geranium

This very large section is recognised by the method of discharge of seeds in which the awn remains attached to the central column of the fruit after the seeds are released. Immediately before the seeds are discharged, they are retained within the mericarp by bristles at the lower end of the mericarp. The section is divided into groups.

Endressii group

These are perennials with erect funnel-shaped flowers, petals usually notched and leaves with diamond-shaped divisions.

G. endressii Gay: This is an evergreen perennial measuring up to about 45 cm with rhizomes near or on the surface. Flowers are bright pink about 3.5 cm across with barely notched petals. Leaves are light green, somewhat wrinkled, 5–10 cm wide deeply divided into five.

This is a summer flowering species native to the western Pyrenees with a number of cultivars such as 'Wargrave Pink' with numerous salmon pink flowers.

G. versicolor Linnaeus: Similar to *G. endressii*, this species has less divided leaves and flowers, which are white, conspicuously veined, reddish-purple and distinctly notched petals. It is native to southern Europe from Italy to Greece.

G. × *oxonianum* Yeo: This is a vigorous fertile hybrid between *G. endressii* and *G. versicolor* reaching 80 cm in height with pink-veined flowers to 4 cm diameter and notched petals. A large number of garden plants are cultivars of this hybrid such as 'A.T. Johnson' with silvery pink flowers. In f. *thurstonianum* the flowers have very narrow strap-shaped purple flowers.

G. nodosum Linnaeus: This is a deciduous perennial, up to 50 cm, with leaves divided into 3–5 toothed but not lobed divisions, 2–5 cm wide. Flowers are up to 3 cm diameter and are bright purple-pink with darker veins.

It is a native of southern Europe from central France to Yugoslavia, flowering in the summer.

Sylvaticum group

These are perennial plants, leaves with broad much lobed divisions and relatively small sepals.

G. sylvaticum Linnaeus: The plant is up to about 70 cm tall with glandular hairs and leaves to over 20 cm wide deeply divided into seven or nine deeply cut divisions. Flowers are usually purplish-blue, but sometimes pink or white, with a white centre 2–3 cm across, produced in early summer.

This plant grows wild in most parts of Europe to Turkey. A number of cultivars include 'Album' with white flowers and 'Mayflower' with a bright violet-purple flower but a smaller white centre.

G. psilostemon Ledebour: This is a tall erect perennial measuring to well over 1 m. Leaves are up to 30 cm across, deeply divided into 7 divisions and lobes are divided and toothed. Flowers are shallow cupped, up to 3.5 cm diameter and petals very bright purplish-red, black at the base and with black veins.

This is a summer flowering species native to North Turkey and the southern Caucasus.

Pratense group

This includes perennials with large blue or pink flowers, petals not notched and usually with glandular hairs at least on inflorescence.

G. himalayense Klotzsch: This is a deciduous spreading perennial to about 50 cm with leaves up to 12 cm across, divided into seven sections, each more or less three-lobed and toothed. Flowers are large, saucer shaped, up to 6 cm wide, blue or deep blue, with a calyx which is not inflating after flowering.

Long flowering through the summer, this species is found wild in the Himalayas.

G. pratense Linnaeus: This is a deciduous perennial which may reach well over 1 m in height. Leaves are divided almost to the base into seven or nine divisions, each deeply pinnately divided, 10–25 cm wide. Flowers are 3.5–4.5 cm, saucer-shaped, blue to violet-blue or sometimes white, with a calyx which is inflated after flowering.

Flowering in summer, this species is native from Europe to western China and many well known garden plants are cultivars of this species such as 'Mrs Kendall Clark' a light pearl gray-blue and the double flowered 'Plenum Caeruleum'. This plant has been used to produce a dye and the roots used in tanning.

Erianthum group

This includes perennials with few stem leaves but dense inflorescence of rather flat flowers with petals unnotched, native to north-east Asia.

G. erianthum De Candolle: This is a deciduous perennial up to about 50 cm with leaves 5–20 cm wide divided nearly to base into seven or nine-lobed, overlapping divisions. Flowers are 3 cm or more in diameter, light to dark purplish-blue and conspicuously veined. Native to Japan, this is a very hardy species flowering in early summer, also found in Siberia and north-western North America.

Wallichianum group

Plants with scrambling or trailing stems and leaves divided to 75 per cent or less. The species are native to the Himalayas.

G. wallichianum D. Don: This is a deciduous, trailing perennial plant without a basal rosette of leaves. Stem leaves are divided into five. The flower is saucer-shaped or more or less flat up to 3.5 cm across, purple or pinkish-purple, veined and sometimes white at the base.

This is a variable species from western Himalayas from which a number of cultivars have been selected such as the dense creeping, 'Buxton's Variety' with smaller marbled leaves and blue flowers. The roots of this plant have been used for dying and tanning.

G. procurrens Yeo: This is a perennial plant with long trailing red stems to 1 m, rooting at the nodes. Leaves are 5–10 cm, divided into five, each division being shallow-lobed. Flowers are 2.5–3.5 cm across, pinkish-purple with a distinct black v-shaped mark at the base and veined with black.

This species is found in the Himalayas in Sikkim and Nepal.

Palustre group

These are bushy perennial plants with main leaves deeply divided, hairy but not glandular and flowers are pink to purple.

G. wlassovianum Link: This is a deciduous, perennial, densely hairy plant up to about 40 cm in height with softly hairy, gray-green leaves normally divided into seven sharply toothed divisions. Flowers are 3–4 cm across, saucer-shaped, pale pink to pale reddish-purple with conspicuous darker veins.

This is a very hardy mid-summer flowering plant from eastern Russia, Siberia and northern China.

Pylzowianum Group

These are low growing alpines from western China with small underground tubers. Flowers are few.

G. orientalitibeticum Knuth: This is a dwarf plant reaching 35 cm in height with distinct strongly marbled, almost round leaves to about 7 cm across, deeply divided into 5-lobed segments. Flowers are more or less flat about 2.5 cm diameter, purplish-pink, white in centre.

This is a summer flowering species native to western China.

G. pylzowianum Maximowicz: This is a low growing perennial up to 25 cm tall and individual leaves raising directly from the tubers, up to 6 cm across, divided into five or seven narrow divisions, each further divided into narrow segments. Flowers are few, about 3 cm in diameter, trumpet-shaped, deep pink fading to white at the base.

This is a species flowering in early summer from western China.

Sanguineum group

This is made up of perennial plants with rhizomes and few basal leaves but leafy flowering stems. Leaves are deeply divided into narrow segments, and flowers are large and petals usually notched.

G. sanguineum Linnaeus: This is a low growing deciduous, perennial to about 30 cm in height, with leaves 2–8 cm across, deeply cut almost to the base into narrow segments. Flowers are saucer-shaped on one-flowered peduncles, to over 4 cm across, petals being notched and white to deep purplish-red.

This is a summer flowering species native to Europe and Turkey. A large number of cultivars are available such as the white flowered 'Album', the rose pink 'Glenluce' and 'Elspeth' with large bright purple flowers. In var. *striatum* the flowers are white with pale pink veins giving the impression of flesh pink flowers.

Sibiricum group

This includes perennial plants with leaves variously divided and small flowers.

G. thunbergii Lindley & Paxton: This is a spreading perennial, hairy plant with trailing stems up to about 25 cm, sometimes rooting at the nodes, with leaves of light green, 5–10 cm wide, often blotched purple between the leaf divisions in winter. Flowers are small to 5 cm across white to dark purple-pink.

This species is native to northern China and Japan and flowers in summer.

Sessiliflorum group

This includes perennial plants with a rosette of leaves but trailing inflorescence stems. Flowers are in one-flowered clusters with the pedicel exceeding the peduncle.

G. sessiliflorum Cavanilles: This is an evergreen perennial plant with leaves 1.5–4.5 cm across and small white flowers barely 1 cm in diameter.

It is a species from Australasia and is represented in gardens by ssp. *novae-zelandiae* and more frequently by the cultivar 'Nigricans' with dark bronze-leaves, which turn orange as they fade.

Incanum group

This is a group of perennial species from South Africa with divided stipules, finely and deeply divided leaves and notched petals.

G. incanum N.L. Burman: This is an evergreen perennial up to over 50 cm tall with aromatic leaves which are finely divided into linear segments rarely exceeding 1 mm in width, with long white hairs on underside. It flowers in early summer to early autumn and these are white or very pale pink in var. *incanum* but deep pink and larger 2.5–3.5 cm in diameter in var. *multifidum*.

Maculatum group

This includes annual or perennial plants from North America with glandular hairs and sepals with long mucronate points.

G. maculatum Linnaeus: This is a deciduous perennial plant, up to about 70 cm tall with leaves up to 20 cm across, divided almost to the base into normally five sections, sharply toothed in the upper part. Flowers are up to 3 cm in diameter, shallow-cupped, pale to deep pink with a white centre, sometimes white sepals, with mucronate apex of 2–3.5 mm long.

This species grows wild in eastern North America and flowers in summer.

Section Dissecta

This includes annual, biennial or perennial plants. The seeds are retained within the mericarp until the moment of release by a part of the wall which extends like a prong and the awn remains attached to the central fruiting column after the seeds are released.

G. aspholdeloides N.L. Burman: This is a perennial species with leaves up to 8 cm across, deeply divided into five or seven lobes. Flowers are up to 3.5 cm diameter, light to dark pink with deeper coloured veins. This species is native to Europe and the Middle East.

Section Tuberosa

This includes annual, biennial or perennial plants with notched petals. The awn of the mericarp does not remain attached to the central fruiting column after the seeds are released.

Tuberosum group

This includes perennial plants with underground tubers, becoming dormant in summer after flowering; leaves are deeply divided. Flowers are usually pinkish to purplish-pink.

G. tuberosum Linnaeus: This is an erect plant up to about 40 cm with leaves up to 10 cm across, deeply divided into seven notched petals.

It is native to the Mediterranean region, eastwards to Iran. The tubers have been used for food.

G. malviflorum Boissier: This is a large perennial, leaves usually over 10 cm wide less finely divided than those of *G. tuberosum*, and the flower is large, bluish-purple with darker veins up to 4.5 cm across, notched at the apex.

It is native to North Africa and southern Spain, flowering in spring.

Platypetalum group

This includes annual, biennial or perennial plants with leaves rarely divided to the base. Flowers are usually bluish-purple.

G. platypetalum Fischer and Meyer: This is a perennial up to about 40 cm, conspicuously hairy with leaves up to about 15–20 cm diameter, divided to about half way into seven or nine segments, each deeply toothed. Flowers are rather flat in form, up to 4.5 cm across, dark purplish-blue with notched petals and darker veins.

Native to Turkey and Iran, this species first flowers in early summer and may produce a further flush later in the season.

G. × *magnificum* Hylander: This is a vigorous sterile hybrid between *G. ibericum* and *G. platypetalum* reaching up to about 70 cm in height, with leaves divided to about 75 per cent into seven to nine deeply toothed divisions which are slightly overlapping. Flower are a rich bluish purple.

G. renardii Trautvetter: This is a low growing perennial plant with rounded gray-green felted leaves up to 10 cm across, divided to half way into five or seven broad toothed divisions. Flowers are white, or very pale purple, with notched petals and conspicuous bluish-purple veins.

This is a summer flowering species native to the Caucasus region.

Subgenus *Robertium*

This is composed of about 30 species. Both annual and perennial species of the subgenus have a fruit discharge described as *carpel projection*. In this subgenus, the mericarp containing the seed is projected some distance from the plant and the awn is separated from the mericarp at the time of the explosive discharge. The mericarp base is rounded and has no hardened point.

Section Batrachiodes

This includes annual and perennial hairy plants with notched petals and a very small mucronate apex to the sepals are included in this section. The mericarps are ribbed, sometimes slightly.

G. pyrenaicum N.L. Burman: This is a perennial up to about 50 cm or so with rounded leaves up to 10 cm across divided into seven or nine divisions. Flowers are up to 2 cm wide on long trailing stems, petals being dark pinkish-purple, white at the base with darker veins and are produced from late spring to early autumn.

It is native to south-western and southern Europe and a white flowered form is also grown.

Section Unguiculata

This includes perennial plants with glandular hairs. Sepals are erect, petals clawed but not notched and mericarps are ridged.

G. macrorrhizum Linnaeus: This is an aromatic rather sticky perennial up to 50 cm tall with leaves up to 20 cm across divided to 75 per cent into seven shallowly divided lobes. It flowers in summer, with a reddish swollen calyx and clawed white or pink petals in a dense inflorescence. Stamens are about twice as long as sepals.

This species is native to southern Europe and a number of cultivars with darker ('Bevan's Variety') or paler ('Album') flowers or variegated leaves ('Variegatum') are grown.

G. dalmaticum (Beck) Rechinger: This is a low growing aromatic perennial up to 15 cm tall. Stems are trailing with small rosettes at the ends of hairless glossy leaves up to 4 cm wide. Flowers are bright pink up to about 2 cm across.

This is a summer flowering plant native to south-western Yugoslavia and Albania.

G. × *cantabrigiense* Yeo: This is a hybrid between *G. macrorrhizum* and *G. dalmaticum* and this plant is aromatic and low growing up to 20 cm or more tall with light green leaves intermediate between the parents and pink or white flowers up to 2.8 cm wide.

Section Lucida

This includes plants which are annual with mostly glandular hairs. Sepals are erect, petals clawed and mericarps glandular hairy with longitudinal ridges.

G. lucidum Linnaeus: This is an annual plant up to about 50 cm with red stems and somewhat succulent glossy leaves up to 5 cm across, deeply divided into five divisions.

Flowers are up to 1 cm across, erect and deep pink. It is found wild from Europe and North Africa to Asia.

Section Ruberta

This includes annual, biennial or perennial plants with mostly glandular hairs and leaves divided to the base. Petals are clawed, barely notched and the mericarps are ridged.

G. robertianum Linnaeus: This is an annual, usually overwintering, unpleasant smelling, with red to reddish-brown hairy stems and leaves. Leaves are over 10 cm across, divided to the base into five-stalked, deeply pinnately divided sections. Flowers are small up to 1 cm wide, pink, flowering over a long period from spring to autumn.

It is found wild across northern hemisphere from Europe to China and also in North America where it may have been introduced. Several cultivars with differing flower colours are grown, especially ones with pure white flowers such as 'Celtic White'. It has been used medicinally in the past.

Section Anemonifolia

This includes short lived perennials forming very large rosette plants with leaves divided to the base. Petals are clawed and mericarps ridged. They originate from Madeira.

G. maderense Yeo (Figure 3.1): This is the largest species reaching up to 1.5 m in height in flower, with leaves up to 60 cm across, each divided to the base into deeply divided lobes. The stem is over 50 cm tall and produces a rosette of leaves from the centre of which the tall flowering stem emerges bearing numerous purplish-pink to magenta flowers, up to 3.5 cm wide, in winter to early spring.

G. palmatum Cavanilles: This is also from Madeira and is not dissimilar but smaller with less divided leaves up to 35 cm across and slightly larger flowers, with stamens about twice as long as the sepals. This species flowers in summer.

Figure 3.1 Geranium maderense.

Subgenus *Erodioideae*

In the third subgenus, the plants are perennial. The fruit discharge is termed *Erodium type* where the awn remains attached to the mericarp, containing the seed, as it is released and becomes coiled but not plumed. The mericarp has a horny point and is covered with bristles on the sides. There are about 15 species in this subgenus.

Section Erodioideae

This section includes tall branching plants with nodding flowers and spreading or reflexed petals. The mericarp apex has three to five ridges.

G. phaeum Linnaeus: This is a perennial up to about 80 cm, flowering in late spring to early summer. Basal leaves are up to 20 cm wide, toothed but not deeply divided, persistent in winter. Flowers are rather flat, over 20 mm diameter, nodding in a one-sided inflorescence, with slightly reflexed, deep purple to pale purple to grayish lilac petals.

A native to southern and central Europe, this species is naturalised in many countries further to the North where it may be found in semi-shaded areas such as wood margins, or in meadows.

A number of cultivars may be found in cultivation which have been selected for their flower colour from white to very dark purple and at least one has variegated foliage.

Section Subacaulia

This includes mostly low growing alpine plants with silvery or gray green foliage. Flowers are erect and petals not reflexed. The mericarp apex has one to three ridges.

G. cinereum Cavanilles var. *cinereum*: This is a rosette forming perennial with a stout rootstock and gray-green basal leaves up to 5 cm across deeply divided into five to seven three-lobed divisions. It flowers in summer, the flowers being 2.5 cm across, petals notched, white or pink with bright pink netted veins.

This is a species from Central Pyrenees, which has resulted in a number of cultivars such as 'Ballerina' with gray leaves and purplish pink flowers.

G. cinereum Cavanilles var. *subcaulescens* (De Candolle) Knuth: This differs in the dark green leaves and deep purplish-red flowers with a black mark at the base of each petal. This variety is native to Turkey and the Balkan region.

G. × *lindavicum* Knuth: This hybrid, between *G. argenteum* and *G. cinereum*, is more vigorous than either parent, with more deeply divided silky leaves compared to those of *G. cinereum* and pale pink, net veined petals. A number of named cultivars are grown such as 'Apple Blossom' with silver-gray leaves and almost white, less conspicuously veined petals.

REFERENCES

Knuth, R. (1912) Geraniaceae, *Das Planzenreich*, 4, 129.

Yeo, P.F. (1984) Fruit discharge type in *Geranium* (Geraniaceae): its use in classification and its evolutionary implications. *Bot. J. Linn. Soc.*, 89, 1–36.

Yeo, P.F. (1985) *Hardy Geraniums*, Batsford, London (Reprinted below).

Yeo, P.F. (1992) *Hardy Geraniums* (2nd edition), Batsford, London.

4 Phytochemistry of the genus *Geranium*

Jeffrey B. Harborne and Christine A. Williams

INTRODUCTION

The phytochemistry of *Geranium*, a genus of some 300 temperate species (Mabberley, 1997) is reasonably well known today. According to Hegnauer's dictionary of plant chemistry (Hegnauer, 1966, 1989), at least 55 species have been investigated chemically. Furthermore, detailed studies have been carried out on at least three well known species: *G. macrorrhizum*, an essential oil plant; Herb Robert, *G. robertianum*, a European medicinal plant; and *G. thunbergii*, a Japanese medicinal plant.

The most characteristic single chemical in the genus is undoubtedly the ellagitannin geraniin, named as such following its crystallisation from *Geranium thunbergii*. This substance (see below) is universally present in the leaves of *Geranium* species. Geraniin, unlike almost all other hydrolysable tannins, fails to produce an astringent taste when leaf extracts of *G. thunbergii* are taken orally (Okuda *et al.*, 1992). This does not mean that other mammals are not affected by the taste properties of this major chemical. *Geranium* plants are well known to be rabbit-proof and one suspects the reason for this is the high content of leaf tannin.

Another natural compound which presumably takes its name from the genus, is germacrone. This is a sesquiterpene isolated in quantity from the essential oil of *G. macrorrhizum*. This plant oil also yields the monoterpene, geraniol, which is named after 'geraniol oil', which is actually derived from plants of the closely related genus *Pelargonium* (see Chapter 11). Geranium oil itself is so widespread in nature in the essential oil of hundreds of plants, that it cannot be regarded as characteristic of *Geranium* or *Pelargonium*. Furthermore, the derived monoterpene radical, geranyl, occurs attached to a variety of other natural products, including a number of flavonoids and coumarins.

In fact, there is little evidence available at present to suggest that either germacrone or geraniol are widely distributed in *Geranium*. It is the flavonoid constituents and particularly two classes of phenolic tannins that have been most widely investigated. The present account describes the whole range of plant polyphenols that have been encountered in the aerial parts and roots of these plants.

ESSENTIAL OILS

Aerial parts of *G. macrorrhizum* have yielded an essential oil containing two well known monoterpenoids, geraniol (1) and β-citronellol (2), together with several sesquiterpenes. These include germacrone (3), which makes up to 50 per cent of the

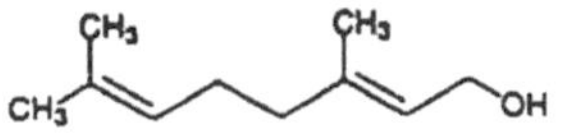

(1) Geraniol

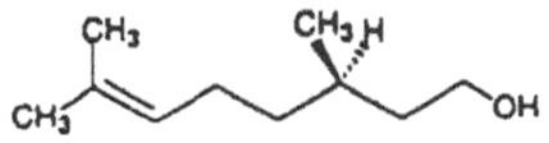

(2) α-Citronellol

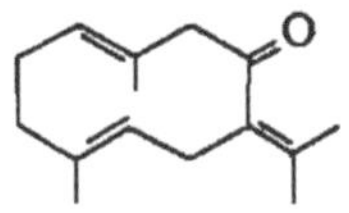

(3) Germacrone

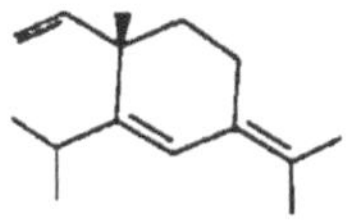

(4) α-Elemene

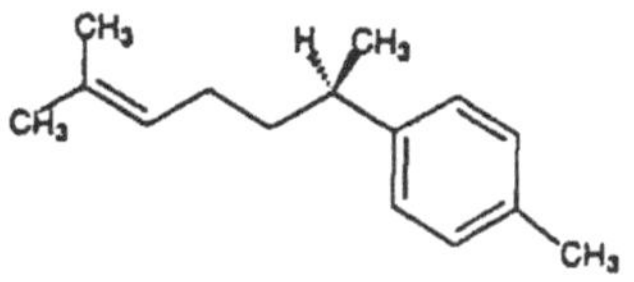

(5) α-Curcumene

Figure 4.1 Essential oil components of *Geranium*.

oils, α-elemene (4) and α-curcumene (5) (Ognyanov *et al.*, 1958) (Figure 4.1). Little has been reported about the essential oils of any other *Geranium* species.

MONOMERIC FLAVONOIDS

Leaf flavonoids

The leaf flavonoids of *Geranium* are typical of the Geraniaceae and related dicotyledonous families. They are predominantly flavonols and the commonly occurring quercetin (6) is universally present. In a survey of acid-hydrolysed leaf tissue of 78 species, Bate-Smith (1972) reported that quercetin is generally accompanied by the lower homologue kaempferol (7) (in 93 per cent of the sample) and by the higher homologue myricetin (8) (Figure 4.2) (in 13 per cent of the sample). Variation in this basic flavonol pattern is to some extent correlated with the geography of the genus. A primitive pattern, including myricetin, predominates in plants from the central Eurasian area, while an advanced pattern, represented by high concentrations of kaempferol, is present in Mediterranean and American species.

Although it is apparent that the above three flavonols occur in *Geranium* in glycosidic combination, relatively little is known of the glycosidic pattern of most species. However, there is one report of an high performance liquid chromatography (HPLC) survey of *Geranium* leaves by Okuda *et al.* (1980). These authors found that quercetin occurs regularly in the genus as the 3-galactoside, called hyperin (9) Figure 4.2. This was detected in direct alcoholic leaf extracts of 12 out of 15 species surveyed (Table 4.1). The content of hyperin varied from 0.03 to 1.6 per cent dry weight, with an average value of 0.43 per cent.

One species apparently lacking in quercetin 3-galactoside is the Japanese *G. thunbergii* (Table 4.1). Instead, leaves of this plant contain either kaempferol 3-rhamnoside

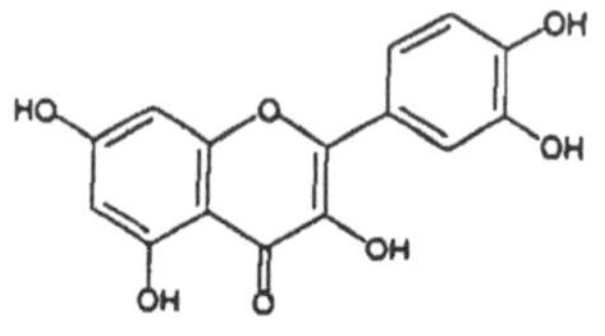

(**6**) Quercetin

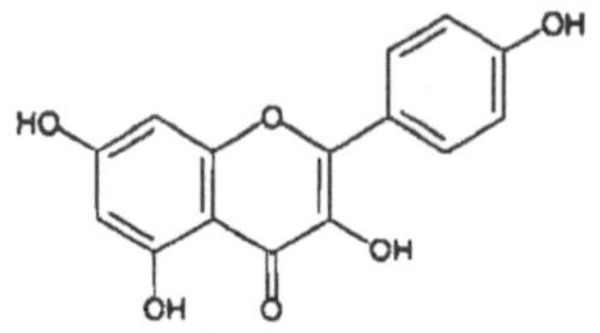

(**7**) Kaempferol

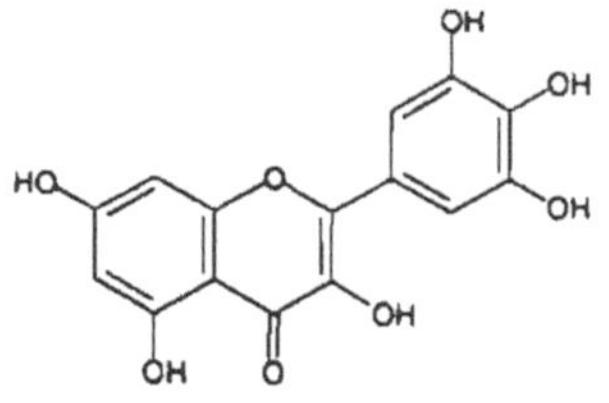

(**8**) Myricetin

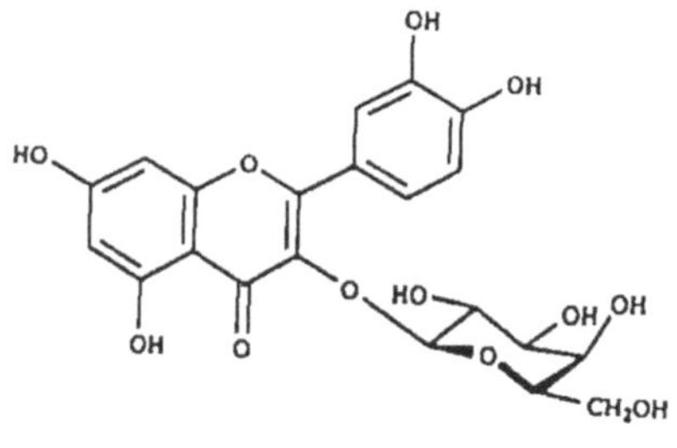

(**9**) Hyperin

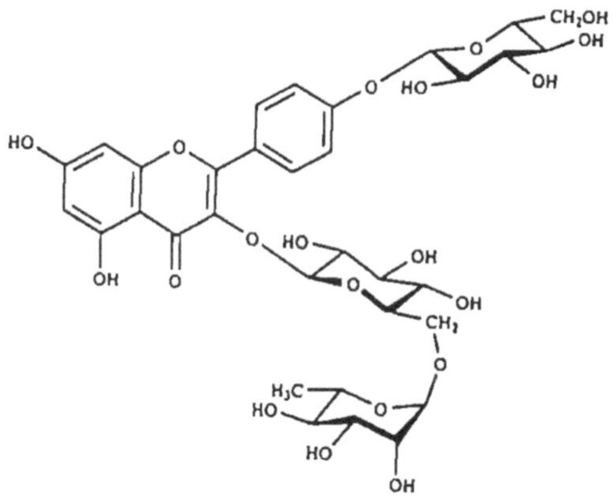

(**10**) Kaempferol 3-rutinoside 4′-glucoside

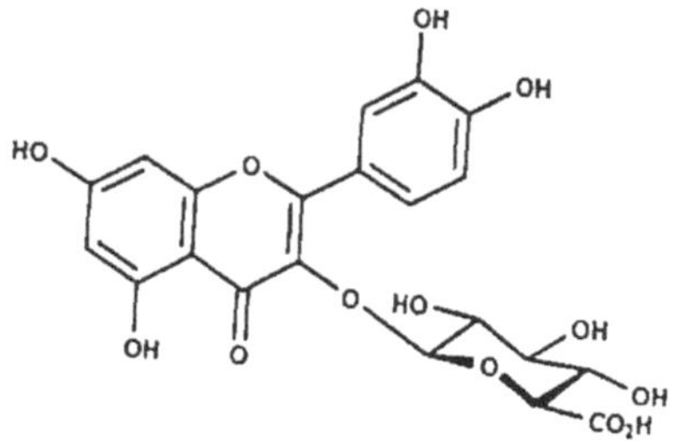

(**11**) Quercetin 3-glucuronide

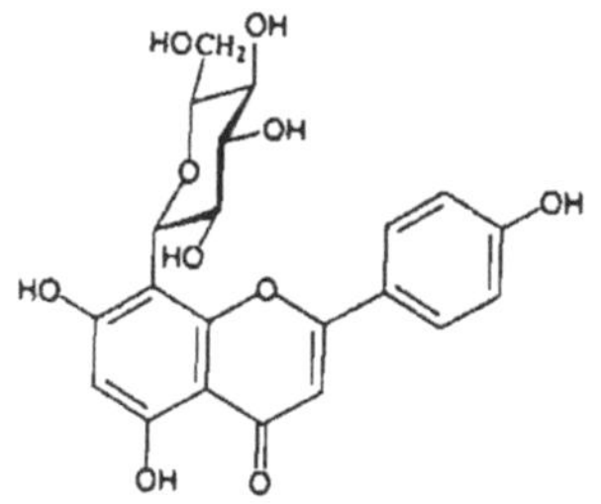

(**12**) Vitexin

Figure 4.2 Leaf flavonoids of *Geranium*.

or a mixture of kaempferol 3-arabinoside-7-rhamnoside and kaempferol 3,7-dirhamnoside (Kawamura *et al.*, 1995). There are thus two chemical races in flavonoid content, but the occurrences of the two races do not correspond with any other variable features of this plant species.

Table 4.1 Geraniin and hyperiin content of the dry leaves of *Geranium* species

Species	*Geranin (%)*	*Hyperin (%)*	*Month of collection*
G. eriostemon Fisch. var *reinii* Maxim	7.5	0.15	July
G. erianthum DC.	7.6	0.13	August
G. soboliferum Komar.	6.8	0.16	October
G. krameri Franch. et Savat.	6.8	0.19	October
G. yoshinoi Makino	9.8	0.55	September
G. yesoense Franch. et Savat.	12	0.18	August
G. yesoense Franch. et Savat. var. *nipponicum* Nakai	12	0.09	October
G. shikokianum Matsum.	6.0	0.59	August
G. sibiricum L. var. *glabrius* Ohwi	8.1	–	October
G. thunbergii Sieb. et Zucc.	12	–	August
G. wilfordii Maxim.	9.5	0.21	September
G. wilfordii Maxim. var. *hastatum* Hara*	0.50	0.03	September
G. tripartitum R. Knuth	12	1.3	September
G. robertianum L.	9.8	–	September
G. carolianum L.	11	1.6	May

Note
* Fresh aerial tissue.

Herb Robert, *G. robertianum*, a plant of medicinal interest, has been examined in some detail for its flavonol glycosides. Aerial parts yield a mixture of six monoglucosides: kaempferol and quercetin 3-galactoside, quercetin 3-glucoside, quercetin 4′-glucoside, quercetin 7-glucoside and quercetin 7-rhamnoside. Accompanying these monoglycosides are seven 3-diglycosides. Only four of the seven were fully characterised as the 3-rutinosides and 3-rhamnosylgalactosides of kaempferol and quercetin (Kartnig and Bucar-Stachel, 1991). Whether Herb Robert varies in its flavonol glycoside content is not yet clear, but it may be noted that Okuda *et al.* (1980) failed to find the quercetin 3-galactoside reported by Kartnig and Bucar-Stachel, 1991) in their particular sample (Table 4.1).

A further four flavonol glycosides, not described so far, have been characterised variously in five *Geranium* species native to Egypt (Table 4.2). The most distinctive is kaempferol 3-rutinoside-4′-glucoside (10), recorded in *G. yemense* and *G. rotundifolium*. The presence of quercetin 3-glucuronide (11) in *G. dissectum* is noteworthy (Saleh *et al.*, 1987) Figure 4.2. The apparent absence of quercetin 3-galactoside from these five Egyptian species should also be noted.

Table 4.2 Flavonol glycosides of five Egyptian species of *Geranium*

Species	*Major flavonols present*
G. yemense	Kaempferol 3-rutinoside-4′-glucoside
G. rotundifolium	Kaempferol 3-rutinoside-4′-glucoside
	Kaempferol 3-rhamnoside
G. trilophum	Quercetin 3-glucoside
G. mascatense	Myricetin 3-rutinoside
G. dissectum	Quercetin 3-glucuronide
	Quercetin 3-glucoside
	Quercetin 3,7-diglucoside

Source: Data from Saleh *et al.* (1983).

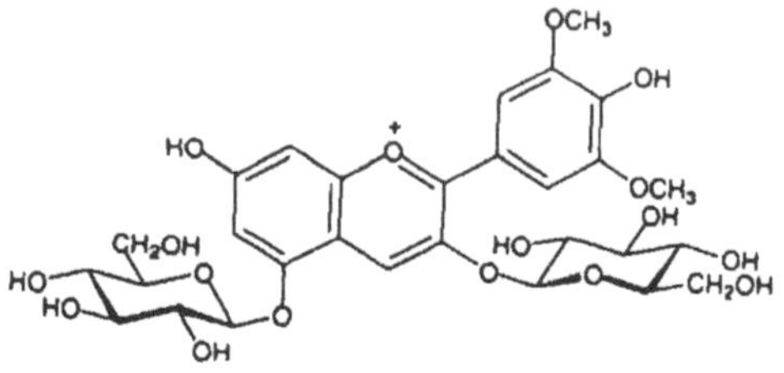

(**13**) Malvidin 3,5-diglucoside

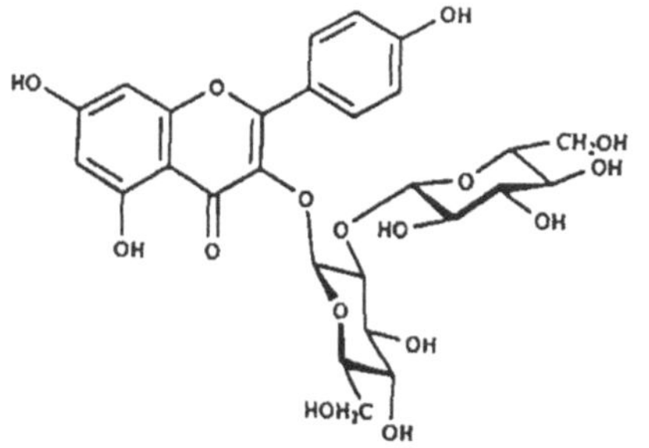

(**14**) Kaempferol 3-sophoroside

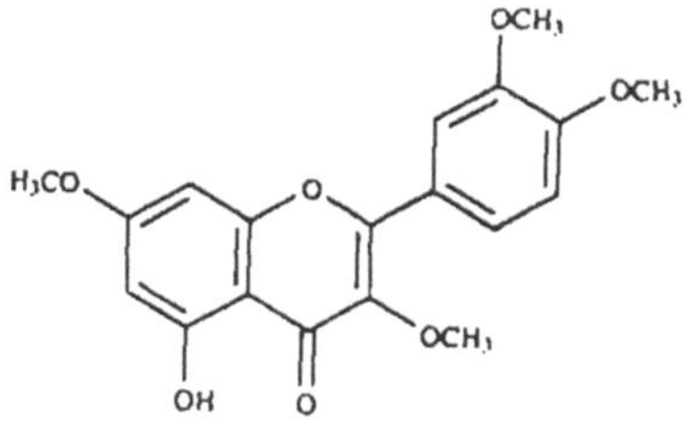

(**15**) Quercetin 3,7,3′,4′-tetramethyl ether

Figure 4.3 Floral and exudate flavonoids of *Geranium*.

Other classes of flavonoid such as glycosylflavones are also present in *Geranium* according to Bate-Smith (1977) but they have not in general been investigated further. They have been found to dominate in the case of *Geranium phaeum*. Here, the flavonol glycosides based on quercetin are minor components, compared to the five glycosyl flavones: vitexin (12) (Figure 4.2), isovitexin, orientin, iso-orientin and vicenin (Boutard and Lebreton, 1975).

Floral flavonoids

Most *Geranium* species have attractive flowers, with colours ranging from blue, purple and red to pink and white. Anthocyanins, together with co-occurring flavonol glycosides are presumably responsible for those flower colours, but surprising little work has been carried out on these pigments in the genus. A major study has, however, been devoted to the bluish-purple or bluish-magenta flowers in *G. pratense* and *G. sanguinea* and in the cultivar 'Johnsons Blue', a hybrid derived from *G. himalayense* × *G. pratense*. The same major anthocyanin is present in the flower of all three plants. It is malvidin 3,5-diglucoside (13), with a labile acetyl substituent at the 6-position of the glucose residue attached to the 5-hydroxyl. Thus, it is malvidin 3-glucoside-5-(6-acetylglucoside) (Markham *et al.*, 1997).

In the petals of two of these plant species, the anthocyanin co-occurs with four flavonol glycosides, namely the 3-glucosides and 3-sophorosides of kaempferol and myricetin. Colour tests *in vitro* indicate that kaempferol 3-sophoroside (14) is the most important co-pigment, shifting the mauve colour of the malvidin glycoside (13) Figure 4.3 towards the blue region. An additional and unusual feature of flower colour production in these

Table 4.3 Exudate flavonol methyl ethers of *G. macrorrhizum* and *G. lucidum*

Kaempferol	*Quercetin*	*Myricetin*
3-methyl ether	–	–
4′-methyl ether*	–	–
3,7-dimethyl ether	3,7-dimethyl ether†	–
3,4′-dimethyl ether	3,3′-dimethyl ether	–
7,4′-dimethyl ether	7,3′-dimethyl ether	–
3,7,4′-trimethyl ether	3,7,3′-trimethyl ether	7,3′,4′-trimethyl ether
	7,3′,4′-trimethyl ether	
	3,7,3′,4′-tetramethyl ether	3,7,3′,4′-tetramethyl ether

Notes
*Lacking in *G. macrorrhizum*;
† lacking in *G. lucidum*; otherwise all compounds present in both species.

Geranium petals is the presence of a cell sap pH of between 6.6 and 6.8, instead of the more usual pH at 5.6. This appears to be a very rare feature in nature but is important for the full colour intensity observed in these flowers (Markham *et al.*, 1997).

Exudate flavonoids

Geranium species regularly have glandular hairs or trichomes on the upper leaf surface. The chemical constituents of these trichomes can be examined separately from the internal leaf components by brief rinsing of leaf surfaces in a solvent such as acetone. Besides the terpenoids and hydrocarbons that are commonly present at the surface, leaf washes occasionally yield mixtures of lipophilic flavonoids, usually flavonol methyl ethers. Such compounds have been obtained from leaf surfaces of *G. macrorrhizum* and *G. lucidum*. The structures present are almost identical in both plants and consist of some 14 kaempferol, quercetin or myricetin methyl ethers (Table 4.3) (Ivancheva and Wollenweber, 1989).

The above report supercedes an earlier paper by Ognyanov and Ivantcheva (1972) in which a so-called novel flavonol, 3,5,7,2′,4′,6′-hexahydroxyflavone and kaempferol 3-methyl ether were reported at the surface of *G. macrorrhizum*. Re-examination of the evidence for the 'new' flavonol suggests that this was a mistaken identification of a known flavonol. It may also be noted that quercetin 3,7,3′,4′-tetramethyl ether (15) was independently identified as a major lipophilic constituent of *G. macrorrhizum* by Nakashima *et al.* (1973). This compound (15) Figure 4.3 crystallised out in 0.4 per cent yield from the essential oil of this plant.

TANNINS

Hydrolysable tannins

The major hydrolysable tannin of *Geranium* is the compound geraniin (16) Figure 4.4, first crystallised from leaf extracts of *G. thunbergii*. This plant has been used extensively in folk medicine in Japan. A boiling water extract of *G. thunbergii* has been taken by numerous people over many years as an antidiarrhetic and for controlling intestinal function (Okuda *et al.*, 1992). Geraniin makes up more than

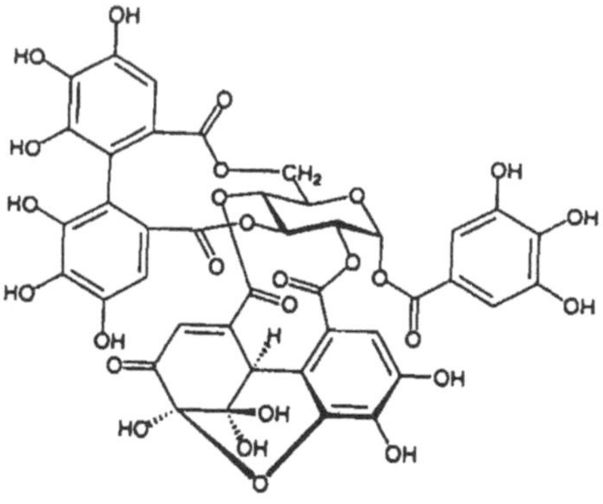

(**16**) Geraniin

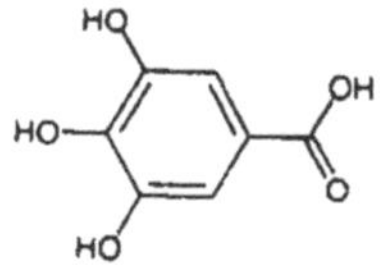

(**17**) Gallic acid

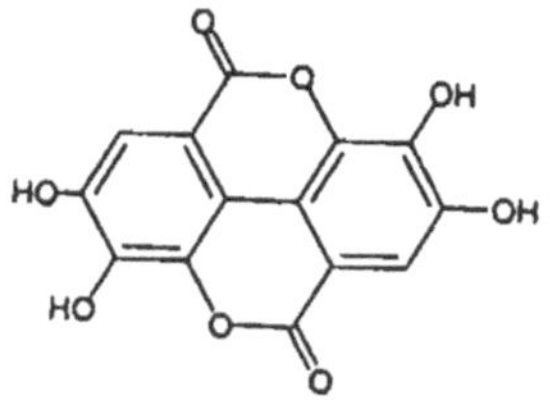

(**18**) Ellagic acid

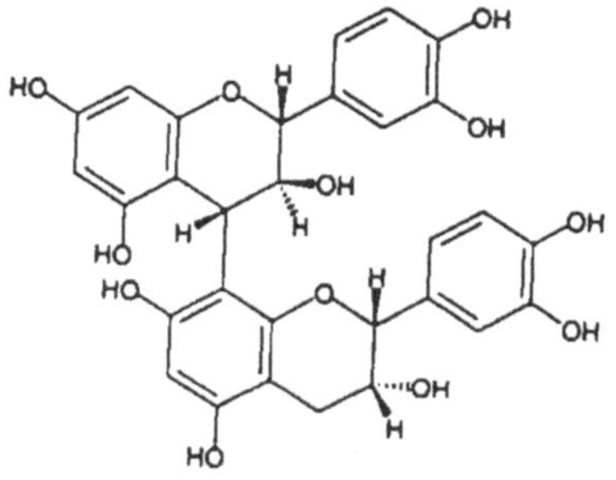

(**19**) Procyanidin

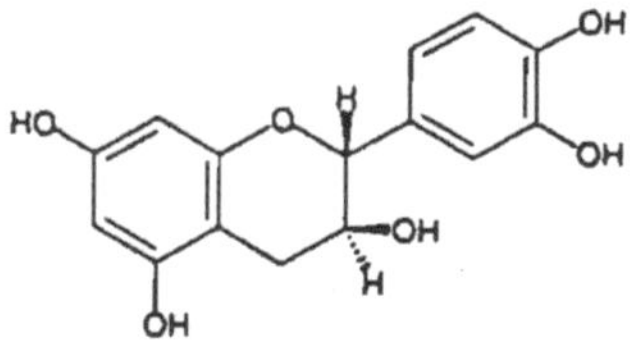

(**20**) (+)-Catechin

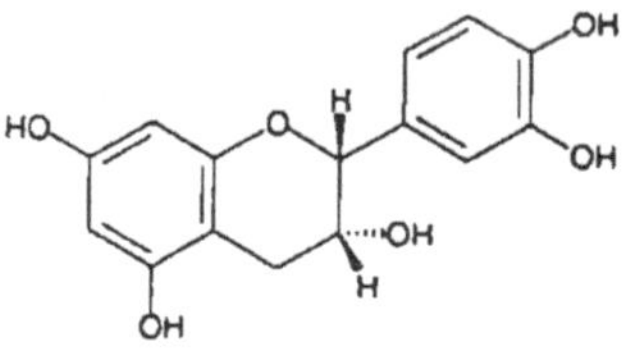

(**21**) (−)-Epicatechin

Figure 4.4 Tannins and their precursors in *Geranium*.

10 per cent of the weight of the dried leaf. It forms yellow crystals and, remarkably for an ellagitannin, completely lacks the astringency normally associated with plant tannins.

In its chemical structure, geraniin (16) is based on a molecule of glucose, which is disubstituted in the 3,6- and 2,4-positions by two hexahydroxygallic acid residues. Additionally, there is a galloyl ester group linked at C-1 of the sugar. Biosynthetically, geraniin is derived from gallic acid (17) via pentagalloylglucose as an intermediate (Haslam, 1989).

Geraniin would appear to be the characteristic hydrolysable tannin of the genus *Geranium*, since it has been detected by HPLC in every one of the 15 species surveyed (Okuda *et al.*, 1980) (Table 4.2). The richest source is *G. thunbergii*, with over 12 per cent of its dry weight made up by geraniin. Other species range from 0.5 per cent up to

12 per cent with an average of about 10 per cent dry weight. By comparison with the leaves, the stems of these plants have only 1–2 per cent dry weight.

In an earlier survey based on a colour test for ellagitannin developed with nitrous acid in the absence of oxygen, Bate-Smith (1972) found that some 30 species, representing 16 sections of the genus, contain ellagitannin (presumably geraniin) in amounts ranging from 1.3 to 20 per cent dry weight. It seems likely that most, if not all, known *Geranium* species contain geraniin or an ellagitannin of similar structure. The same compound, incidentally extends its distribution to other members of the Geraniaceae and to other families in the order Geraniales. It has been detected, for example, in 28 plant species of the Euphorbiaceae. It also occurs in the cocaine-containing *Erythroxylon coca* (Erythroxylaceae). Geraniin is however apparently absent from the closely related genus, *Pelargonium*, in spite of the fact that ellagitannins are also abundantly present in *Pelargonium* species (Okuda *et al.*, 1980) (see Chapter 11).

Co-occurring with geraniin in the leaves of *Geranium* species is the related structure, ellagic acid (18). Bate-Smith (1962) records ellagic acid in the leaves of 4 out of 6 surveyed, in *G. meeboldii*, *G. phaeum*, *G. robertianum* and *G. sylvaticum*. A richer source of ellagic acid, is however the plant root and rhizome. Here, it has been recorded in some 61 species (Hegnauer, 1966). Moreover, so much is present in the roots that ellagic acid can be isolated in crystalline form. Gallic acid (3,4,5-trihydroxybenzoic), which is a presumed precursor of ellagic, has also been recorded regularly in the leaf (Bate-Smith, 1962) and the root (Hegnauer, 1966). Gallic acid has been isolated, for example, from roots of *G. maculatum*, *G. nepalensis* and *G. pratense*.

Condensed tannins

Plants of *Geranium* contain both hydrolysable and condensed tannin, but the distribution in the different organs varies considerably. The main occurrence of condensed tannin (or proanthocyanidin) is in the root stock, according to Bate-Smith (1973). There is apparently a suppression of this chemical character in the leaves, where ellagitannins dominate (see above). Only a handful of 60 species of *Geranium* surveyed have significant amounts of proanthocyanidins in the leaves. These are: *G. polyanthes* (Eurasia), *G. platypetalum* (Armenia), *G. renardii* (Caucasus), *G. sinense* (China), *G. incanum* (South Africa) and *G. lindenianum* (Venezuela).

The proanthocyanidins in *Geranium* are based on either procyanidin or prodelphinidin or a mixture of the two. No detailed chemistry has yet been carried out on the condensed tannins of these plants. However, it is likely that the procyanidins are of a common type (e.g. (19)), since the two procyanidin precursors, (+)-catechin (20) and (−)-epicatechin (21) Figure 4.4 have been detected in roots of *G. pratense* and *G. palustre* (Hegnauer, 1966).

The content of procyanidin in fresh rhizomes, as compared to the amount of ellagitannin, has been shown to be about the same in *G. sylvaticum*. By contrast, there is only one-seventh the amount of procyanidin, compared to six-sevenths ellagitannin, in *G. pratense* (Hegnauer, 1966). The high content of tannins in the roots has meant that *Geranium* species have been employed in the past as good sources of tanning material for the leather industry. At least two species, *G. nepalense* and *G. wallichianum* have been used in this way.

MISCELLANEOUS CONSTITUENTS

Aerial parts of *G. richardsonii* and *G. viscosissimum* characteristically accumulate the organic acid, tartaric acid. This acid occurs regularly in members of the Geraniaceae (Stafford, 1961). However, it is not always present in every *Geranium* species. Thus, *G. robertianum* and *G. sanguineum*, when analysed, showed the presence of malic and citric acids, but tartaric acid was missing (Kinzel, 1964).

CONCLUSION

The chemistry of *Geranium*, as is clear from the above summary, is dominated by phenolic constituents. Not only are these two classes of plant tannin – proanthocyanidin and ellagitannin – widely distributed in both aerial and underground tissues. But also there are a wealth of monomeric flavonoids, variously present in the leaf and petal. We are only just beginning to appreciate the chemical complexity bound up in the phenolic fraction of *Geranium* plants and much further work is required to establish the precise range of structures that are present in any given species.

Since both the monomeric flavonoids (Rice-Evans, 2000) and the various tannins (Haslam, 1989) have long been considered to be active components of many medicinal plants, it is likely that the useful properties in terms of human medicine associated with *Geranium* plants may be due to the type and quantity of particular phenolics present. However, there is no doubt that much further biological experimentation is required before we can adequately explain the curative properties of these plants.

REFERENCES

Bate-Smith, E.C. (1962) Phenolic constituents of plants and their taxonomic significance. I Dicotyledons. *J. Linn. Soc. (Bot.)*, 58, 39–54.

Bate-Smith, E.C. (1972) Ellagitannin content of leaves of *Geranium* species. *Phytochemistry*, 11, 1755–1757.

Bate-Smith, E.C. (1977) Chemotaxonomy of *Geranium*. *Bot. J. Linn. Soc.*, 67, 347–359.

Boutard, B. and Lebreton, P. (1975) The presence of C-glycoflavones in *Geranium phaeum*. *Plantes Med. Phytotherapi.*, 9, 289–296.

Haslam, E. (1989) *Plant Polyphenols. Vegetable Tannins Revisited*, 230pp., Cambridge University Press.

Hegnauer, R. (1966) *Chemotaxonomie der Pflanzen*, vol. 4, pp. 195–197, Birkhauser Verlag, Basel.

Hegnauer, R. (1989) *Chemotaxonomie der Pflanzen*, vol. 8, pp. 511–516, Birkhauser Verlag, Basel.

Ivancheva, S. and Wollenweber, E. (1989) Leaf exudate flavonoids in *Geranium macrorhizum* and *G. lucidum*. *Indian Drugs*, 27, 167–168.

Kartnig, T. and Bucar-Stachel, J. (1991) Flavonoide aus den oberirdischen Teilen von *Geranium robertianum*. *Planta Med.*, 57, 292–293.

Kawamura, T., Hisata, Y., Noro, Y., Nishibe, S., Sakai, E. and Tanaka, T. (1995) *Polyphenols*, 94, *Palma de Mallorca*, INRA, Paris, pp. 301–302.

Kinzel, H. (1964) Organic acids in the leaves of some plants. *Ber. Deut. Botan. Ges.*, 77, 14–21.

Mabberley, D.J. (1997) *The Plant Book*, Cambridge University Press.

Markham, K.R., Mitchell, K.A. and Boase, M.R. (1997) Malvidin 3-glucoside-5-(6-acetylglucoside) and its colour in Geranium 'Johnson's Blue' and other 'Blue' Geraniums. *Phytochemistry*, 45, 417–423.

Nakashima, R., Yoshikawa, M. and Matsuura, T. (1973) Quercetin 3,7,3′,4′-tetramethyl ether from *Geranium macrorrhizum*. *Phytochemistry*, 12, 1502.

Ognyanov, I.V. and Ivantcheva, S. (1972) A new hexahydroxyflavone and isokaempferide in *Geranium macorrhizum*. *Dokl. Bulg. Akad. Nauk.*, 25, 1057–1059.

Ognyanov, I., Ivanov, D., Herout, V., Hovak, M., Pliva, J. and Sorm, F. (1958) Structure of germacrone. *Chem. Listy*, 52, 1163–1173.

Okuda, T., Mori, K. and Hatano, T. (1980) The distribution of geraniin and mallotusinic acid in the order Geraniales. *Phytochemistry*, 19, 547–551.

Okuda, T., Yoshida, T. and Hatano, T. (1992) Pharmacologically active tannins from medicinal plants. In: *Plant Polyphenolsi* (Hemingway, R.W. and Laks, P.E., eds), pp. 539–569, Plenum Press, New York.

Rice-Evans, C. (ed.) (2000) Wake up to Flavonoids, 74 pp., Royal Society of Medicine Press, London.

Saleh, N.A.M., El-Karemy, Z.A., Mancour, R.M. and Fayed, A.A. (1987) A chemosystematic study of some *Geraniaceae*, *Phytochemistry*, 22, 2501–2505.

Stafford, H.A. (1961) Distribution of tartaric acid in the *Geraniaceae*. *Amer. J. Bot.*, 48, 699–701.

5 Cultivation and harvesting of *Geranium macrorrhizum* and *Geranium sanguineum* for medicinal use in Bulgaria

Tatyana Stoeva

INTRODUCTION

Species belonging to the genus *Geranium* are known as hardy geraniums and a number of them are broadly used as ornamental, medicinal and melliferous plants. The most economically important for their healing and/or aromatic properties are *G. macrorrhizum* L. and *G. sanguineum* L., which are cultivated in Bulgaria.

GERANIUM MACRORRHIZUM L. (SYN. BIGROOTED *GERANIUM*, LONGROOTED CRANESBILL)

Introduction

G. macrorrhizum is a perennial herbaceous plant with a long, nodulous, horizontal, superficial over- and under-ground rhizome, covered with dry brown scales. Basal leaves are 4–10 cm wide, fragrant, divided into 5–7 lobes on 25 cm long petioles growing in clumps; stem leaves are opposite, on short petioles or cauline. Stems are 15–50 cm high, erect, straight with magenta flowers on top. All leaves and stems are covered with simple and glandular hairs (Webb and Ferguson, 1968; Petrova and Kozhuharov, 1979; Assenov *et al.*, 1998). The species is distributed in temperate zones of Europe: Southern Alps, Apennines, south and central Carpathians and Balkan Peninsula. The plant has become naturalized in Central and Northern Europa, where it is grown for ornamental or medicinal purposes and for the feeding of bees (Hegi, 1924; Webb and Ferguson, 1968; Petrova and Kozhuharov, 1979).

The plant is highly valued in Bulgaria for its ornamental, aromatic, melliferous and variety of healing properties which makes it a garden favourite. It is an emblematic plant in the Bulgarian folk tradition: a symbol of health, wealth and longevity: it's folk name *Zdravetz* means 'health' (Ivancheva, 1998). *G. macrorrhizum* is the source for production of the Bulgarian geranium essential oil (EO) called 'Zdravetz oil', which is distilled from the overground parts of the plant and possesses a very pleasant odour reminiscent of clary sage, orris, and particularly of rose (Guenther, 1950; Gildemeister and Hoffman, 1959).

Cultivation

The first experiments on the industrial cultivation of *G. macrorrhizum* in Bulgaria were carried out during 1924–1931 at the State Experimental Field for Roses, Essential oil and Medicinal Plants in the town of Kazanlik (Rose Valley). Rhizomes of wild geranium were transfered to the field conditions of the Rose Valley but survived for 3 years only, because of the hot summers and low air and soil humidity in this area (Georgieff, 1933). The abundance of natural high qualitative resources for the plant all around the southern slopes of the Central Balkan Mt., as well as the northern ones of the Sredna Gora Mt. which were in the immediate vicinity of the distillers in the Rose Valley, made the oil production from wild sources more profitable than those from cultivated plants (Georgieff, 1933; Irinchev, 1956). Recently a large part of the commercially valuable populations of *G. macrorrhizum* in the region, were protected within the 'Central Balkan' National Park, which forced the development of primary technology for large-scale cultivation for EO production. Experiments were carried out at the Institute of Botany in Sofia and various soil-types were assessed under different environmental conditions.

Climatic and soil requirements

In the wild, the species inhabits humid, shady, shrubby and rocky places, among rock fragments where soil is scanty, preferentially on steep slopes facing north, generally on limestone. The species is sciophylic and cold-resistant and takes part in the herbaceous canopy of forestal coenoses in mountains and lower mountain slopes from 300 to 2500 m (Webb and Ferguson, 1968; Petrova and Kozhuharov, 1979; Genova, 1995; Assenov *et al.*, 1998).

Choice of the correct place for plantation is of great importance. The species has no special requirements as to soil-type but it prefers semi-permeable to well-drained soils. It needs high humidity and its cultivation on open and sunny places have to be compensated by irrigation, spraying or a showering system. This could be of great advantage in the utilization of terrains in mountain regions which are unsuitable for main crops (near the forests, slopes), and out of crop rotation. As an intercrop in orchard gardens it could be used as a melliferous plant as well. The crop performs best in areas with an average daily temperature of 18–22 °C (Yankulov *et al.*, 1993).

Propagation

Plant multiplication in nature occurs both vegetatively, by rhizome sprouts (suckers), and by seeds. Plants form large clumps and a common garden practice is to divide them into single main rhizomes (Genova, 1995). Despite the almost unlimited rooting capacity of the plant and possibility to produce single-node cuttings, conventional vegetative propagation using the overground part of rhizomes from the previous crop, still remains the only method used at the horticultural level.

Planting and aftercare

Preliminary soil-cultivation: deep ploughing, 30–35 cm deep, after a previous crop (cereals, earthed-up crops), harrowing and cultivation. Suckers (rhizome tips with 3–6 leaves) 10–15 cm long are planted at 4–6 cm depth in rows and covered with soil (this could be performed mechanically). The recommended spacing is 50 cm between rows and 30 cm between plants. The rate of planting material is 1500–2200 kg/ha, which yields 66 000–70 000 plants per hectare. The best time to plant is late autumn from

15th October to 15th November, which ensures sufficient soil humidity, 90 per cent rooting occurs and plants spend the winter successfully (Yankulov *et al.*, 1993).

Aftercare during the first and second year includes: 3–5 times of earthing up (weeding) and abundant watering (7–8 times). From the third year, plants form wide clumps merged into continuous compact rows and suppress weeds. Irrigation remains to be the main aftercare during the further exploitation of the plantation, which is up to 7–8 years (Yankulov *et al.*, 1993).

Fertilization

G. macrorrhizum is responsive to fertilization. Incorporation of 600–800 kg/ha of superphosphate and 150–200 kg/ha K_2O at the time of ploughing and land preparation or mixing half of these amounts with 30–40 t per hectare of cattle manure is the usual method. In the first spring, when rows are well outlined by developed plants, 200–300 kg/ha of N fertilizer has to be applied with the first earthing up. Annually, in the spring, plants require full mineral fertilization (Yankulov *et al.*, 1993).

Harvesting and yields

Harvesting begins in the second year after planting and a regular yield could be obtained after the third. The above-ground mass is harvested twice a year: before flowering (end of May to the beginning of June) and at the end of vegetation in late autumn (end of September to the beginning of October) when accumulation of phytomass is highest. The EO content varies from 0.02 to 0.13 per cent depending on the plant's origin and the vegetative stage.

Different Bulgarian populations were studied for their EO productivity and a high oil-yielding selection (from Rila Mt.) was used for estimating the possible harvest of EO from the species under conditions of large-scale cultivation; however, breeding has still not been carried out. Herbage has to be cut (reaped) with a sickle 4–5 cm above the overground rhizome, so as to ensure new growth. Fresh herbage is used for distillation and it withstands pressing, wilting but should not be stored for more than 3 days. After the first harvest, the plantation needs irrigation and earthing.

The annual yield (two harvests) of phytomass varies from 4000 to 6000 kg/ha, which at an average EO content of 0.05 per cent could yield 2–3 kg/ha of EO (Yankulov *et al.*, 1993).

Use of *G. macrorrhizum*

Phytotherapy of some Balkan countries makes use of both herbage and rhizomes of *G. macrorrhizum* mainly for their hypotensive, sedative, astringent, cardiotonic and antiatheromatous effects. This biological activity is due to: EO with about 0.1 per cent of the main compound, the sesquiterpenic ketone germacrone, which makes up to 50 per cent of the EO; there are also tannins, flavonoids, phenolic acids and waxes in the herbage and tannins (about 16 per cent), flavonoids and phenolic acids in the rhizomes (Mihailov and Tucakov, 1974; Petkov, 1982; Assenov *et al.*, 1998). Phytotherapy prescriptions involve the water-soluble extract (cool extraction of dry rhizomes at a ratio of 1:10 for 8 h) per os: 400 ml daily split into four doses, as well as externally as compresses (Petkov, 1982).

Pharmacological effect

The extracts and aromatic products of *G. macrorrhizum* have been studied in detail in Bulgaria. The main pharmacological effects of the two herbal drugs was proved experimentally to be due to their water-soluble, ethanolic and methanolic extracts (Manolov *et al.*, 1977; Petkov, 1982). Isolated flavonoid fractions possess hypotensive and spasmolytic activity as well (Ivancheva *et al.*, 1976). Different fractions of the rhizome methanolic extract demonstrate sedative and moderate central myorelaxing effects (Manolov *et al.*, 1977; Petkov, 1982). After the EO extraction, a polymeric phlobaphen complex was obtained from the herbage as a durable (stable) product which clinical tests showed it to have a hypotensive effect (Petkov, 1982). The antimicrobial activity of polyphenols was also reported (Ivancheva *et al.*, 1992).

GERANIUM SANGUINEUM L. (BLOODY CRANESBILL)

Introduction

Geranium sanguineum is a perennial plant with stout, horizontal rhizomes. Stems are diffused, branched, erect to decumbent, with long white, patent hairs and sessile glands. Leaves are mostly cauline, 3–5 cm wide, divided into 5–7 lobes, each with 1–3 linear, oblong, acute segments on each side. Flowers are single on a 7–15 cm peduncle and stick out from nodes (Webb and Ferguson, 1968; Petrova and Kozhuharov, 1979; Assenov *et al.*, 1998). The species is distributed in most of Europe and the Caucasus, southwards from *c.* 60°N (Hegi, 1924; Webb and Ferguson, 1968; Petrova and Kozhuharov, 1979).

The species is a valuable ornamental, but in Bulgaria it is valued mostly for its medicinal properties. Because of the red colour of its rhizomes, the plant was considered to be a blood-strengthener and in folk medicine, the water-soluble extracts of rhizomes are prescribed as a remedy for malignant illness of the blood-generated organs (Assenov *et al.*, 1998). Some aspects of large-scale cultivation, for production of both rhizomes and herbage, were examined in the field, in the Sofia Plain, where plant samples of wild origin were introduced and propagated for the field studies.

Cultivation

Species occur on rocky or sandy ground or on well-drained soils, in herbaceous places, among shrubs and sparse woods in lowlands, mountains and base of mountains up to 1700 m (Webb and Ferguson, 1968; Petrova and Kozhuharov, 1979; Assenov *et al.*, 1998).

The species have no special requirements of soil under conditions of cultivation. The crop is sensitive to waterlogging, so terrains for plantations have to be well-drained and sunny. It grows well on sandy to clayey and alluvial-meadow soils as well.

Propagation

In nature, plant propagation occurs both by seeds and vegetatively by rhizome sprouts (suckers). As an ornamental, which could be kept at the same place for years, the plant forms a large round clump (up to 50–60 cm in diameter and up to 50 cm in height) which is due both to new suckers and seedlings growing up inside the clump. Seedlings were used in the studies, produced under green-house conditions from the end of January/early February to mid-April.

Planting and aftercare

Normal preliminary soil cultivation and fertilization conditions for *G. macrorrhizum* were applied. Seedlings were planted in rows spaced 70 cm apart and with a 30 cm distance between individuals and irrigation was supplied. A 1 ha plot was covered by 50 000 seedlings. The plantation needs 1–2 times of irrigation to be applied and weed control (regular earthing up) during the growth period.

Harvesting and yields

There are two approaches to the utilization of plantations depending on which of the herbal drug is to be produced: herbage or roots and rhizomes.

Roots and rhizomes: The crop is ready for harvest in the third year after the end of vegetation. Yield of fresh rhizomes reaches 20 000 kg/ha (4000 kg/ha dry rhizomes). During these years, overground parts could be used for seed production.

Herbage: Experimental data shows that on the third year of cultivation, the yield of fresh above-ground mass of 12 000 kg/ha (300–350 kg/ha dry weight) could be achieved.

Possibilities for simultaneous utilization (rotation) of the plantation for production of both sources and duration of exploitation have to be studied.

Use of *G. sanguineum*

Bulgarian phytotherapists apply Rhizoma et Radix *Geranii sanguinei* for their astringent, vasodilatant, anti-inflammatory, hypotensive, cardiotonic and slight sedative effect (Petkov, 1982; Staneva *et al.*, 1982; Assenov *et al.*, 1998). It contains catechin tannins (up to 29 per cent), leucoanthocyanidins, the bitter substance geraniin, resins, etc. (Assenov *et al.*, 1998). Phytotherapists prescribe water-soluble extracts (cool extraction of dry rhizomes at a ratio of 1:10 for 8 h) per os: 400 ml daily split in four doses. It is also applied externally (Petkov, 1982). Numerous investigations have been reported about the healing effect of the plant extracts and the polyphenolic complex obtained from herbage and rhizomes of this species, which possess antimicrobial (Ivancheva *et al.*, 1992) antiviral, against *Herpes simplex* (Zgorniak-Nowosielska *et al.*, 1989; Serkedjieva, 1995, 1997; Serkedjieva and Ivancheva, 1999) and influenza (Manolova *et al.*, 1986; Serkedjieva *et al.*, 1986; Serkedjieva and Manolova, 1987, 1988; Serkedjieva, 1997; Serkedjieva and Hay, 1998) viruses and also its radio-protective activity. Three Bulgarian patents were taken out on the basis of *G. sanguineum* extracts, concerning the main pharmacological effects of the plant: i.e. its hypotensive, antiinfluenza and radio-protective (immunostimulative) functions (Ivancheva, 1998).

REFERENCES

Assenov, I., Gussev, C.H., Kitanov, G., Nikolov, S. and Petkov, T. (1998) *Bilkosabirane*. BILER, Sofia.

Georgieff, C. (1933) Annual Report of the State Experimental Field for Roses, Etherial and Medicinal Plants in Kazanlik (for 1929–1932), Publ. house 'Gutenberg' Kazanlik.

Genova, E. (1995) *Geranium macrorhhizum* L. In: *Chorological Atlas of Medicinal Plants in Bulgaria*, I. Bondev (ed.), Acad. Press 'Prof. M. Drinov', Sofia, 94–95.

Gildemeister, E. and Hoffmann, Fr. (1959) *Die ätherischen Öle*. Bd. 5, Academie Verlag Berlin.

Guenther, E. (1950) *The Essential Oils*, Vol. 4, D. Van Nostrand Company, Inc., Toronto, New York, London.

Hegi, G. (1924) *Ilustrierte Flora fon Mittel-Europa*, IV Bd., 3 Teil, J.F. Lehmanns Verlag, München, p. 1708.

Irinchev, I. (1956) Zdravetzovo maslo. *Priroda*, 5(2), 65–67.

Ivancheva, S. (1998) *Zdravetzut – bilka i poema*. Integral G, Sofia.

Ivancheva, S., Zapessotchnaya, G. and Ognyanov, I. (1976) Flavonoids and other substances in roots of *Geranium macrorrhizum* L. *Compt. rend. de l'Acad. bulg. des Sci.*, 29(2), 205–208.

Ivancheva, S., Manolova, N., Serkedjieva, J., Dimov, V. and Ivanovska, N.(1992) Polyphenols from Bulgarian medicinal plants *Geranium macrorrhizum* and *G. sanguineum* with anti-infectious activity. *Basic Life. Sci.*, 59, 717–728.

Manolov, P., Petkov, V. and Ivancheva, S. (1977) Studies on the central depressive action of methanol from *Geranium macrorrhizum* L. *Compt. rend. de l'Acad. bulg. des Sci.*, 30(11), 1657–1659.

Manolova, N., Gegova, G., Serkedjieva, J., Maksimova-Todorova, V. and Uzunov, S. (1986) Antiviral action of a polyphenol complex isolated from the medicinal plant *Geranium sanguineum* L. I. Its inhibiting action on the reproduction of the influenza virus. *Acta Microbiol. Bulg.*, 18, 73–77.

Mihailov, M. and Tucakov, J. (1974) Pharmacognostic study of long-rooted geranium (*G. macrorrhizum*) in Yugoslavia. Preliminary communication. *Glas Srp. Akad. Nauka (Med.)*, 25, 9–19.

Petkov, V. (1982) Ed. *Modern Phytotherapy*, Medicina i Fizkultura, Sofia.

Petrova, A. and Kozhuharov, S. (1979) *Geranium* L. In: *Flora Republicae popularis bulgaricae*, vol. VII, (B. Kuzmanov, Ed.), Aedibus Academiae Scientiarum bulgaricae, Sofia, 31–34.

Serkedjieva, J. (1995) Inhibition of influenza virus protein synthesis by a plant preparation from *Geranium sanguineum* L. *Acta Virol.*, 39(1), 5–10.

Serkedjieva, J. (1997) Antiinvective activity of a plant preparation from *Geranium sanguineum* L. *Pharmazie*, 52(10), 799–802.

Serkedjieva, J. and Ivancheva, S. (1999) Antiherpes virus activity of extracts from the medicinal plant *Geranium sanguineum* L. *J. Ethnopharmacol.*, 64(1), 59–68.

Serkedjieva, J., Manolova, N., Gegova, G., Maksimova-Todorova, V. and Ivancheva, S. (1986) Antiviral action of a polyphenol complex isolated from the medicinal plant *Geranium sanguineum* L. II. Its inactivating action on the influenza virus. *Acta Microbiol. Bulg.*, 18, 78–82.

Serkedjieva, J. and Manolova, N. (1987) Antiviral action of a polyphenol complex isolated from the medicinal plant *Geranium sanguineum* L. V. Mechanism of the anti-influenza effect *in vitro*. *Acta Microbiol. Bulg.*, 21, 66–71.

Serkedjieva, J. and Manolova, N. (1988) Antiviral action of a polyphenol complex isolated from the medicinal plant *Geranium sanguineum* L. VI. Reproduction of the influenca virus pretreated with the polyphenol complex. *Acta Microbiol. Bulg.*, 22, 16–21.

Serkedjieva, J. and Hay, A.J. (1998) *In vitro* antiinfluenza virus activity of a plant preparation from *Geranium sanguineum* L. *Antiviral Res.*, 37(2), 121–130.

Staneva, D., Panova, D., Raynova, L. and Assenov, I. (1982) *Bilkite vuv vseki dom*. Medicina i Fizkultura, Sofia.

Webb, D. and Ferguson, A. (1968) *Geranium* L. In: *Flora Europaea*, Univ. Press, Cambridge, p. 195.

Yankulov, Y., Taleva, R. and Stoeva, T. (1993) Obiknovenniyat zdravetz e eterichnomaslena kultura. *Zemedelie*, 91 (11/12), 6.

Zgorniak-Nowosielska, I., Zawilinska, B., Manolova, N. and Serkedjieva, J. (1989) A study on the antiviral action of a polyphenol complex isolated from the medicinal plant *Geranium sanguineum* L. VIII. Inhibitory effect on the reproduction of *Herpes simplex* virus type 1. *Acta Microbiol. Bulg.*, 24, 3–8.

6 Essential oil of *Geranium macrorrhizum* L. production, extraction, distillation and use

Tatyana Stoeva

INTRODUCTION

The essential oil (EO) of *Geranium macrorrhizum* is produced solely in Bulgaria and the industry dates back to the end of the nineteenth century (Guenther, 1950; Gildemeister and Hoffman, 1959; Irinchev, 1956). The above-ground herbage is harvested in the autumn (September–October). In post-flowering plants, the EO content was found to be higher (0.083–0.13 per cent) than during flowering (0.05 per cent). Herbage consists of 43–49 per cent leaf blades, 39–43 per cent leafstalks, 4–8 per cent yellowing or withered leaves and flowers and 5–10 per cent rhizomes (Irinchev, 1956; Georgiev, 1995). In recent years, the availability of the wild source of *G. macrorrhizum* became scarce and is unavailable for industrial usage due to the new biodiversity protection laws and conservation strategies. The only way to solve the lack of resource was to amend the cultivation and increase plant productivity.

ESSENTIAL OIL DISTILLATION

Essential oil distillation is done in the same distillation vessels as those where peppermint and lavender are processed. The main methods used are: water distillation, steam and water distillation and steam distillation.

During steam distillation, the herbage is preventing from immersion by supporting it, without compaction, or by the use of 1–2 perforated grids, and the process proceeds at 120 kg/cubic meter.

In water plus steam distillation, it requires 80–85 kg herbage per cubic meter, the boiling water being poured in at a rate of 1:5–6. Distillation is continued for 3.4–4.5 h at a speed of 5–6 per cent and a temperature of the distillation of 45–50 °C as the EO components crystallise out at lower temperatures. About 48 per cent of the oil is obtained during the first hour, with 17–22 per cent obtained after the second and third hour. Since the distillate waters contain 15–20 per cent of the total oil, they are cohabated for 1.5–2 h and 80 per cent of the EO in the water is isolated during the first hour (Irinchev, 1956; Georgiev, 1995).

Primary and secondary oils are rectified separately. They are left in a warm place for precipitation of admixtures to occur and then are filtered while hot. The commercial EO is prepared by combining the primary (raw) and secondary (cohabated) oils in a proportion of 8:1. According to the temperature, the EO is a yellow-green to dark green liquid (above 35 °C) or a semi-solid mass which consists of a mixture of colourless to pale-yellow crystals (stearoptene) with a liquid (eleoptene) at room temperature (Irinchev, 1956; Georgiev, 1995; Ivanov *et al.*, 1952).

The odour is typically 'zdravetz': it is a pleasant, reminiscent of clary sage, orris and rose (Guenther, 1950). The Physico-chemical characteristics of *G. macrorrhizum* are as shown in Table 6.1.

ESSENTIAL OIL COMPOSITION

The EO is typically sesquiterpenic, as the major component is germacrone (50–55 per cent) which constitutes the stearoptene. The other terpenes are mainly ketones: α- and β-elemenone, germazone etc. Hydrocarbons account for 11–13 per cent, comprising: α- and β-selinene; β-elemene, ar-curcumene, α-santalene, caryophyllene, α-humulene, γ-muurolene, δ-cadinene, calamenene etc. Alcohols make up 10–20 per cent, including: juniperic camphor, junenol, β-eudesmol, elemol. Monoterpenic compounds amount to 7–10 per cent, consisting of γ-terpinene, terpinolene, ρ-cymene, α-pinene, δ-3-carene, α-phellandrene, limonene and borneol. The green colour of the oil is due to azulenes (Georgiev, 1995; Georgiev *et al.*, 1989; Ivanov *et al.*, 1952; Ivanov and Ognyanov, 1955; Ognyanov and Ivanov, 1958a,b; Tsankova and Ognyanov, 1972). The EO should be stored in 5 kg copper canisters (kumkuma) under standard conditions for unlimited periods. The oil is usually semi-crystalline at room temperature.

The zdravetz oil is used for its excellent fixative properties in fougeres, chypres, oriental bases, colognes and fantasy fragrances. It blends well with oakmoss, labdanum, olibanum, sandalwood, clary sage, lavender and bergamot (Arctander, 1960).

CONCRETE AND ABSOLUTE OF *G. MACRORRHIZUM*

Concrete

This is obtained by extraction using petroleum ether at room temperature through a three-step extraction: the first lasts for 40–60 min, the second lasts 20–30 min and the third lasts for 5–15 min at a rate of 120 kg zdravetz herbage per cubic litre

Table 6.1 Physico-chemical characteristics of *G. macrorrhizum*

	*Commercial sample 1**	*Sample 2***
Relative density at 20 °C	0.930–0.952	0.9380–0.9680
Refractive index at 20 °C	1.5024–1.5135	1.5003–1.5189
Optical rotation	−4.8 to −8.6	not given
Melting point	34–50 °C	34–50 °C
Ester value	4.97–7.91	7–15
Ester after acetylation	20.8–49.8	25–55
Free alcohols (as geraniol)	5.8–11.9	2–7.5
Total alcohols (as geraniol)	7.7–14.5	6.8–20.7
Stereopten content	not given	41.7–52.8

Notes
*Ognyanov and Ivanov (1958);
** Georgiev *et al.* (1989).

Figure 6.1 Geranium macrorrhizum.

solvent. The herbage is distributed on 4–6 perforated grids. The concrete yields 0.35–0.40 per cent. It looks like a vaseline-like green to yellow-green mass with a strong typically zdravetz odour but much milder, fuller and more lasting than the EO produced by distillation.

The concrete contains: geraniol, germacrol, mono and bicyclic sesquiterpenes; sesquiterpene alcohol, free acids, paraffins, dyes etc. It should be stored in 5 kg tin-plated containers for up to 2 years (Guenther, 1950; Gildemeister and Hoffman, 1959; Georgiev, 1995; Georgiev *et al.*, 1989).

Applications in perfumery: as a rose-scented aromatic with green and fresh notes, exhibiting good fixative values.

Absolute

The absolute is applied in different perfume compositions (Zhelev, 1971; Nikolov *et al.*, 1971). It is a dark green thick liquid obtained from zdravetz concrete by extraction with 95 per cent ethyl alcohol. The absolute has strong, penetrative powers but a gentle zdravetz odour with strong fixative properties.

The remaining mass after the absolute is extracted consists of a dark green solid matter containing mainly *n*-triacontane. It is fatty, tasteless, with a pleasant zdravetz odour and possesses strong photoprotective properties (Georgiev, 1995; Toleva *et al.*, 1971; Toleva and Tolev, 1956). The zdravetz oil and concrete also had antibacterial and antimycotic activity. (Zhelev, 1971; Nikolov *et al.*, 1971; Toleva *et al.*, 1971).

Figure 6.1 shows a typical example of *Geranium macrorrhizum*.

REFERENCES

Arctander, S. (1960) *Perfume and Flavor Materials of Natural Origins*, Elisabeth, New Jersey.

Georgiev, E. (1995) *Technologia na estestvenite I sintetichni aromatni producti*, Zemizdat, Sofia.

Georgiev, E., Dimitrov, D. and Angelakova, M. (1989) *Spravochnik na specialista po aromatichnata I kozmetichna promishlenost*, Technika, Sofia.

Gildemeister, E. and Hoffmann, Fr. (1959) *Die ätherischen Öle*. Bd. 5, Academie Verlag Berlin.

Guenther, E. (1950) *The Essential Oils*, vol. 4, D. Van Nostrand Company, Inc., Toronto, New York, London.

Irinchev, I. (1956) Zdravetzovo maslo. *Priroda*, 5(2), 65–67.

Ivanov, D., Ognyanov, I. and Nikolov, N. (1952) Sur l'huile de zdravetz bulgare, *Compt. Rend. Acad. Bulg. Sci.*, 5, 33–36.

Ivanov, D. and Ognyanov, I. (1955) Bulgarian zdravetz oil. II. Sesquiterpenic hydrocarbons from oil, *Compt. Rend. Acad. Bulg. Sci.*, 8, 45–48.

Nikolov, N., Blagoeva, I. and Boyadzhiev, P. (1971) Aromatnie produkti iz *Geranium macrorrhizum*. In: Congress International des huilles essentielles, Tbilissi (USSR), Sept. 1968, vol. 1, *Chimie et Technologie des huilles essentielles et des substances odorantes*. Pishchevaya promishlenosti, Moscow, p. 248

Ognyanov, I. and Ivanov, D. (1958a) Bulgarian zdravetz oil III. Its terpene compounds, *Compt. Rend. Acad. Bulg. Sci.*, 11, 379–382.

Ognyanov, I. and Ivanov, D. (1958b) Bulgarian zdravetz oil IV. Sesquiterpene oxygen compounds in the oil is eleoptene. *Compt. Rend. Acad. Bulg. Sci.*, 11, 469–472.

Toleva, P. and Tolev, I. (1965) Des possibilities d'utilizer l'huile essentielles et la concrete de Geranium dans la preparation de produits de cosmetique. Travaux Scientifiques de ITSIA, 12, 133–136.

Toleva, P., Hristov, D., Georgieva, A. and Surtalova, T. (1971) Issledovanie sostava voskov iz konkretov zdravetz I akatzii I vozmozhnosti ih ispolizovaniya. In: Congress International des huilles essentielles, Tbilissi (USSR), Sept. 1968, vol. 1, Chimie et Technologie des huilles essentielles et des substances odorantes. Pishchevaya promishlenosti, Moscow, pp. 362–367.

Tsankova, E. and Ognyanov, I. (1972) Sesquiterpene hydrocarbons in the Bulgarian zdravetz (*Geranium macrorrhizum* L.) oil. *Compt. Rend. Acad. Bulg. Sci.*, 25, 1229–1231.

Zhelev, Z. (1971) Aromatischeskie produkti iz zdravetza In: Congress International des huilles essentielles, Tbilissi (USSR), Sept. 1968, vol. 1, *Chimie et Technologie des huilles essentielles et des substances odorantes*. Pishchevaya promishlenosti, Moscow, pp. 97–99.

7 Use of *Geranium* species extracts as herbal medicines

Elizabeth M. Williamson

INTRODUCTION

Geranium species are common throughout temperate regions and are used in many different parts of the world in traditional systems of medicine. 'Geranium oil' is also widely used, however this normally refers to the essential oil (EO) distilled from *Pelargonium* species rather than *Geranium*. In general, most species are used for similar disorders, although some of the different local indications for each are given in Table 7.1. Geraniums used medicinally usually contain high levels of tannins, which are responsible for the folklore use as haemostatics and astringents. They are used internally for haemorrhage and diarrhoea, and externally for wounds, grazes, sores and fissures (British Herbal Pharmacopoeia, 1983; Brendler *et al.*, 1999). More recently various tannin-containing drugs, including *Geranium* species, have been used as antiinfective agents particularly for viral diseases (Serkedjieva and Hay, 1998) and antioxidant activity, which would be expected from the content of polyphenolic compounds, has also been demonstrated (Lamaison *et al.*, 1993).

EUROPEAN AND AMERICAN HERBAL MEDICINE

Geranium maculatum L.

The species most commonly used worldwide is probably *G. maculatum* L., variously known as American cranesbill, storksbill, spotted cranesbill, crowfoot and others. This species is found in shady and moist ground throughout the whole of Europe and in North America from Newfoundland to Manitoba and as far south as Missouri. Both the rhizome and the herb have been used for medicinal purposes since antiquity to treat fever, including malaria, abdominal and uterine disorders, inflammation and as an external application for wounds, excessive bleeding, and sores. Internally it was used as a styptic for metrorrhagia, menorrhagia, haematuria and haemorrhoids; and particularly for diarrhoea, peptic and duodenal ulcers and dysentery. These are still the most important indications. In the middle ages, Paracelsus described it as having cardiotonic and antidepressive activity, and there are other uses for which *Geranium* has been suggested, including worms, leucorrhoea and as a mouthwash (Brendler *et al.*, 1999).

Table 7.1 Species of *Geranium* used medicinally

Geranium sp.	*Local names*	*Locality*	*Internal use*	*External use*	*References*
G. aconitifolium	Palto	Ladakh		Insect bites, ulcers	Srivastava *et al.* (1982)
G. core-core	Core-core	Mexico	Shock, fever, cataract	Toothache, inflammation	Rodriguez *et al.* (1994)
G. japonicum	Cranesbill Storksbill	Japan, China	Rheumatism, numbness of limbs, pain		Perry (1980)
G. macrorrhizum	Bulgarian Geranium, Zdravetz	Bulgaria, Romania, Poland	See G. maculatum; antiviral	Oil used in aromatherapy	Serkedjieva, 1997; Tisserand *et al.* (1995)
G. maculatum	American or Spotted Cranesbill, Storksbill, Crowfoot	North America, Europe	Diarrhoea; dysentery; GI ulcers; styptic in menorrhagia and haematuria	Haemorrhoids; wounds; sores; bleeding; as a mouthwash	Brendler *et al.* (1999); British Herbal Pharmacopoeia, (1983)
G. mexicanum	Mano de Leon, Pata de Leon, Agujas	Mexico, Venezuela	Laxative in infants, antispasmodic	Infant bathing to treat rashes and wounds	Morton (1981)
G. nepalense		N. India, Nepal	Diarrhoea, endometriosis		Perry (1980)
G. niveum		Mexico	Infectious diarrhoea		Calzada *et al.* (1998)
G. pratense	Meadow Cranesbill, Cao-Yuan-Lao-Guan-Cao	China, Japan, Europe, temperate regions worldwide	Acute bacillary dysentery		Yau *et al.* (1999)
G. sanguineum	Bloody geranium	Eastern Europe	Haemorrhage, diarrhoea		Serkedjieva (1996, 1997)
G. robertianum (*Geranium foetidum* Gilib)	Herb Robert, Herb Robin, Dragon's Blood, Wild Cranesbill, Storkbill.	Europe, USA, China, Japan, North Africa, USA, India, South America	Diarrhoea, haemorrhage, jaundice, dispersal of milk, kidney and gall stones	Gargle, mouthwash, burns, wounds	Brendler *et al.* (1999) Ambasta (1986)
G. thunbergii	Gen-No-Shoko	China, Japan	Inflammation of GI system, diarrhoea, haematological and liver disorders	Cataract	Kimura *et al.* (1994)
G. wallichianum	Wallich Cranesbill, Laljhari	Northern India, Punjab, Kashmir	Astringent, diarrhoea	Toothache and eye disorders	Ambasta (1986)
G. wilfordii	Lao Kuan Tsao	China	Chronic rheumatism		Pei-gen (1989)

Constituents and preparations

G. maculatum contains gallotannins. The dried herb and the root are used at the same dose, which is normally 1–2 g powdered drug taken 3 times daily. The herb may be taken as an infusion, and the root as a decoction, liquid extract 1:1 (45 per cent ethanol) or tincture 1:5 (45 per cent ethanol). *G. maculatum* is often found in preparations combined with other herbs such as *Geum urbanum*, *Agrimonia eupatoria* for duodenal ulcer, and *Bidens pilosa* for bleeding of the digestive tract (British Herbal Pharmacopoeia, 1983). Few side effects have been observed, although patients with sensitive stomachs may feel nauseous because of the high tannin content, and theoretically it should not be taken with certain other medicines because of the possibility of the formation of tannin-drug complexes. Homoeopathic preparations are also used, including drops, tablets or injections, to treat stomach ulcers and bleeding of the mucus membranes (Brendler *et al.*, 1999).

Geranium robertianum L.

G. robertianum L., also known as Herb Robert, Herb Robin, Dragon's Blood and others (see Table 5.1) is indigenous to a large area covering Europe to China and Japan; most of North America; Africa as far south as Uganda, and temperate parts of South America. Historically, it was used as an astringent in diarrhoea and excessive bleeding from various causes in much the same way as *G. maculatum*, but it also had a reputation for resolving or dispersing milk in women after parturition, and as a coolant and pain reliever. It was even recommended as a gargle for angina. The Indians of western North America are said to have used the root as a cure for syphilis. It is now used as a mouthwash for inflammation of the mouth and pharynx; topically for burns and wounds, and internally for haemorrhage, kidney and bladder stones, and for liver and gall bladder problems. It is used in India for similar purposes and particularly for jaundice (Ambasta, 1986) and has been used in veterinary medicine for dysentery and haematuria. Modern usage also centres on the antiviral effect of extracts of the plant, which has been confirmed in some studies but not in others. No activity was shown against polio type 1, measles, coxsackie-B2, adeno- or Semliki forest viruses, but a mild effect was demonstrated against a stomatitis virus. Extracts also have inhibitory effects on some types of pathogenic microbial growth *in vitro* (Brendler *et al.*, 1999) The effects of geraniin will be discussed later in connection with the effects of *G. thunberghii*.

Constituents and preparations

G. robertianum contains tannins including geraniin; isogeraniin; β-penta-O-galloyl glucose; and flavonoids such as rutin, hyperoside and others. The dried aerial parts of the herb are used at a dose of 1.5 g, taken as an infusion up 3 times daily. This may be used internally or externally. Fresh cleaned leaves have been chewed to relieve inflammation of the oral cavitiy. Occasionally the herb may be found adulterated with *G. palustre*, or *G. pratense*.

Geranium macrorrhizum L. and *G. sanguineum* L.

Bulgarian folk medicine and that of other Eastern European countries such as Poland and Romania, use *G. macrorrhizum* (known as zdravetz in Bulgarian) and *G. sanguineum* for similar astringent purposes in diarrhoea and haemorrhage to those outlined for

G. maculatum. *G. sanguineum* is known to contain a number of polyphenolic compounds including gallotannins, catechins and flavonoids, and has been shown to inhibit replication and synthesis of the influenza virus *in vitro* and *in vivo* in mice (Serkedjieva, 1996, 1997) and Herpes, Vaccinia and HIV viruses *in vitro* (Serkedjieva, 1996, 1999). It appears to work by inhibiting influenza virus neuramidase enzyme (Serkedjieva, 1995a) and protein synthesis (Serkedjieva, 1995b). Other susceptible organisms included *Staphylococcus aureus* and *Candida albicans* (Serkedjieva, 1997). The EO of 'Zdravetz' (*G. macrorhizum*) is occasionally used in aromatherapy and contains about 50 per cent of germacrone (Tisserand and Balacs, 1995).

ORIENTAL MEDICINE

Geranium thunberghii Sieb et Zucc.

G. thunberghii, known as 'Gen-no-shoko' in Japanese, is used in Kampo (traditional Japanese medicine) particularly as a remedy for diarrhoea induced by inflammation of the small intestine and also for liver and haematological disorders (Kimura *et al.*, 1984). It also has been used to treat cataract (Fukaya *et al.*, 1988). The biological activity is usually ascribed to the tannin content, particularly geraniin. Extracts of the plant have been shown to reduce gastro-intestinal motility in isolated rat intestine (Kan and Taniyama, 1992); have inhibitory effects on cholera toxin-induced secretion from rat gastric mucosa and other astringent effects (Ofuji *et al.*, 1998), giving support to the traditional usage of *Geranium* for diarrhoea. They also exhibit strong antimutagenic effects against direct acting mutagens (Okuda *et al.*, 1984). Like other tannin-containing drugs, extracts reduced levels of ureamic toxins, including creatinine, methylguanidine and guanidinosuccinic acid, in rats with renal failure (Yokozawa *et al.*, 1995).

A considerable amount of research has been done on isolated geraniin, which has been shown to protect against experimental hepatic injuries in rats induced by galactosamine, carbon tetrachloride and thioacetamide (Nakanishi *et al.*, 1998a). Geraniin also prevented the accumulation of liver triglycerides and lipid peroxide levels and lowered serum levels of hepatic enzymes associated with liver damage in the rat (Nakanishi *et al.*, 1999a). A protective effect on oxidative damage to mouse ocular lens supports the traditional use for cataract to some extent (Fukaya *et al.*, 1988). Haematological effects of geraniin in rats have also been studied, and a decrease in serum lipid levels observed as well as a reduction in erythrocytes and leucocytes (Nakanishi *et al.*, 1998b). The effects of geraniin on aminonucleoside nephrosis in rats included a reduction in proteinuria, and in serum and cholesterol levels produced by puromycin (Nakanishi *et al.*, 1999b). Macrophages are known to be affected by geraniin, and marked increases in phagocytosis observed in *in vitro* cultures (Ushio *et al.*, 1991). The clinical significance of some of these findings is not clear, and few human studies on *Geranium* herb are to be found in the literature, however the results of these various experiments show that the tannins it contains have pharmacological effects which may support its therapeutic use.

Constituents and preparations

G. thunberghii contains tannins, the most important being geraniin, corilagin, dehydrogeraniin, furosin and furosinin, ellagic and gallic acids, geraniic acids B and C and phyllanthusiin F (Kan and Taniyama, 1992; Okuda *et al.*, 1982; Ito *et al.*, 1999).

Geranium wilfordii Maxim, *G. pratense* L. and *G. nepalense* Sweet

G. wilfordii, or 'Lao-kuan-tsao' is used in Chinese medicine to treat chronic rheumatism, often steeped in wine and taken orally, either alone or in combination with other antirheumatic drugs. *Erodium stephanium*, also known as 'Lao Guan Cao', is considered interchangeable and used in a similar way (Pei-gen, 1989). The closely related Meadow Cranesbill, *G. pratense* ('Cao Yuan Lao Guan Cao') is used in traditional Chinese medicine to treat acute bacillary dysentery (Yau *et al.*, 1999). *G. nepalense* is used in a similar manner to *G. thunberghii*, which is sometimes considered a variety of it (Perry, 1980).

ASIA

Geranium aconitifolium L'Herit.

This plant is known as 'palto' by the traditional medical practitioners of Ladakh in the Himalayas. The Tibetan system of medicine is known as 'Amchi' and is a form of Ayurveda. The flowers of *G. aconitifolium* are used in the form of a paste to soothe insect bites, and the root paste is used for ulcers, and to treat cattle wounds, especially when an insect repellent action is needed (Srivastava and Gupta, 1982).

Geranium wallichianum D. Don.

The Wallich Cranesbill (locally called Laljhari or Liljhari, and in Kashmir, Kao-ashud) is used in Uttar Pradesh, India; and in Kashmir and Punjab as an astringent, particularly for eye problems, and toothache. The root has been substituted for *Coptis teeta* (Ambasta, 1986) and like many other species of *Geranium* is also used for tanning. Other species used in a similar way in India include *G. robertianum* (see above), *G. pusillum*, *G. rotundifolium* and *G. sibirica*; all are used as astringents and for wound dressing (Ambasta, 1986).

SOUTH AND CENTRAL AMERICA

Geranium core-core L.

An infusion of the leaf of this plant, known as 'core-core', is used in Chile by the Mapuche Amerindians to treat cataract, shock and fever. The root, considered astringent, was used to treat toothache and the whole plant used as an antiinflammatory agent (Rodriguez *et al.*, 1994). It is not normally used for diabetes, but the hypoglycaemic effects of an extract of the whole plant was assessed in normoglycaemic and alloxan diabetic rats and the activity found to be significant, but less than that produced by tolbutamide (Rodriguez *et al.*, 1994).

Geranium mexicanum HBK.

In Mexico, the juice of this plant is used as a laxative in infants although it is reputed to have no effect in adults. A decoction is used to bathe babies to prevent and treat rashes and wounds. In parts of Venezuela, the leafy stems are sold by herb vendors to make a decoction, which is drunk for digestive spasms (Morton, 1981).

Geranium niveum

This plant is used in Mexican traditional medicine to treat infectious diarrhoea, which is usually amoebic dysentery, and has recently been tested as for antiprotozoal activity against *Entamoeba histolytica* and *Giardia lamblia*. Activity was confirmed to support this usage, although the plant extract was less effective that metronidazole (Calzada *et al.*, 1998). Bioassay-guided fractionation led to the isolation of two new antiprotozoal constituents, identified as A-type proanthocyanins and named geranins A and B (Calzada *et al.*, 1999).

CONCLUSION

The use of *Geranium* species throughout the world demonstrates reassuring evidence that even in a wide variety of systems of medicine, with different original philosophical principles, traditional use shows very similar indications and supports the pharmacological studies carried out more recently. There is very little discrepancy between usage in different parts of the world and these depend usually upon the astringent effect of the tannins contained in the herb, present in whichever species grows locally. The more recent studies on the antiviral effects and other antimicrobial actions will add to the supporting data for the traditional use of this species.

REFERENCES

Ambasta, S.P. (ed.) (1986) *The Useful Plants of India*, CSIR, New Delhi, India.

Brendler, T., Gruenwald, J. and Jaenicke, C. (eds) (1999) *Heilpflanzen/Herbal Remedies*. Deutcher Apotheker Verlag, Stuttgart, Germany.

British Herbal Pharmacopoiea (1983) British Herbal Medicine Association, Keighley, UK.

Calzada, F., Meckes, M., Cedillo-Rivera, R., Tapia-Contreras, A. and Mata, R. (1998) Screening of Mexican medicinal plants for antiprotozoal activity. *Pharmaceutical Biology*, 36(5), 305–309.

Calzada, F., Cerda-Garcia-Rojas, C.M., Meckes, M., Cedillo-Rivera, R., Bye, R. and Mata, R. (1999) Geraniins A and B, new antiprotozoal A-type proanthocyanins from *Geranium niveum*. *J. Nat. Prod.*, 62(5), 705–709.

Fukaya, Y., Nakazawa, K., Okuda, T. and Iwata, S. (1988) Effect of tannin on oxidative damage of ocular lens. *Jap. J. Opthalmol.*, 32(2), 166–175.

Ito, H., Hatano, T., Namba, O., Shironon, T., Okuda, T. and Yoshida, T. (1999) Constituents of *Geranium thunbergii* Sieb. et Zucc. XV. Modified dehydroellagitannins, geraniic acids B and C and phyllanthusiin F. *Chem. Pharm. Bull.*, 47(8), 1148–1151.

Kan, S. and Taniyama, K. (1992) Mechanism of inhibitory actions of *Geranium thunbergii*, tannic acid and geraniin on the motility of rat intestine. *Jap. J. Pharmacognosy*, 46(3), 246–253.

Kimura, Y., Okuda, H. and Mori, K. (1984) Studies on the activities of tannins and related compounds from medicinal plants and drugs IV. Effects of various extracts of Geranii Herba and geraniin on liver injury and lipid metabolism in rats fed peroxidised oil. *Chem. Pharm. Bull.*, 32(5), 3755–3758.

Lamaison, J.L., Petitjean-Freitet, C. and Carnat, A. (1993) Tannin in polyphenols and antioxidant activity in French Geraniaceae. *Plantes Medicinales et Phytotherapie*, 26(2), 130–134.

Morton, J. (1981) *Atlas of Medicinal Plants of Middle America*. Charles C. Thomas, USA.

Nakanishi, Y., Kubo, M., Okuda, T., Oruda, M. and Abe, H. (1998a) Hematological effects of geraniin in rat. *Natural Medicines*, 52(2), 179–183.

Nakanishi, Y., Orita, M., Okuda, T. and Abe, H. (1998b) Effects of geraniin on the liver in rats I. Effects of geraniin compared to ellagic acid and gallic acid on hepatic injuries induced by CCl_4, D-galactosamine and thioacetamide. *Natural Medicines*, 52(5), 396–403.

Nakanishi, Y., Kubo, M., Okuda, T. and Abe, H. (1999a) Effect of geraniin on aminonucleoside nephrosis in rats. *Natural Medicines*, 53(2), 94–100.

Nakanishi, Y., Okuda, T. and Abe, H. (1999b) Effects of geraniin on the liver in rats III. Correlation between lipid accumulation and liver damage in CCl_4-treated rats. *Natural Medicines*, 53(1), 22–26.

Ofuji, K., Hara, H., Sukamoto, T. and Yamashita, S. (1998) Effects of an antidiarrhoea containing an extract from *Geranium* herb on astringent action and short-circuit current across jejunal mucosa. *Folia Pharmacol. Jap.*, 111(4), 265–275.

Okuda, T., Hatano, T. and Yazaki, K. (1982) Dehydrogeraniin, furosinin and furosin, dehydroellagitannins from *Geranium thunbergii*. *Chem. Pharm. Bull.*, 30(1), 1110–1112.

Okuda, T., Mori, K. and Hayatsu, H. (1984) Inhibitory action of tannins on direct acting mutagens. *Chem. Pharm. Bull.*, 32(9), 3755–3758.

Pei-gen, X. (1989) Excerpts from the Chinese Pharmacopoeia. In: L.E. Craker, J.E. Simon (eds) *Herbs, Spices and Medicinal Plants*, Vol. 4. Oryx, USA, pp. 43–85.

Perry, L.M. (1980) *Medicinal Plants of East and South East Asia*, MIT Press, USA, p. 159.

Rodriguez, J., Loyola, J.T., Maulen, G. and Schmeda-Hirschmann, G. (1994) Hypoglycaemic activity of *Geranium core-core, Oxalis rosea* and *Plantago major* in rats. *Phytother. Res.*, 8(6), 372–374.

Serkedjieva, J. (1995a) A polyphenolic complex isolated from *Geranium sanguineum* L. inhibits influenza virus neuramidase. *Fitoterapia*, 63(2), 111–117.

Serkedjieva, J. (1995b) Inhibition of influenza virus protein synthesis by a plant preparation from *Geranium sanguineum* L. *Acta Virologica*, 39(1), 441–443.

Serkedjieva, J. (1996) A polyphenolic extract from *Geranium sanguineum* L. inhibits influenza virus protein expression. *Phytother. Res.*, 10(5), 441–443.

Serkedjieva, J. (1997) Antiinfective activity of a plant preparation from *Geranium sanguineum* L. *Pharmazie*, 52(10), 799–802.

Serkedjieva, J. (1999) Anti herpes virus activity of extracts from the medicinal plant *Geranium sanguineum* L. *J. Ethnopharmacol.*, 64(1), 59–68.

Serkedjieva, J. and Hay, A.J. (1998) *In vitro* anti-influenza activity of a plant preparation from *Geranium sanguineum* L. *Antiviral Res.*, 37(2), 121–130.

Srivastava, T.N. and Gupta, O.P. (1982) Medicinal Plants used by Amchies in Ladakh. In: C.A. Atal and B.M. Kapoor (eds) *Cultivation and Utilization of Medicinal Plants*, Regional Research Laboratory, CSIR, Jammu-Tawi, India, 519–526.

Tisserand, R. and Balacs, T. (1995) *Essential Oil Safety*. Churchill Livingstone, Edinburgh, UK.

Ushio, T., Okuda, T. and Abe, H. (1991) Effect of geraniin on morphology and function of macrophages. *Int. Arch. Allergy Appl. Immunol.*, 96(3), 224–230.

Yau, X., Zhou, J. and Xie, G. (1999) *Traditional Chinese Medicines*. Ashgate Publishing, UK.

Yokozawa, T., Fujioka, K., Oura, H., Tanaka, T., Nonaka, G. and Nishioka, I. (1995) Confirmation that tannin-containing crude drugs have a uraemic toxin-decreasing action. *Phytother. Res.*, 9(1), 1–5.

Part 2

Pelargonium

8 The taxonomy of *Pelargonium* species and cultivars, their origins and growth in the wild

Diana M. Miller

CHARACTERISTICS OF THE GENERA WITHIN THE FAMILY *GERANIACEAE*

The genera *Geranium* and *Pelargonium* are classified in the family *Geraniaceae* and like the three other genera included in *Geraniaceae* they also have a similar elongated fruit with five mericarps, each containing a single seed. When the seeds mature, the mericarps split apart and the plumed seed is released. *Monsonia* is found mainly in Africa with a few species in Asia. The fleshy stemmed *Sarcocaulon* is limited to southern Africa where it is called Bushman's candle. The remaining three genera were included by Linnaeus in the one genus *Geranium* named from the resemblance of the fruit to a crane's bill. The true *Geranium* has a wide distribution throughout the temperate regions of the world but the genus of plants with irregular flowers was later separated from *Geranium* and named *Pelargonium* for the Greek word for stork, 'pelargos'. The third genus, *Erodium*, mainly found around the Mediterranean, was later separated and renamed from the Greek for 'heron'. The differences between *Geranium* and *Pelargonium* are compared in Table 8.1.

The majority of *Pelargonium* species are found in the Southern hemisphere mainly in South Africa with a few in Australia, eastern Africa and on some islands including St. Helena, Tristan de Cuhna and Madagascar. On the other hand *Geranium* grow wild in the Northern hemisphere in North America, Europe and Asia. Although, as Table 8.1 indicates, the species of each genus should be readily distinguished, confusion has arisen over the nomenclature of *Pelargonium*. When first introduced into Europe in the seventeenth century, all pelargoniums were called geraniums, presumably because of the similar fruit structure. However, later botanists noted distinct differences and the two genera were separated about 200 years ago. However, the name 'geranium' has become re-established during the twentieth century as a vernacular name for pelargoniums especially for the zonal cultivars so widely used in bedding.

HISTORY OF DISCOVERY

As with the majority of plants, the discovery and introduction of the first *Pelargonium* species into the great gardens of Europe is closely connected with the history of exploration and the opening up of trade routes. Once the route around the southern tip of Africa to the East had been discovered and the spice trade was established, the East India Companies of Britain and Holland were set up to create permanent trading posts

Table 8.1 The distinctions between *Pelargonium* and *Geranium*

Pelargonium	*Geranium*
Flowers irregular, the upper two petals different in shape and size to the lower three	Flowers regular, all petals more or less the same in shape and size
Hypanthium present	No hypanthium present
Fertile stamens less than ten	Fertile stamens ten
Habit various	Mostly herbaceous
Mostly frost tender	Mostly hardy
Mostly from the Southern hemisphere	Mostly from the Northern hemisphere

in southern Africa. Naturalists on board collected plants for possible food or medicine for the sailors and plants were brought back to Europe to botanic gardens such as the one at Leiden, established in 1577.

The earliest plants from South Africa were bulbs and the first recorded *Pelargonium* was the tuberous rooted, *P. triste*. It is found wild in the regions near the early settlements on the coast. Other pelargoniums grow in similar areas with much more showy flowers, but *P. triste* with a tuber would have been more able to survive the long journey back. It may have been collected as a medicine as it has been used as such in its native land, or as a possible food source, or simply for the night-scented flowers. It was first recorded in 1631 by John Tradescant but it was initially considered to be a native of India, presumably because it was brought back on a ship returning from the East.

Paul Hermann in 1672 was perhaps the first serious collector who pressed examples of the South African Flora on his way to Ceylon (now Sri Lanka). Seeds were sent to Jacob Breyne who illustrated one or two *Pelargonium* species in 1678. On his return to Holland, Hermann published a catalogue of the plants of the gardens in Leiden which listed nine pelargoniums, all of which are still grown today. Jan Commelin at the new botanic garden in Amsterdam, also received plant material and 2 years later was listing one or two new species and, by 1710, about ten more new species had been added to this collection.

Exchanges of plants between Britain and Holland were frequent, and records around the turn of the century show that in London, James Sherard, an apothecary with a large estate at Eltham in Kent and the Bishop of London, Henry Compton in his famous contemporary gardens at Fulham included pelargoniums. In South Africa at about this time, explorations inland to the north and east resulted in the discovery of more new species of *Pelargonium* several of which were recorded in European gardens. During the next 50 years there was a slow but steady introduction of species, but in 1772 a young collector from Kew, Francis Masson was sent to South Africa and was responsible for about 20 years, for the introduction of a large number of new species from inland areas of the southern and south-western Cape.

Another Kew gardener was Anthony Hove. On his way to India in 1795, he collected 17 pelargoniums from near the Bay of Angra, on the south-western coast of Africa, about 500 miles north of the Cape. Unfortunately only three survived.

The explorer William Paterson in 1777 also collected plants as he travelled northward along the coast where he discovered the Orange River as well as *P. sibthorpifolium* and *P. klinghardtense*. The first Australian species such as *P. australe* found as early as 1792, were collected following the expeditions of Captain Cook and Joseph Banks.

Towards the end of the eighteenth century and during the nineteenth century, a large number of hybrids were grown, at first as accidental seedlings and later as the process was understood, more deliberately produced. Zonal, regal, scented-leaved and ivy-leaved cultivars were well known by the late nineteenth century and an explosion of interest in these groups throughout the world has occurred during the twentieth century. New species have been discovered and a number, especially from the section *Hoarea* have been named as recently as the 1990s.

GENERAL CHARACTERISTICS OF THE GENUS *PELARGONIUM*

The flowers of *Pelargonium* are arranged in an umbel-like inflorescence of one to about 50 individual flowers. Whereas in a true umbel, the youngest flowers are at the centre of the flower head, in pelargoniums, the oldest flowers are at the centre, so the inflorescence is normally termed a pseudoumbel.

Each flower has five sepals with the posterior one modified so that a nectary is formed at the base. The length of this nectary is characteristic for each species and in some is a mere indentation but in others may be several centimetres long. The nectary is fused to the pedicel and known as the hypanthium. At the end of the hypanthium is a hump, sometimes quite pronounced, and the relative lengths of the hypanthium and pedicel is often important in the classification of species. The five free petals are often clawed and arranged so that the upper two are separated from the lower three usually with a distinct difference in size, shape and markings. This distinction becomes less obvious in the modern zonal and regal cultivars, deliberately bred in Victorian times to bear regular flowers with rounded petals. The relative sizes, arrangement and positioning of the petals, together with the differing lengths of pedicel and hypanthium, create the wide range of flower formation in the genus, each having evolved alongside the appropriate pollinating insect found in its native habitat.

Of the ten stamens, seven or fewer bear fertile pollen and this number is, another significant feature in the recognition of the sections. The remaining infertile stamens are known as staminodes, which may be curved. The filaments are joined at the base, occasionally for most of their length. The style divides into five stigmas, which open after the anthers have dehisced, thus avoiding self pollination. Once the flower is fertilised, the mericarps develop and the plumed seeds are dispersed.

In habit, the species of *Pelargonium* may be woody, succulent or herbaceous, annual or perennial, evergreen or deciduous, and some have tubers. All have alternate petiolate leaves with stipules. The leaf venation may be palmate or pinnate, simple or compound and of almost any shape. In some sections, the petioles, the stipules or both are persistent, remaining on the stems for several seasons often hardening to form spines.

The presence of aromatic oils in many species is often a useful tool for identification. A few species have scented flowers but this scent is usually only emitted at night.

NATIVE HABITATS

The genus *Pelargonium* includes about 270 distinct species and the majority of these are to be found in southern Africa. In southern Africa, itself, over 80 per cent of these are associated with the Mediterranean type of climate and vegetation found in the south-western part of Cape Province in an area of winter rainfall. The remaining species are found in southern and eastern Cape where the rain falls mainly in summer. Approximately 20 or so grow in East Africa. About a dozen species are to be found wild outside Africa, some in Australia, one or two in New Zealand, two in the Middle East and one or two are to be found on the islands of Madagascar in the Indian Ocean and St. Helena and Tristan da Cunha in the southern Atlantic Ocean (van der Walt, J.J.A., 1977).

The characteristic feature of all these localities is the relatively low rainfall which in most cases falls mainly in one season of the year, either summer or winter. No species grow in areas of high humidity and those that are found nearer the equator live at higher altitudes where the temperatures are not excessive. However, in other respects, all types of habitat are occupied from desert to tall grasslands, on rock faces or amongst shrubs.

The genus is almost unequalled in the plant kingdom for the enormous diversity of form created to adapt to these different habitats. The genus includes shrubs, annuals, succulents and tuberous plants; plants of a few centimetres tall or shrubs reaching several metres; plants with fragrant flowers or with scented foliage, some of which are used in the commercial production of aromatic oils; plants which have spines and a number valued for their medicinal properties.

CLASSIFICATION HISTORY

The first pelargoniums were grouped with geraniums. As early as 1732, Dillenius suggested that the African species of *Geranium* with unequal and irregular flowers might be called 'Pelargonium' but he did not utilise his own epithet. Six years later, in 1738, Johannes Burman of Amsterdam again brought up the name and did employ '*Pelargonium*' for at least some of the species he depicted. Not long after this date, in 1753, Linnaeus published his *Species Plantarum* which although it established the binomial system of naming, did not recognise *Pelargonium* as a distinct genus and retained the generic name *Geranium* for the 20 *Pelargonium* species known at that time as well as those known today as *Erodium* and *Geranium*. Such was the stature of Linnaeus at the time, that the name *Geranium* was retained for another 40 years before the name *Pelargonium* was finally approved.

Arguably, the most important taxonomic works were the illustrations and unpublished descriptions of Charles-Louis L'Héritier, who is given the credit for publishing the name *Pelargonium* in *Geraniologia* 1787–1788 where he clearly distinguished between the three genera, *Pelargonium*, *Erodium* and *Geranium*. Although the generic names were disputed for a few years, they were eventually established and have remained since.

By the latter part of the eighteenth century, the number of new *Pelargonium* species and hybrids had risen so dramatically that a means of classifying them became imperative. Cavanilles in 1787 subdivided the species of *Geranium* into groups. He separated the African geraniums, on the basis of their flower shape and further subdivided these on leaf characters such as the presence of a zone, and then on the degree of lobing. Aiton in 1789 placed species with similar morphology together but did not attempt to formalise the groupings with names. The first serious attempt was by Sweet in 1820, in the first of his five volumes of *Geraniaceae*. He separated many of the more distinct species by creating ten new genera. These differences were obviously recognised, but not necessarily considered sufficient to merit the status of unique genera, and de Candolle in 1824 soon reduced Sweet's genera to sections within the original genus *Pelargonium*. Although his classification included many new names for sections which were then subdivided into series, it made the recognition of a plant easier because each grouping was smaller to include far fewer species. Different classifications were published in the following years such as that by George Don in 1831.

Most of the classification systems which followed were variations or amalgamations of the systems produced by Sweet and de Candolle. Ecklon and Zeyher in 1835, prolific plant collectors in South Africa, reinstated the generic status for several sections and created other new ones as well but they seem to have little influence on classification as these names were soon discarded by Harvey in *Flora Capensis*. William Harvey, Colonial Treasurer at the Cape in 1835 produced his '*Flora Capensis*' in 1860 where he separated the genus into 15 sections, many of which were combinations of those proposed by Sweet and de Candolle. Reinhard Knuth, who published an important work on *Geraniaceae* in 1912, basically followed Harvey's naming but included a number of new species.

Harvey's division of the genus into sections was followed in principle for over a 100 years until the last 20 years when the taxonomists in South Africa began a major study of the genus. This investigation is still being carried out by botanists in several parts of the world looking at all aspects of taxonomy. These modern methods investigate all aspects of the species, including location, habitat, chromosome size and number, pollen characteristics, alkaloid and protein content, external morphology and internal anatomy. In many cases, the original classifications have remained very similar although the recent work has solved many anomalies and is able to show evolutionary trends and relationships. Even more recently a group or researchers have made a detailed phylogenetic study, the results of which are not completely published, and their latest classification has been followed in this publication.

SYNOPSIS OF SECTIONS

The most recent classification is based on the molecular work but has initially been divided into two subgenera based on the chromosome size. Each subgenus is further divided into a number of sections. These divisions into sections in most cases matches the morphological-based classification of the older botanists and closely relates to areas of similar climatic conditions with a characteristic type of vegetation. It does solve a number of apparent past anomalies. All species of the section *Cortusina*, for example, are to be found in southern Namibia and the north-western part of Cape Province in desert or semi-desert regions where the winter rainfall may not exceed 2–3 cm a year and plants may be found protected by rocks or in loose sand. However, the species of section *Reniformia*, until relatively recently classified in the same section, are to be found towards the southern and eastern part of South Africa in areas of summer rainfall, up to 20–30 cm a year, some in long grass and others at the edges of forest. The adaptation of each species to the microclimate in which it is found, explains the enormous variation in habit. Even within a single species in the wild, the variation in the habit of the plant collected from different localities can reflect the conditions under which it was growing, especially availability of water and light. This adaptability can cause some problems with identification. Hybrids between subgenera are not known and hybrids between the sections are extremely rare.

Subgenus *Ciconium*

This subgenus representing about 23 per cent of the species in the genus has large chromosomes and is divided into six sections. Most of the species are found in eastern and southern Cape Province and East Africa, Yemen to the Middle East.

Section Ciconium

This extremely important section for horticulture, includes several species involved in the ancestry of both the zonal and ivy-leaved pelargoniums. Earlier, botanists separated species with the trailing habit of *P. peltatum* into a distinct genus, *Dibrachya*. Nowadays, this section is merged with *Ciconium* (van der Walt and Vorster, 1988) and more recently species have been added from other sections such as *Ligularia*. All have large chromosomes with a basic number x = 9. If not pruned back for the winter, nearly all are evergreen in cultivation, with simple, palmately lobed, but not repeatedly divided, leaves.

The flowers tend to be in shades of red and pink or white but less frequently in the purplish shades associated with species in other sections. Unusually the lower petals may be equal in size or sometimes even larger than the upper. There are seven or fewer fertile stamens of which two or more may have extremely short filaments or all the filaments may be united for almost their whole length. The hypanthium is long, mostly over five times the pedicel length.

The majority of the species grow wild in the eastern or southern parts of Cape Province. Two important species are *P. zonale* and *P. inquinans*, considered to be the major parents of the zonal cultivars. Both are among the early introductions of the genus at the beginning of the eighteenth century but the main development of the zonals did not begin until well over a 100 years later. Sports and selections of *P. peltatum* on the other hand resulted in the ivy-leaved pelargoniums with probably little or no influence from other species.

P. acetosum (Linnaeus) L'Héritier: A branching, subshrub with hairless brittle stems and fleshy, glaucous leaves, sometimes with a reddish crenate margin. The pale to salmon-pink flowers have very narrow petals and five fertile stamens. This was one of the very early species of *Pelargonium* from eastern Cape Province which was included in illustrations by Casper Commelin of the plants in the medicinal plant garden in Amsterdam in 1703. Perhaps it was thought that the sorrel tasting leaves had some curative properties although they have no known medicinal value.

P. articulatum (Cavanilles) Willdenow: A plant with a very short stem above ground but an underground rhizome which is alternately thick and thin. The more or less orbicular leaves usually have a narrow dark zone towards the margin with deeply cordate base and large cream to white flowers have upper petals veined with purple, much narrower lower petals and seven fertile stamens. This species has recently been used in Australia with zonal cultivars to produce a range of interesting hybrids. It is thought that Thunberg was the first to collect this species in the early 1770s from the western part of Cape Province.

P. frutetorum R.A. Dyer: A spreading, branching plant with thick reddish-brown stems and shallow five-lobed, green leaves with a very distinct dark brown-purple ring-shaped zone towards centre of lamina. The pale salmon pink flowers have seven fertile stamens.

This species from the eastern Cape Province has been used in hybridisation to create zonal type cultivars with attractive foliage. It is less robust than *P. zonale* but has a much more clearly defined zone on the leaf. 'The Boar' is almost certainly derived from this species.

P. inquinans (Linnaeus) L'Héritier: An erect, branching, subshrub to about 2 m tall with red glandular hairs and unmarked, almost circular shallow lobed leaves. The flowers are scarlet, occasionally pink or white, almost regular in shape, the upper petals slightly smaller than lower. The stamens and style are barely exserted and the filaments with seven fertile anthers are joined for most of their length.

This species and *P. zonale* are generally considered to be the main parents of the modern zonal pelargoniums, and both were very early introductions from the Eastern Cape into the great gardens of Europe in the early eighteenth century. Many plants found under this name are early hybrids but the true species is easily recognised by the red glandular hairs found on the stems and leaves which leave a red-brown stain on the hands. The plant is another of those that has been used in the past for its medicinal properties.

P. peltatum (Linnaeus) L'Héritier: A trailing or climbing, variable perennial with somewhat fleshy, slightly aromatic leaves, usually peltate, with five triangular lobes, either glossy and glabrous or with a short soft velvety pubescence, often with a darker circular zone. The flowers are strongly zygomorphic, white, pink or pale purple, up to 4 cm across, the upper petals larger and veined darker; fertile stamens seven, two of which are very short.

Typically *P. peltatum* is found scrambling through other shrubs, able to reach to 2 m or more. It grows wild over a large area of the winter rainfall zone in Cape Province but the distribution extends much further east into the regions of summer rainfall of Natal and eastern Transvaal. It was apparently introduced into Europe in 1700. The leaves have been used as an antiseptic and a deep blue dye, which has been used in painting, may be obtained from the petals but the species is most important as the parent of the ivy-leaved pelargoniums.

P. quinquelobatum Hochstetter: A herbaceous, hairy, perennial plant with dull green, bluish-tinged deeply lobed leaves which are broadly triangular in outline. The flowers are pale yellowish-green to grayish green-blue, the upper petals faintly lined with pink, and three of the seven fertile stamens have very short filaments.

The unusual colour of the flowers of this species from Tanzania and Kenya to Ethiopia, has encouraged the use of this plant for hybridisation although few have proved successful so far.

P. salmoneum R.A. Dyer: A weakly erect, branching subshrub to about 1.5 m or more with semi-succulent stems and thick, green to glaucous, unmarked leaves which are somewhat folded upwards. The narrow petals of the salmon pink flowers are almost equal and there are five fertile stamens.

P. salmoneum was first found in a garden in Port Elizabeth and described in 1932. For many years it was considered to be a species but later, because it was only known in cultivation, it was listed as a cultivar.

P. zonale (Linnaeus) L'Héritier: An erect usually hairy subshrub but often found scrambling through bushes. The orbicular leaves, often have a darker brownish-purple horseshoe-shaped mark. The inflorescence has up to 50 pink, sometimes white or red, flowers with more or less equal petals, marked with darker veins and seven fertile stamens, two of which are very short. In the wild, this species is widespread from eastern to western Cape Province and Commelin, records how it was grown in 1700 in the medicinal garden of Amsterdam.

It has lent its name to the enormous group of modern cultivars, the zonals, named on account of the zoned leaf, and it was certainly involved in their parentage.

Section Chorisma

Old references, especially in German, placed *P. tetragonum* in the genus *Chorisma* which was later referred to as a section by de Candolle. Harvey transferred the species to section *Jenkinsonia* but recent studies indicate that the section *Chorisma* should be reinstated and include three other less well-know species, until recently classified in the section *Ligularia* (Albers *et al.*, 1995). These are all shrubs or subshubs with herbaceous or semi-succulent stems. The simple leaves have petioles usually longer than the lamina. The inflorescence has up to six flowers on short pedicels, each with two very large upper petals and two or three very much smaller lower ones. There are five or seven fertile stamens. All have very large chromosomes with a basic number $x = 11$. The species may be found wild in the southern part of Cape Province from the western to eastern Cape.

P. tetragonum (Linnaeus f.) L'Héritier: A sprawling or upright tufted plant with brittle green, three or four-angled, succulent stems jointed at the nodes. The rounded or reniform dark green, often marked with a darker zone leaves are quickly deciduous. The flowers are large, cream or pale pink, one or two at the ends of the branches with the upper petals veined dark red, a hypanthium up to 6 cm long and seven strongly exerted fertile stamens, the filaments bent sharply upwards.

It was collected by Francis Masson on his second expedition around the Cape Peninsula and introduced to Kew in 1774 and also grows wild in dry rocky ground from the west coast to Grahamstown.

Section Myrridium

Most species in this section are herbaceous plants with pinnately divided leaves and rather long straggly stems sometimes woody at the base. The flowering stems become very elongated, with smaller leaves along their length, while long peduncles hold the flowers above the foliage. The inflorescence is few flowered and each flower is very irregular in shape with two very large upper petals, usually with long claws, but three or sometimes only two, small lower ones. The sepals have a membranous texture and are conspicuously veined and ribbed. The hypanthium may be quite long and the fertile stamens five or seven. The fruits enlarge quite considerably as they ripen (van der Walt and Boucher, 1986). The plants are often found as sprawling roadside weeds and many are pioneer plants colonising disturbed ground. Most grow wild in the winter rainfall areas of south-western Cape Province, though some species extend further east and north into the summer rainfall regions. The basic chromosome number is $x = 11$.

P. longicaule Jacquin: These are low growing, spreading, herbaceous plants with dark green leaves, often tinged red, deeply pinnately divided to bi-pinnatisect. The white to pale pink flowers, often tinged yellow in bud, have upper petals with conspicuous red feathering and seven fertile stamens. The first reference to this plant in 1767 described this species as a form of *P. myrrhifolium* but there has been much confusion about the nomenclature of members of this section. It grows in mountainous regions of south-western Cape Province.

P. caucalifolium Jacquin: Superficially this species might be mistaken for *P. longicaule* but in *P. caucalifolium* the stems tend to be woodier and the flowers rarely have more than four petals and there is usually only one flower, rarely two on each flower stem.

P. myrrhifolium (Linnaeus) L'Héritier: This is a low growing, short lived subshrub with pinnatifid or bi-pinnatisect leaves. Flowers are white to pale purplish-pink, each peduncle bearing up to five flowers, the upper petals with long claws and deep red feathered markings and five or seven fertile stamens. This species is very variable in the wild being found from Namaqualand through south-western Cape Province to Port Elizabeth in southern Cape Province. It was one of the earliest pelargoniums to be introduced and it was first illustrated in 1678. The variety *coriandrifolium* has more finely divided leaves and larger flowers but many plants found in cultivation under this name are the larger flowered *P. longicaule*.

P. candicans Sprengel: This species shows some resemblance to *P. myrrhifolium* var. *myrrhifolium* with rather small pinkish coloured flowers. It is however distinguished by leaves which are a gray almost silvery green, often with a darker blotch in the centre and with two lobes at the base. The flowers always have four petals and five fertile stamens.

P. suburbanum Clifford ex Boucher: This is a sprawling herbaceous perennial, and the leaves have long conspicuous hairs. The flower buds are large and the flowers bright pink or white to pale yellow held well above the foliage, the upper petals with dark red feathered markings and with seven fertile stamens. Two subspecies are easily distinguished in flower as ssp. *suburbanum*, which has bright pink flowers with a shorter hypanthium, while the flowers of ssp. *bipinnatifidum* are cream with the hypanthium exceeding 15 mm. Whereas ssp. *suburbanum* is found in eastern Cape Province, ssp. *bipinnatifidum* grows in the south-west.

Section Jenkinsonia

In his first volume of *Geraniaceae* in 1820–1822, Robert Sweet proposed a new genus, *Jenkinsonia*, which he named after Robert H. Jenkinson, one of the well known collectors of members of the family *Geraniaceae* in the early nineteenth century. He included in this genus, species now reclassified in the related sections *Myrrhidium* and *Chorisma* (van der Walt *et al.*, 1997). Most of the species in this recently redefined section are subshrubs, some with tubers although two species, *P. senecioides* and the recently described *P. redactum* are annuals, rarely seen in cultivation. The leaves, sometimes soon deciduous, are palmately or pinnately divided and in several species, aromatic. The length of the hypanthium may be shorter or longer than the pedicel, the petals four or five, the upper usually larger and sometimes rolled to form a tube-like structure. There are two, three, five or seven fertile stamens. The species of this section with a basic chromosome number x = 9, are distributed through most of southern Africa including Namibia and Botswana but the majority may be found in the north-western area.

Figure 8.1 P. praemorsum. (See Colour plate 1)

P. antidysentericum (Ecklon and Zeyher) Kosteletzky: This is an erect branching shrub with a very large, partly underground tuber and thin branches becoming woody with age and often with hard curved spines formed from persistent stipules. Rounded, five-lobed, aromatic leaves are produced in clusters on short branches. The flowers have five purplish-pink or white petals veined with purple, a long hypanthium up to 30 mm and seven prominently exserted fertile stamens, the filaments slightly curved upwards. This unusual species is found growing in very arid regions of Namibia and Namaqualand. The enormous tubers were used by the Hottentots as a medicine in cases of dysentery and anaemia. First described in 1835, it was known to be in cultivation a few years later but is only seen in a few specialist collections today.

P. praemorsum (Andrews) Dietrich: This is a branching subshrub up to 30 cm or so in height with rather thin brittle stems (Figure 8.1). The leaves are almost round, deeply divided into five, sometimes clustered at the nodes, soon deciduous but the stipules persistent and rigid. The flowers are cream or pink to purple, the upper petals are very large, veined with deep red or purple, the lower petals two or three and the seven exserted fertile stamens have filaments strongly curved upwards.

It is thought to have been first introduced into cultivation in England from Namaqualand at the end of the eighteenth century and although not common is spectacular in flower.

P. trifidum Jacquin: This is a scrambling herbaceous plant woody at base with bright green trifoliate to trifid leaves, strongly scented when touched. Flowers large, cream to almost white, the larger upper petals with conspicuous red veining, the hypanthium long and seven fertile stamens (Figure 8.2).

This freely seeding plant is found wild in arid regions from the south-western to the eastern regions of the Cape Province, and was first introduced into cultivation at Kew

Figure 8.2 P. trifidum. (See Colour plate 2)

in 1790. Several forms, differing slightly in leaf and flower size are seen in cultivation, some with a more pleasant scent than others.

Section Subsucculentia

The four species of this recently proposed section with large chromosomes and a basic number x = 10, have been separated from the section *Ligularia* (van der Walt *et al.*, 1995). All are rare in cultivation but in the wild may be found on the west coast of southern Africa from Namibia to the western Cape. They form deciduous, shrubs or subshrubs with woody or somewhat succulent stems, often covered with spines formed from the persistent remains of the petioles. The inflorescence of up to ten flowers, each with five petals, the upper and lower more or less similar in size, five or seven fertile stamens and the hypanthium long, usually exceeding the pedicel.

Section Quercetorum

As a result of recent investigations, it has been proposed that a number of species from the Sections *Ciconium*, *Ligularia* or *Jenkinsonia* are included in this new section Although this grouping reflects their phylogenetic identity, the four species are geographically and morphologically very diverse. Two more or less hardy species from the Middle East in Iraq and Turkey, (*P. endlicherianum* and *P. quercetorum*), have swollen roots with two very large magenta coloured upper petals but with minute or absent lower petals and seven fertile stamens. One species from the south-western Cape Province (*P. karooicum*) is a small woody plant with long semi-succulent branches becoming woody with age, deciduous leaves, white pink or pale yellow flowers, the upper petals slightly larger than the lower and five fertile stamens (Figure 8.3). The fourth species from Madagascar (*P. caylae*) is a tall evergreen shrub with pale purple-pink flowers with somewhat undulate petals and five fertile stamens. All are extremely rare in cultivation in Europe.

Figure 8.3 P. karooicum. (See Colour plate 3)

Subgenus *Pelargonium*

All members of this subgenus have small chromosomes and it is now separated into eight sections, several of which have been subdived into subsections. About 77 per cent of the recognised species belong in this subgenus. The majority are found in the winter rainfall areas of the south-western Cape and Western areas of southern Africa although members of the Section Peristera are geographically spread from Australia or several Oceanic islands and through southern and eastern Africa.

Section Campylia

Initially, the name for this section was used by Sweet in 1820 for a new genus to include *P. ovale*, on account of the two bent staminodes. Others of the present section such as *P. tricolor*, with erect staminodes and shiny raised bumps at the base of the two upper petals, he assigned to yet another genus, *Phymatanthus*, meaning wart-flower. These patches are false nectaries to attract insects for pollination (MacDonald and van der Walt, 1992). The two genera were then amalgamated into one section, *Campylia*, by de Candolle a few years later and the recent work in South Africa has added the newly described species, but as yet rarely cultivated, *P. ocellatum*, as well as some from other sections (van der Walt and Roux, 1991).

The plants are small, with short stems and all have an extensive root system enabling the plants to tap water from well below the surface. The leaves have long petioles with conspicuous membranous stipules at the base. The flowers are often rather open or 'flat' in shape with five, occasionally four, usually broad petals, a long pedicel and five or seven fertile stamens. The hypanthium is usually very short and in some cases the only sign is a slight depression at the base of the upper sepal. In some species such as *P. ovale*, the upper petals have constricted or auricled bases. The basic chromosome number is x = 10. There appears to be some affinity between this section and the section *Pelargonium*. *Campylia* includes some of the most difficult species of the genus to grow. In the wild, this section is centred in south-western and southern parts of Cape Province where

Figure 8.4 'Splendide' a hybrid between *P. tricolor* and *P. ovale*. (*See Colour plate 4*)

the plants tend to grow in mountainous regions with very freely draining soils. Of about ten species in the genus, about three are found in general cultivation.

P. elegans (Andrews) Willdenow: This is a small erect tufted plant with green, hairy or glabrous, broad ovate to almost rounded leaves, the margin often reddish. The flowers large, to 3 cm across, pale to deep pink, the upper petals with darker markings, and with seven fertile stamens. *P. elegans* introduced by the British nursery firm of Lee and Kennedy in 1795, is found in south-western Cape Province, often near the coast.

P. ovale (Burman filius) L'Héritier: This is a low growing plant with hairy, often gray-green, almost rounded to lanceolate leaves. The flowers, about 4 cm across, are white to deep pink, the upper petals with darker markings and have five fertile stamens. This is a variable species and several subspecies have been named (van der Walt, J.J.A. and Van Zyl, L., 1988). In the wild, it is found in mountainous regions of Cape Province and Francis Masson collected slightly different plants for Kew on at least three of his expeditions in 1774, 1790 and 1794. It is one of the easier of the section to cultivate.

P. tricolor Curtis: This is a small plant to about 30 cm high. The leaves are gray-green, hairy, narrow with sharp irregular teeth and often two larger lobes at the base. The flowers *c.* 2 cm across, resemble a wild pansy, the upper petals deep red with a glossy black raised spot and the lower petals white. Fertile stamens are five in number. *P. tricolor* is found growing wild, often under other shrubs, in dry sandy soils in the foothills of the mountains in south-western Cape Province.

It was first collected by Francis Masson on his expedition of 1791. Plants are occasionally found with unicoloured white or mauve-pink flowers still bearing the dark patches at the base of the petals. The true species is rare in cultivation and difficult to grow but the cultivar 'Splendide', a hybrid with *P. ovale*, is easier and may be distinguished by the long petioled, gray-green, hairy leaves which are broadly ovate and more regularly serrated, the slightly larger flowers and dark purple stamens (Figure 8.4).

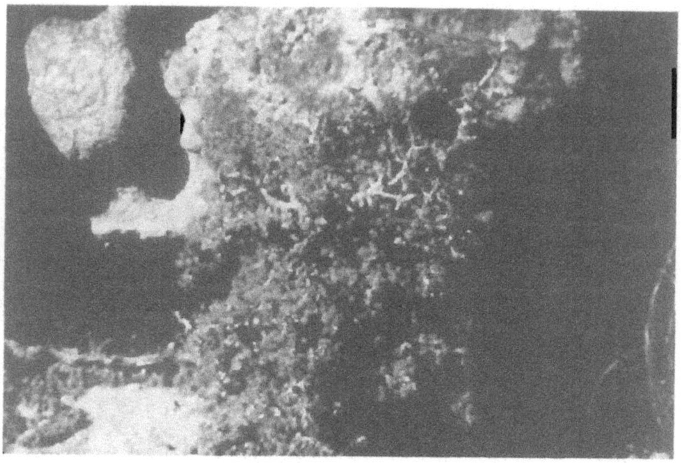

Figure 8.5 P. alternans. (*See Colour plate 5*)

Section Otidia

This small section is well defined by the succulent habit of the species, totally unlike other pelargoniums, showing some resemblance to members of the families *Crassulaceae* or *Euphorbiaceae*. For this reason Sweet in 1822, adopted the genus *Otidia* proposed earlier by Lindley for *P. carnosum*. This was later demoted to a section within *Pelargonium*. All but one of these species are found in the arid regions of the western side of South Africa and Namibia in areas where the winter rainfall may be under 100 mm a year. The plants shed their leaves, becoming more or less dormant during the hot, dry summers. Some grow near the coast and have to withstand salt-laden winds from the sea. The plants grow in rocky areas but do survive in shifting sand dunes.

The plants have succulent stems, fleshy pinnate, often compound leaves and very small stipules but the habit may be variable, especially so in cultivation when water is less scarce. The leaves are usually clustered towards the top of stems as the lower ones fall in previous seasons. The inflorescence is often branched, each branch bearing a few white or cream flowers. The petals are not markedly dissimilar in size and shape, although in most species, the upper two have reddish marks while the lower three are plain. The flowers have five fertile stamens and a short hypanthium. The basic chromosome number is x = 11.

P. alternans Wendland: This is a succulent branching plant with deeply lobed, hairy, pinnate leaves (Figure 8.5). The white flowers *c.* 1.5 cm across, have narrow reflexed petals, the upper marked with red lines. Plants in cultivation tend to be more vigorous whereas in the wild the old plants are stunted and woody. The species may be found in the dry and mountainous regions of south-western Cape Province and is said to have been introduced to Kew by F. Masson in 1791.

P. carnosum (Linnaeus) L'Héritier: This is a succulent plant 30 cm or more tall with very variable oblong, more or less succulent, gray-green to green leaves, to *c.* 15 × 5 cm. The white, sometimes greenish-tinged flowers are almost regular in shape, about 1.5 cm across with slightly reflexed upper petals marked with reddish-purple lines.

This, the first known member of the section first illustrated by Dillenius as early as 1732, is still the most widely grown today. With its very wide natural distribution in southern and south-western Africa, it is variable in its morphological characteristics and has been collected on many expeditions.

P. ceratophyllum L'Héritier: This is a much smaller plant with thinner stems and almost rounded succulent leaf segments and small petals barely longer than the sepals. This is one of several species which grows wild in south-western Namibia and north-western Cape Province. It was first collected in 1786 by A.P. Hove, a collector for the Royal Botanic Gardens of Kew.

P. dasyphyllum Meyer, E.: This species forms a branching succulent which has been mistaken for *P. alternans*, but is less visibly hairy with a less woody appearance. The narrow, unmarked, strongly reflexed petals give the flowers an asymmetric appearance. It grows in the north-western part of Cape Province and was first collected in about 1830.

P. laxum (Sweet) Don, G.: Apart from the much more irregular appearance of the flowers, this species might superficially be mistaken for *P. carnosum*. It is the only species of the section to be found in the summer rainfall zones of central and eastern Cape Province. It is said to have been cultivated from seed sent to Colvill in 1821.

Subsection Cortusina

This section has recently been split and several species have been transferred to the new section, *Reniformia* (Dreyer *et al.*, 1992). However the most recent work by Bakker *et al.* (2000) from the most recent phylogenetic analyses, has proposed that as this group of plants are so closely related to members of the section *Otidia* that a subsection is more appropriate. Many distinct morphological characteristics as well as the karyological evidence and very widely separated geographical locations, made this an obvious division.

In 1824, de Candolle created a subseries for species such as *P. echinatum*. Ecklon and Zeyher and raised it to the status of genus in 1835, but Harvey in Flora Capensis in 1859 used the name as a section. The species are from the desert and semi-desert areas of north-western part of Cape Province and Namibia where the winter rainfall is exceedingly low, often under 10 cm annually. The stems of all except one, are succulent for the storage of water and often covered with the remains of persistent hard stipules or petiole bases, giving extra protection. The long petioled, simple, usually almost rounded leaves, are shed in periods of drought. Several species have the added advantage of tuberous or thickened roots. The inflorescence is rarely branched, but each peduncle bears several flowers. Each flower has an almost regular appearance with six or seven fertile stamens and, except for *P. desertorum*, the hypanthium is always conspicuously longer than the pedicel. The basic chromosome number is $x = 11$.

P. cortusifolium L'Héritier: This is a branching plant to 30 cm with thick succulent stems and almost rounded hairy leaves with a silvery sheen. Flowers are white to pale pink or lilac in inflorescence of about ten flowers, the lower petals often darker in colour than the upper; fertile stamens six. This species is found along the coastal areas of southern Namibia in desert conditions. It was first collected by A.P. Hove in 1786 during an expedition to the south west coast of Africa.

Figure 8.6 P. echinatum. (*See Colour plate 6*)

P. crassicaule L'Héritier: This is a low growing plant with thick brown knobbly succulent stems and very broad ovate gray-green leaves with a soft velvety texture. It has white to pale pink or lilac flowers in inflorescence of about eight flowers which are slightly scented. The reflexed upper petals are often blotched and lined with purple; fertile stamens six or seven. Like *P. cortusifolium*, this species was introduced by Hove in 1786 from southern Namibia.

P. desertorum Vorster: This species is very similar to *P. xerophyton*, but with strongly aromatic foliage and the hypanthium shorter than the pedicel.

P. echinatum Curtis: This is an erect subshrub with thick spines on stems from persistent petioles and ovate, shallow 3–5 lobed, gray-green, leaves paler and hairy below, up to 6 cm across. Flowers are usually white in cultivation but also pink to purple in wild, 1.5–2 cm across; ovate upper petals each have a dark red blotch and lines and 6–7 fertile stamens (Figure 8.6).

The species grows wild in the north-western regions of Cape Province and a white flowered specimen was first collected by F. Masson for Kew in 1789. 'Miss Stapleton' which has bright purplish pink flowers spotted on each petal with longer straighter spines, was raised from seed collected at Colvill's nursery and first flowered in 1823.

P. xerophyton Schlechter ex Knuth: This is a low growing plant with numerous thick and almost woody branches with dull to blue-green broad ovate leaves. It has white, usually solitary flowers, the upper petals are narrow and spotted red; there are seven fertile stamens. This species, first recorded in 1897, is native to southern Namibia and north-western Cape Province, surviving in the shelter of rocks.

Section *Pelargonium*

This is a large section which contains some very important plants for both the horticultural industry and as a crop for the commercial production of Geranium oil. It includes the parents of most of the scented leaved cultivars, the angel pelargoniums and the Uniques, as well as the species, *P. cucullatum*, from which all the regal pelargoniums have been developed. These species may usually be found in areas where there is at least some water, either in the subsoil or near streams or rivers, even though these are often dry for part of the year. The majority are to be found in the winter rainfall regions of south-western or southern Cape Province, many near to the coast. Unpruned, most will become large and eventually, quite woody shrubs. The foliage is often aromatic and sometimes viscid. The flowers, in few flowered inflorescences are white, pink or purple but none are a true red or yellow. The upper petals are larger in size and often a different shape compared to the lower petals and are marked with darker spots or lines. There are seven fertile stamens and the hypanthium may be longer or shorter than the pedicel (van der Walt, 1985). The basic chromosome number is x = 11. They appear closely related to the subsection *Glaucophyllum* but do not grow in such arid situations.

The cultivars grown for rose-scented Geranium oils have been developed from species such as *P. capitatum* and *P. radens*, and several other species have scented foliage which have been considered as possible sources of aromatic oils, may be used in pot pourri and in some cases for culinary flavouring.

P. betulinum (Linnaeus) L'Héritier: This is an upright or decumbent branching woody plant with rather stiff, toothed leaves *c.* 20 × 15 mm. Flowers are pink or purplish-pink, sometimes white, 3 cm or more across, upper petals heavily veined with purplish-red, lower petals unmarked or with faint markings, narrower than upper. This species, which is found near the south-western and southern coast line of South Africa, was known in European gardens in the mid-eighteenth century. Natural hybrids with *P. cucullatum* are found in the wild and it is almost certainly one of the ancestors of some of the regal and Unique pelargoniums, as a result of accidental crosses in the early days of the eighteenth century. It has been used for the treatment of chest complaints in its native land.

P. capitatum (Linnaeus) L'Héritier: This is a decumbent, somewhat spreading, or weakly erect, softly hairy plant and densely hairy, strongly rose-scented foliage to 2–8 cm across. Flowers measure to 15 mm across, usually mauve-pink in very dense compact heads of 10–20 flowers, upper two petals narrow obovate, veined darker pink. *P. capitatum* found in many areas along the coasts of South Africa, was one of the earlier species imported to Europe and records indicate that it was brought into England from Holland in 1690. The true species is quite rare in gardens but is represented by the cultivar 'Attar of Roses' with a more upright habit and rougher but strongly aromatic foliage and pinker flowers.

P. citronellum J.J.A. van der Walt: This is a rough, hairy, strongly lemon-scented shrub to 2 m with irregularly toothed leaves (Figure 8.7). Flowers are pale pinkish-purple in

Figure 8.7 P. citronellum. (*See Colour plate 7*)

an open branching inflorescence, upper petals reflexed, marked with dark purple lines and blotches. Until recently, this plant was not recognised as a distinct species because of its similarity to *P. scabrum*. It normally grows near streams in a small area of southern Cape Province. The popular, lemon-scented cultivar 'Mabel Grey' is very similar.

P. crispum (Bergius) L'Héritier: This is a very upright plant, the base of the stems becoming woody with small strongly lemon-scented, rough leaves up to about 7 mm across with crisped margins. The large flowers are usually pink, the upper petals with darker markings and notched at the apex. The species was first described in 1767 and grows wild in south-western Cape Province. Garden plants show a wide range of leaf size and a number of cultivars are grown for their lemon-scented foliage such as the smaller leaved, 'Minor', the more vigorous 'Major' and 'Variegatum' with leaves edged with creamy white.

P. cucullatum (Linnaeus) L'Héritier: This is a large, erect, branched, hairy, shrub up to over 2 m in height. Leaves rounded to triangular in shape, toothed and sometimes shallow lobed, usually hood- or cup-shaped with long hairs, sometimes aromatic, *c.* 4.5 × 50 cm. Flowers measure to over 4 cm or more across, and are bright purplish pink in a large branching inflorescence, with the upper two petals veined deeper pink, lower petals slightly smaller, narrower.

P. cucullatum grows wild in Cape Province and was introduced into cultivation in the early part of the eighteenth century. Three subspecies, for many years treated as distinct, are found in geographical areas that rarely overlap (Volschenk *et al.*, 1982). It is almost certain that ssp. *cucullatum* and ssp. *strigifolium* are parents of the regal or fancy pelargoniums. *P. cucullatum* ssp. *tabulare* with softly hairy, distinctly cup-shaped and strongly scented leaves was involved in early hybrids such as 'Purple Unique'. In cultivation, it is sometimes difficult to decide which subspecies to assign a plant, probably because of the hybridisation between the subspecies.

P. denticulatum Jacquin: This is an erect branched shrub up to over 1.5 m, with strongly balsam-scented, sticky dark green leaves which are bi-pinnatisect, the ultimate segments being irregularly toothed. Flowers are purplish-pink, up to 2 cm across; the peduncle measures up to 5 cm, bearing about six flowers; upper petals are narrow and spathulate, veined darker pink, with the apex usually notched, *c.* 18 × 6 mm; lower petals are slightly smaller, unmarked; the hypanthium is *c.* 8–9 mm; the pedicel up to 2 mm; there are seven fertile stamens. It comes from south-western and southern Cape Province growing in the moister areas especially beside streams and in deep valleys and was introduced into cultivation in 1789. The cultivar 'Filicifolium' with very finely divided leaves is a form of this species.

P. glutinosum (Jacquin) L'Héritier: *P. glutinosum* also has the viscid, strongly scented foliage of *P. denticulatum* and *P. quercifolium* but has a far wider distribution in the wild. The leaves are less rough than in these species. The flowers are pale to dark pink, *c.* 1.5 cm across. It shows a range of variation in both its leaf and flower characteristics. Some scented-leaved cultivars such as 'Viscosissimum' are probably forms of this species.

P. '*Graveolens*': It is a vigorous, branching, rose-scented, shrub to over 1.5 m in height with somewhat rough, deeply divided green leaves *c.* 4 × 6 cm, with the ultimate segmented margins slightly curved under. Flowers are pink to pale purplish-pink, up to 15 mm across in inflorescence of 5–10 flowers, the upper petals are narrow with rounded or notched apex, veined dark purplish-pink, and have stamens with stunted anthers rarely bearing pollen.

The naming of this old and exceedingly well known rose-scented *Pelargonium* grown on the window sills of so many houses in this country, Europe and North America, is not easy to resolve. In 1792, a plant was illustrated by L'Héritier which is more or less identical morphologically with the plant most commonly grown today as 'Graveolens' and it was almost certainly a hybrid.

'Rosé', the cultivar widely grown for the production of rose-scented Geranium oil has been shown to be a hybrid between *P. capitatum* and *P. radens* (Demarne and van der Walt, 1989) and is very similar morphologically to 'Graveolens', 'Radula', 'Little Gem' and the variegated 'Lady Plymouth'. The species *P. graveolens* found wild in the northern part of South Africa, has white flowers and deeply divided, softly hairy, somewhat peppermint-rose scented leaves. The flowers are larger, fewer with longer pedicels and hypanthia giving the inflorescence a more open appearance also bears seven fertile anthers.

P. quercifolium (Linnaeus f.) L'Héritier: This is an erect, branching viscid shrub with a strong balsam scent. Leaves are rough with long glandular hairs, usually up to 5 × 5 cm, pinnately lobed to pinnatisect. Flowers are pale pink to darker purplish-pink, in inflorescence of 3–6; the upper petals have a notched apex and darker purplish-pink lines and blotches (Figure 8.8). This species, from southern Cape Province, was first introduced to Kew about 1774. It is variable in leaf and flower characteristics in its native habitat and some forms resemble cultivars seen in gardens such as 'Royal Oak'.

Figure 8.8 P. quercifolium. (See Colour plate 8)

P. radens H.E. Moore: Erect branched aromatic shrub to over 1 m in height. The very rough grayish-green leaves are deeply divided into narrow segments, the margins of each segment rolled under giving an even narrower appearance. The pale or pinkish-purple flowers are in a loose inflorescence of about five, the upper two petals with darker markings. This species from south-western Cape Province to Transkei, was first introduced to Kew by Masson in 1774. It is not common in cultivation and should not be confused with the rose-scented cultivar 'Radula'. *P. radens* has very rough leaves and the segments are exceedingly narrow as the margins are rolled under. This species is considered with *P. capitatum* to be a parent of the cultivar 'Rosé' grown in great quantities for the commercial production of Geranium oil on the island of Reunion.

P. tomentosum Jacquin: This is a low growing, but wide-spreading, branched subshrub strongly scented of pepperment. Leaves with three or five round lobes 4–6 × 5–7 cm, have long, soft, velvety hairs. The white flowers in a loosely branched inflorescence are small, the upper two petals obovate, with purple lines. This species is a popular ornamental plant and has also been used as a culinary flavouring and for the production of peppermint oil. It has been cultivated continually since its introduction by Francis Masson in 1790. Unlike many *Pelargonium* species, it will grow in more slightly shaded and moist situations.

Subsection Glaucophyllum

This is rather a small section and is very close to section *Pelargonium* and hybrids between the two sections may even be found occasionally in the wild. Bakker *et al.*, 2000 have proposed that the section *Glaucophyllum* be demoted to a subsection within section *Pelargonium*. The leaves are usually glaucous, as the name suggests, and often rather leathery, simple or divided. The white, cream or purplish-pink flowers in a few flowered inflorescence, are very irregular in shape with a hypanthium usually considerably longer than the pedicel. Five to seven fertile stamens are present. The

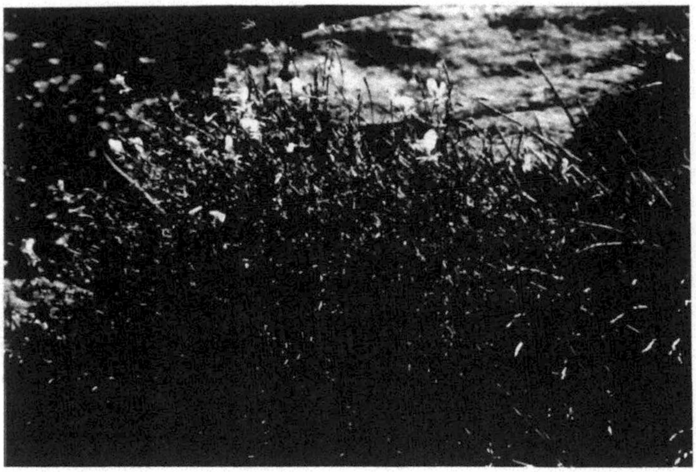

Colour plate 1 P. *praemorsum*. (*See page* 58)

Colour plate 2 P. *trifidum*. (*See page* 59)

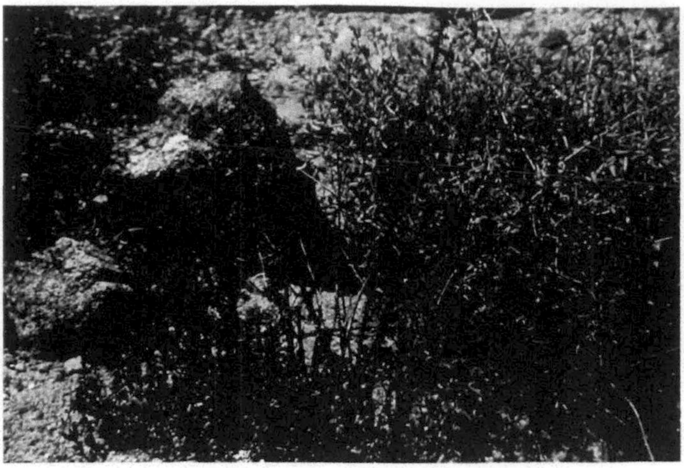

Colour plate 3 P. *karooicum*. (*See page 60*)

Colour plate 4 'Splendide' a hybrid between *P. tricolor* and *P. ovale*. (*See page 61*)

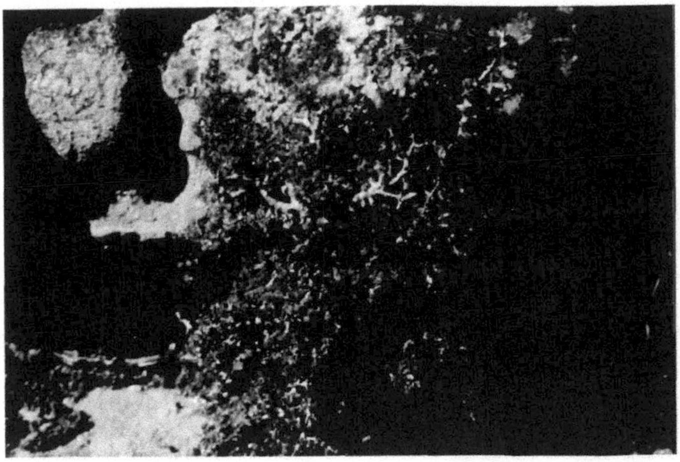

Colour plate 5 *P. alternans*. (See page 62)

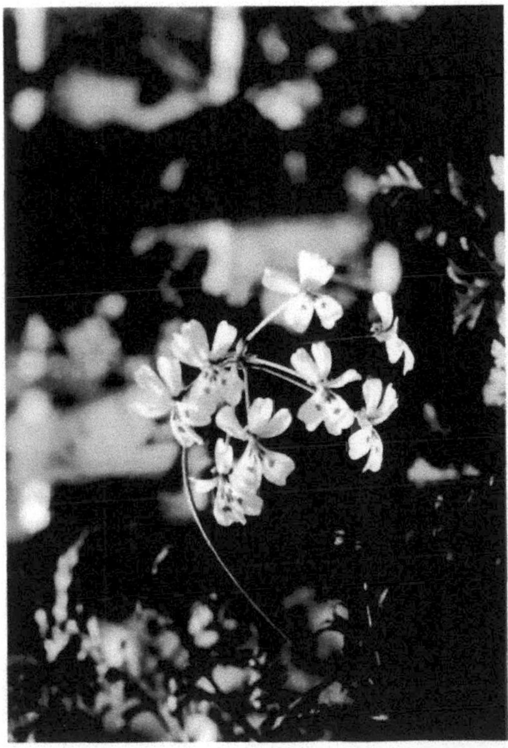

Colour plate 6 *P. echinatum*. (See page 64)

Colour plate 7 P. citronellum. (*See page* 66)

Colour plate 8 P. quercifolium. (*See page* 68)

Colour plate 9 P. grandiflorum. (*See page 69*)

Colour plate 10 P. lanceolatum. (*See page 70*)

Colour plate 11 P. appendiculatum. (*See page 71*)

Colour plate 12 P. fulgidum. (*See page 72*)

Colour plate 13 P. reniforme. (*See page* 76)

Reniform leaves ('reniforme')

Flower head with distinctive stripes and dots

Morphology of *Pelargonium reniforme* ssp. *reniforme*

Colour plate 14 P. reniforme CURT. (*See page* 264)

Cordate leaves

Flower head

Morphology of *Pelargonium sidoides*

Colour plate 15 P. sidoides DC. (*See page 273*)

plants are subshrubs, either upright or trailing and the lower parts of the stems are woody. The basic chromosome number is x = 11. The species are found in the rocky, mountainous regions of southern and south-western regions of Cape Province in the winter rainfall region (van der Walt *et al.*, 1990).

P. grandiflorum (Andrews) Willdenow: This is an erect, almost hairless plant with glaucous deeply 5–7 palmately lobed, sometimes zoned leaves. Flowers 2–3, pink to pale purple-pink or sometimes creamy white, the upper petals marked with darker blotches and streaks (Figure 8.9). It is thought to have been introduced in 1794 by Francis Masson and was used in the early nineteenth century in hybridising.

P. tabulare (Burman filius) sensu J.J.A. van der Walt and *P. patulum* Jacquin.

These are similar but with smaller flowers. *P. tabulare* is an erect plant but shorter than *P. grandiflorum* whereas *P. patulum* has a trailing habit.

P. laevigatum (Linnaeus filius) Willdenow: This is a small, very variable, often rather straggling, usually glabrous plant with variable, usually glaucous, slightly fleshy leaves divided into three linear or rounded segments, themselves three parted, with almost spine-like apices. The white or pale pink flowers are normally single, the upper petals reflexed and marked with red. First collected by Thunberg around 1773, described in 1781, illustrated in several early books and easy to cultivate, this species is not widely grown.

P. lanceolatum (Cavanilles) Kern: This is an upright branching plant with lanceolate, untoothed, glaucous leaves (Figure 8.10), Flowers one to two in number, are cream to pale yellow with the upper petals marked with red and strongly reflexed. In leaf, this species is very atypical but the flower is quite typical of the section.

Section Hoarea

In the wild, the species of this section grow mainly in the arid inhospitable areas of winter rainfall of west and south-west Cape Province with a few extending into

Figure 8.9 P. grandiflorum. (See Colour plate 9)

Figure 8.10 P. lanceolatum. (See Colour plate 10)

southern Cape Province. This is the largest section within the genus containing about 80 species. The plants die down immediately after flowering and pass most of the year underground in a dormant state. The foliage appears after the winter rain but the flowers do not develop until the leaves have died down, although in cultivation the leaves may remain during flowering. All are geophytes with one or more tubers covered with papery sheaves which help to prevent the tuber drying out in summer. The inflorescence, often branched, usually has many flowers and each flower may have either two, three, four or five petals and five, sometimes two, three or four fertile stamens. The flowers are very irregular and have a hypanthium exceeding the pedicel in length. The basic chromosome number is x = 11.

P. appendiculatum (Linnaeus filius) Willdenow: This is a plant with a large underground tuber and gray-green, densely hairy, aromatic, feathery leaves, to 10 × 5 cm. Yellow flowers about 4 cm across, borne on a long stem, have very narrow petals, the upper petals spotted with red (Figure 8.11). It was collected by Thunberg in the late eighteenth century. Although not common in cultivation, it is worth growing for its tall many-flowered inflorescence.

P. auritum (Linnaeus filius) Willdenow: This is a rather small tuberous plant with very variable leaves from simple to bi-pinnatifid, the lamina up to 12 cm or more long. The flowers may be very dark purple, pink or white, in many flowered heads with linear petals and long exserted stamens giving a star-like appearance. It may be found growing wild in south-western Cape Province and has been known in Europe since 1697.

P. incrassatum (Andrews) Sims: This is a plant with a large fleshy underground tuber and deeply pinnately lobed leaves with soft grayish-white hairs. The 20–40 bright magenta flowers are borne terminally on long stems, and have three lower petals with in-rolled edges. It was first collected by Francis Masson in 1791 from the coastal regions of western Cape Province and Namaqualand.

Figure 8.11 P. appendiculatum. (See Colour plate 11)

P. pinnatum (Linnaeus) L'Héritier: This stemless tuberous plant has bluish-green leaves, oblong in outline, pinnate with elliptic pinnae each *c.* 10 × 5 mm. The flowers, each about 2 cm across, are clear pink but pale yellow and white forms may be found in the wild. It is not uncommon in the south-western Cape Province and has been known since the early eighteenth century.

P. rapaceum (Linnaeus) L'Héritier: This is a stemless plant with a large and several smaller tubers. The leaves are hairy, linear, pinnate, the pinnae further divided into linear segments forming whorls along the rachis. The flowers are similar in shape to a legume, pink or yellow, in heads of ten or more. The upper petals are held together erect, while the lower petals are held together enclosing the stamens. It was one of the earliest of the section to be collected with early references of Commelin and Hermann at the end of the seventeenth century and in the wild has a very wide distribution through Cape Province.

Subsection Ligularia

This section as circumscribed by Knuth is one of the most diverse both morphologically and geographically. However a karyological study together with attempts to hybridise species of this section and others, has shown that many of the species should be moved to other sections and for some new sections created (Albers *et al.*, 1992). In most cases, the results were confirmed by morphological similarities and also by their natural geographical locations. The remaining members of the existing section have small chromosomes with a basic number x = 11. All have fleshy stems, often covered with the remains of stipules, sometimes becoming quite hard. The leaves in most species are adapted to survive in regions of quite low rainfall on the western side of South Africa.

It is interesting to note that some of the species included in this section and others originally included but now moved, are of more recent introduction and were not known to the early botanists at the height of the interest in these new Cape introductions. More recently phylogenetic work by Bakker *et al.* (2000) have suggested that the remaining species within this reduced section are classified as a subsection of *Hoarea*.

P. fulgidum (Linnaeus) L'Héritier: This is a spreading or scrambling, softly hairy plant with somewhat succulent stems. The pinnately lobed leaves often have two almost free lobes at base and a silver sheen. Flowers are scarlet in a branched inflorescence with 4–9 flowered heads; upper petals veined darker red, strongly reflexed; the hypanthium is long, often dark brownish-red, very conspicuous and swollen at the base and it has seven fertile stamens (Figure 8.12).

This is one of the older species to be introduced into Holland in the early eighteenth century and from there sent to Italy and England. It appears to be one of the few members of the genus which is able to hybridise with species of several other sections

Figure 8.12 P. fulgidum. (See Colour plate 12)

such as *Polyactium*, and *Pelargonium*. It is involved in the parentage of some red flowered cultivars such as 'Scarlet Unique' and possibly the regal pelargoniums but is not part of the parentage of the zonal or ivy-leaved cultivars. It grows along the west side of South Africa, near to the coast.

P. pulchellum Sims: This is a hairy plant with a very short, rarely branched, thick semi-succulent stem and thickened root. The leaves are softly hairy and pinnately lobed to pinnatifid. Many white flowers are held in a branching inflorescence, the upper petals with faint red lines at base, sometimes red blotched, while unusually the lower petals have a prominent red blotch; the hypanthium is long and fertile stamens seven in number. This is one of the most striking species of the genus and will flower early in the season. It grows wild in Namaqualand.

Section Peristera

The section includes some of the more insignificant short-lived, dwarf herbaceous plants, often with straggly or weedy, trailing habits. Initially the plants are usually quite compact, but at flowering, elongate to produce many spreading branching stems with long internodes. The flowers are usually small, rather regular in shape, the petals often barely longer than the sepals, and borne on very fine pedicels in many flowered heads. The fertile stamens are five or seven, occasionally four and the hypanthium often very short or more or less non-existent. The leaves may be simple or divided, pinnately or palmately veined but usually with rather long petioles. This section representing about 17 per cent of the total number of species of *Pelargonium*, is more extensive than any other with species found in Australia, east Africa, islands such as Madagascar and Tristan da Cunha as well as in South Africa. Many individual species themselves have a very wide geographical distribution which is unusual within the genus as is the variable chromosome number.

P. australe Willdenow: This is a herbaceous perennial with a short erect stem, but with leafy flowering stems extending to about 30 cm long. Leaves are hairy or more or less glabrous sometimes flushed purple below. Flowers are white or pale pink, in a fairly compact inflorescence, the upper petals veined red; fertile stamens seven in number. This very variable species may be found in several parts of Australia, eastern Tasmania and New Zealand growing in a range of different habitats, usually in sandy dunes in coastal areas but also inland. Before flowering, the plants make rather neat rosettes but as the flower stems elongate and branch, the whole plant becomes straggling.

P. drummondii Turczaninow: This is a plant with semi-succulent thick stems covered with persistent brown membranous stipules and dark green, more or less orbicular leaves. Flowers are white to very pale pink, the upper petals veined dark pink; fertile stamens seven in number. This is a little known species that grows in a restricted location in rocky conditions of inland areas of western Australia and was first discovered about 150 years ago.

P. grossularioides (Linnaeus) L'Héritier: This is a spreading, short lived, herbaceous plant often with red-tinged stems. The rounded to reniform shallow 3–5 lobed leaves have a fruity scent variously described as peach, blackcurrant or coconut. Many tiny, usually purplish-red, flowers in a rather tight head, are about 8 mm across, and almost regular in shape; fertile stamens seven in number. It may be found over a wide area of southern Africa as well as Tristan da Cunha and was cultivated as early as 1731.

P. minimum (Cavanilles) Willdenow: This is a prostrate small plant with numerous small tubers, gray green pinnately divided leaves. Minute white flowers are in a cluster of 3–6 and have almost no hypanthium; fertile stamens five in number. It has been known since the late eighteenth century, and in the wild it is found in arid conditions throughout southern Africa where it is used in folk medicine.

P. rodneyanum Mitchell ex Lindley: This is a plant with brown tuberous roots and short aerial stems. Leaves are shallowly lobed on a petiole measuring up to about twice the length of the lamina. Flowers are dark rose or purplish-pink, in a branched inflorescence of about five flowers, the upper petals with darker pink markings; fertile stamens seven in number. This may be found in south-eastern and South Australia, often growing among rocks, and was first collected in 1836.

Subsection Isopetalum

P. cotyledonis (Linnaeus) L'Héritier: This is the only species in the section which Sweet assigned to its own genus and gave it the name *Isopetalum*. However, de Candolle in 1824 reduced the genus to a section. Although the remote geographical location as well as many unique characters of the only species, phylogenetic studies indicate a close relationship to section *Peristera* and section *Isopetalum* it is now considered to be more correctly classified as a subsection (Bakker *et al.*, 2000). It is a short, slightly branched plant with stout, succulent stems covered with scaly rough bark about 30 cm tall. The deciduous leaves, clustered at the ends of branches usually turn red before falling, are rounded 2–5 cm across, leathery, glossy dark green above and, gray hairy beneath. The numerous white unmarked flowers are regular, 1.5 cm across and have five, occasionally six, fertile stamens but the hypanthium is exceedingly short. The plants have 22 chromosomes. This species is endemic to the island of St. Helena in the southern Atlantic where it grows on rocky cliffs often exposed to salt sea spray. Unfortunately like many species of the native flora, it has become very rare in its native country and wild goats appear to have been one of the major causes of its decline. Grown as a curiosity and known as 'old man live forever', it has not lost its appeal since introduced to Kew in 1765 by John Bush.

Subsection Reniformia

The species found wild in the region of summer or all year round rainfall in the central and eastern parts of South Africa were assigned to the section, *Reniformia* (Dreyer *et al.*, 1992). However, the most recent proposed classification suggests that this section is very close to *Peristera* and should be included as a subsection (Bakker *et al.*, 2000). Most grow in areas receiving over 50 cm of rain during the summer and some have additional rain in winter. They are mostly small subshrubs, occasionally with fleshy stems. The stipules or leaf bases are persistent and the leaves simple, lobed and frequently aromatic. The many-flowered inflorescence is branched, each individual branch bearing relatively few small flowers but over quite a long season. The flowers are irregular in shape with the rather narrow upper two petals, held distinctly together and erect while the lower three are usually slightly broader and widespread. The hypanthium is long, the pedicel short and there are seven fertile stamens. The basic chromosome number is x = 8. Most species of this section are valued for their aromatic foliage and have been cultivated for nearly 200 years.

P. abrotanifolium (Linnaeus f.) Jacquin: This is a branching, erect somewhat straggly plant, with stems becoming woody with age and bearing remains of leaf stalks. Aromatic, gray-green leaves are deeply divided more or less to the midrib into linear segments. Flowers in cultivation are usually white but pink flowered plants are common in the wild. The typical white flowered form known in cultivation in Europe for over 200 years, is found over quite a wide area of Cape Province in dry often rocky situations in areas of mainly winter rainfall.

P. album J.J.A. van der Walt: This is an erect herbaceous plant with thick fleshy branching stems and aromatic rather viscid leaves, with whitish tinge. Flowers are white in branched inflorescence, each peduncle bearing about eight flowers, the upper petals are sometimes with small red markings. This recently described species may be found in the eastern Transvaal.

P. exstipulatum (Cavanilles) L'Héritier: This is a branching, woody subshrub with aromatic viscous, gray-green leaves Flowers are pale pink, 2–3 on an unbranched inflorescence, the upper petals with dark red feathered markings. It is known to have been grown by the Countess of Strathmore as early as 1779 and is found wild in southern Cape Province.

P. fragrans Willdenow: This is a small erect branching plant with gray-green, soft velvety leaves and a spicy aromatic scent. Flowers are white in a branched inflorescence, the upper petals erect, feathered with red. There has always been a question about whether this plant is a true species or not and for many years has been considered to be a hybrid between *P. odoratissimum* and *P. exstipulatum*. It was first discovered in Berlin in the early nineteenth century by Willdenow who considered it to be a true species, but has not been found in the wild.

P. odoratissimum (Linnaeus) L'Héritier: This is a low growing herbaceous plant with short thick main stem, the light green, usually apple-scented leaves arising from the top, but with spreading or trailing flowering stems. Flowers are white, inflorescence branching with small leaves, upper petals are finely marked with red. This popular species has probably been grown continually since its first introduction into Europe in 1724. Many have apple-scented foliage but in other plants the scent may be more pungent or spicy. In the wild it grows over a large area in the southern and eastern South Africa.

P. reniforme sensu J.J.A. van der Walt: This is an erect or trailing subshrub with small tuberous roots. Leaves are gray-green with velvety texture, silvery below and slightly aromatic. Flowers are bright pink to magenta; inflorescence branched; fertile stamens seven in number, sometimes six (Figure 8.13). The first introduction was by Francis Masson to Kew in 1791 from southern and eastern Cape Province. The tuberous roots of both this species and *P. sidoides* have been used medicinally for a variety of purposes in their native country and are now developed and used as a drug in Europe against tuberculosis (see Chapter 25).

Figure 8.13 P. reniforme. (*See Colour plate 13*)

P. sidoides de Candolle: This plant, with thick underground roots and a short erect stem, is covered with persistent petioles and stipules, from which arises a rosette of gray-green slightly aromatic leaves with silvery sheen. Flowers are a distinctive deep blackish-purple colour, the upper petals narrow, oblong, often twisted and curled backwards. This species growing from the Transvaal to southern Cape Province, was collected by the Swedish botanist Thunberg in 1772 and also by F. Masson.

Section Polyactium

The section includes species which are found in the winter rainfall area of south-west large, not covered by the papery sheaves typical of members of the section *Hoarea*. The leaves are lobed or pinnately divided and are produced, at least in the wild, at the same time as the flowers. The almost regular yellowish, greenish, dark brown or purple night-scented flowers each with a long hypanthium are formed on a many-flowered scape and have six to seven fertile stamens. The basic chromosome number for the section is x = 11.

P. lobatum (Burman filius) L'Héritier: This is a plant with a very large irregular tuber covered with a rough bark and a very short aerial stem and leaves variable in shape and degree of lobing, up to 30 cm long in the wild. Flowers are very dark purple-black *c.* 2 cm across, sweetly night-scented in umbels of 5–20 on a branched inflorescence, rounded petals, margined yellowish-green; fertile stamens seven in number. Tubers of this species were first sent to Holland from the Cape in 1698 from its natural habitat in south-western and southern Cape Province.

P. triste (Linnaeus) L'Héritier: This is a geophytic plant with a large tuber as well as several smaller ones and a very short succulent aerial stem. Basal leaves are hairy, somewhat carrot-like, deeply divided into narrow segments. Flowers are night-scented, brownish-purple, sometimes yellow or brown, with a broad yellowish margin about 1.5 cm across, upper petals, slightly reflexed; fertile stamens seven in number. This species was the first *Pelargonium* to have been introduced into cultivation in Europe in about 1630. It is found in sandy soils, often in colonies in the south-western and western regions of the Cape Peninsula. The tuber has been used as a remedy for dysentery.

Subsection Gibbosum

The distinct habit of the only species included in this subsection has resulted in the creation of this subsection.

P. gibbosum (Linnaeus) L'Héritier: This is a spreading or scrambling almost hairless plant with succulent stems swollen at the nodes, becoming woody later. Leaves are somewhat succulent, glaucous, pinnately lobed with one or two pairs of unevenly toothed or lobed leaflets. Up to 15, greenish-yellow flowers, 1.5–2 cm across, are sweetly scented at night, the upper petals slightly reflexed; fertile stamens seven in number. In the wild, along the western coast of South Africa, this species may often be found scrambling for several metres through shrubs.

Section Magnistipulacea

This is a small section separated from *Polyactium*, with species found in the southern and eastern Cape Province and east Africa. The flowers have a long hypanthium, seven fertile stamens and the petals are often fringed. The plants are tuberous and the basic chromosome number is x = 11.

P. bowkeri Harvey: This is a perennial plant with an underground tuber and a very short aerial stem with attractive erect, soft, gray-green feathery foliage. Flowers are pale yellowish-green, the lower petals flushed purple, *c.* 4 cm across, with all petals deeply cut into linear segments.

Introduced in 1863 from the grassland of eastern Cape Province and Natal. The leaves are said to have been eaten by some of the local people and it has also been used a medicinal plant.

P. luridum (Andrews) Sweet: This is a stemless plant with an underground tuber and extremely variable hairy leaves, the early ones shallow pinnately lobed becoming more deeply divided into linear segments as the season progresses. The night-scented white, pink, yellow or occasionally red flowers reach 2.5–3 cm across with broad almost equal petals. This is a species that shows considerable morphological variation with a wide geographic distribution from Tanzania to the southern coast of South Africa. It has also been used medicinally and in concoctions for courtship rituals.

P. schizopetalum Sweet: This geophyte has a large tuber covered with brown scales and a very short aerial stem with shallow to deeply pinnately divided and pubescent basal leaves. The pale yellow or yellowish-green flowers, often flushed reddish-purple are about 3 cm across, and are unpleasantly scented at night; all the petals are similar and deeply cut into linear segments. The leaves of plants collected from different locations in eastern Cape Province may show considerable variation. It was first collected in 1821.

REFERENCES

Albers, F., Gibby, M. and Austmann, M. (1992) A reappraisal of *Pelargonium* section *Ligularia* (*Geraniaceae*). *Plant System. Evolution*, 179, 257–276.

Albers, F., van der Walt, J.J.A., Gibby, M., Marschewski, D.E., Price, R.A. and du Preez, G. (1995) A biosystematic study of *Pelargonium* section: 2. Reappraisal of section *Chorisma*. *S. Afr. J. Bot.*, 61, 339–346.

Aiton, W. (1789) *Hortus Kewensis*, Ed 1, Vol. 2., London.

Bakker, F.T., Culham, A. and Gibby, M. (2000) Phylogenetics and diversification in *Pelargonium*. *Adv. Plant Mol. System*. 57, 353–374.

Bakker, F.T., Culham, A., Daugherty, L.C. and Gibby, M. (1999) A trnL-F based phylogeny for species of *Pelargonium* (*Geraniaceae*) with small chromosomes. *Plant System. Evolution*, 216, 309–324.

Burman, J. (1738) *Rariorum africanarum plantarum*, Amsterdam.

Cavanilles, A.J. (1787) *Monadelphiae classis dissertationes decem: Quarta dissertatio botanica de Geranio*, Paris.

De Candolle, A.P. (1824) *Prodramus systematis naturalis regni vegetabili*, Vol. 1., Paris.

Demarne, F. and van der Walt, J.J.A. (1989) Origin of the rose-scented *Pelargonium* cultivar grown on Réunion Island. *S. Afr. J. Bot.*, 55, 184–191.

Dreyer, L.L., Albers, F., van der Walt, J.J.A. and Marschewski, D.E. (1992) Subdivision of *Pelargonium* sect. *Cortusina* (*Geraniaceae*). *Plant System. Evolution*, 183, 83–97.

Harvey, W.H. (1860) *Flora Capensis*, Hedges, Smith & Co., Dublin.

Knuth, R. (1912) *Das Pflanzenreichi*. 4, 129, Berlin.

L'Héritier de Brutelle, C.-L. (1787) *Geraniologia, sue Erodii, Pelargonii, Monsoniae et Grieli historia iconibes illustrata*, Paris.

MacDonald, D.J. and van der Walt, J.J.A. (1992) Observations on the pollination of *Pelargonium tricolor*, section *Campylia* (*Geraniaceae*). *S. Afr. J. Bot.*, 58, 386–392.

Sweet, R. (1820–1830) *Geraniaceae*, vol. 1–5, London.

van der Walt, J.J.A. (1985). A taxonomic revision of the section *Pelargonium* L'Herit. (*Geraniaceae*) *Bothalia*, 15, 345–385.

van der Walt, J.J.A., Albers, F., Gibby, M., Marschewski, D.E. and Price, R.A. (1995) A biosystematic study of *Pelargonium* section: 1. A new section *Subsucculentia*. *S. Afr. J. Bot.*, 61, 331–338.

van der Walt, J.J.A., Albers, F. and Gibby, M. (1990) Delimitation of *Pelargonium* sect. *Glaucophyllum* (*Geraniaceae*). *Plant Syst. Evolution*, 171, 15–26.

van der Walt, J.J.A., Albers, F., Gibby, M., Marschewski, D.E., Hellbrugge, D., Price, R.A. and van der Merwe, A.M. (1997) A biosystematic study of *Pelargonium* section: 3. Reappraisal of section *Jenkinsonia*. *S. Afr. J. Bot.*, 63, 4–21.

van der Walt, J.J.A. and Boucher, D.A. (1986) A taxonomic revision of the section *Myrrhidium* of *Pelargonium* (*Geraniaceae*) in southern Africa. *S. Afr. J. Bot.*, 52, 438–462.

van der Walt, J.J.A. and Roux, J.P. (1991) Taxonomy and phylogeny of *Pelargonium* section *Campylia* (*Geraniaceae*). *S. Afr. J. Bot.*, 57, 291–294.

van der Walt, J.J.A. and Van Zyl, L. (1988) A Taxonomic revision of *Pelargonium* section *Campylia* (*Geraniaceae*). *S. Afr. J. Bot.*, 54, 145–171.

van der Walt, J.J.A (1977) *Pelargoniums of Southern Africa*, vol. 1. Purnell and Sons, Cape Town.
van der Walt, J.J.A. and Vorster, P.J. (1981) *Pelargoniums of Southern Africa*, vol. 2. Juta and Co. Ltd., Cape Town.
van der Walt, J.J.A. and Vorster, P.J. (1988) *Pelargoniums of Southern Africa*, vol. 3. National Botanic Gardens, Kirstenbosch.
Volschenk, B., van der Walt, J.J.A. and Vorster, P.J. (1982) The subspecies of *P. cucullatum* (*Geraniaceae*) *Bothalia*, 14, 45–51.

9 Cultivation and sales of *Pelargonium* plants for ornamental use in the UK and worldwide

Janet James

INTRODUCTION

Historical survey

The *Pelargonium* family first reached Europe some 350 years ago, when sailing ships were beginning to travel further afield. The ship's surgeon was usually the person most interested in botany, and he would explore the new countries in search of unknown plants. The probable incentive was to use the plants for medicinal purposes.

In Europe, it became fashionable for rich people to build glasshouses to accommodate plants from warmer climates that would not have survived frosts. The lifting of the tax on glass in 1851 had an impact on this activity. Pelargoniums became popular subjects for hybridising, because cross-pollination was not difficult, and they have the advantage of growing all the year round, so there was no dormancy to interrupt the work. The length of the flowering period was also a bonus and inspired gardeners to use the results of the breeding work as long lasting summer bedding.

By the time Queen Victoria was on the throne, pelargoniums, or 'geraniums' as they were popularly known, were very fashionable plants. The misnomer of 'geranium' came about because the seed of the pelargonium is very similar to the seed of the geranium, and at first the botanists decided that the pelargoniums were, in fact, closely related to European geraniums, and named them accordingly. As science progressed, this was found not to be true, and it was later on that the term 'pelargonium' was applied to these plants. But by then it was too late – the word 'geranium' was well and truly part of the English language for what we now know as a 'zonal pelargonium'.

Cost of pelargoniums

One indication of the value put on these plants in the nineteenth century days, was the fact that a zonal variety called 'Paul Crampel' was put on the market in 1896 at £1 a plant – compared to the price of bread at that time (about 2d a loaf – £1 = 240d), this was an enormous amount of money. We know from nurserymen's catalogues of the day that pelargoniums were plentiful in a great number of varieties. The *Pelargonium* section of Cannell's Nursery, of Swanley in Kent, catalogue in 1910 listed 810 varieties. A great many mutations occurred during this period of cultivation, and many of the coloured-leaf varieties that we still grow heralded from the nineteenth and early twentieth century. 'Caroline Schmidt', for instance, came from Germany in1898 and is still going strong – on the Continent it is known as 'Wilhelm Langguth'. During the First World War, when anti-German feeling was running high in Britain, the plant was called 'Caroline Smith', and reverted to 'Schmidt' once feelings had calmed down. 'Mrs Parker' is a pink sport from 'Caroline Smith', and has identical leaves and tallish robust growing habit. 'Chelsea Gem' (1880) is paler and more compact in it's growth habit, but is frequently mistakenly labelled as 'Mrs Parker'.

First World War: effects on plants grown

The First World War heralded a disaster for many tender plants. It was the law that glass could only be used to grow food. Therefore many varieties were lost to cultivation, but fortunately many novelty varieties were preserved by amateur growers. After the war all things 'Victorian' went out of fashion, wealth had largely dissipated and labour was short. All these things led to a decline in the popularity of the *Pelargonium* family.

The war did not have such an impact in North America, and breeding work continued there. About the 1960s a new strain was introduced called 'Irene' variety. The original 'Irene' was raised by Charles Behringer and was named after his wife. (Shellard, 1981). These were short-jointed, bushy, floriferous plants and soon set criteria by which all varieties were judged. It did not take long for them to arrive in England, and these became the standard bedding varieties. At this time new plants were obtained by cuttings taken from mother stock, and seed raised varieties were only in the hands of the hybridisers.

F1 hybrid seeds

The drawback to cultivation by cuttings was the cost of maintaining the stock, the cost of labour and the fact that viruses and disease could be perpetuated from one generation to the next. Therefore much breeding work was aimed at producing plants that could be satisfactorily raised from seed. The first series of F1 hybrid seed zonal pelargoniums to be taken up by commercial growers was called the 'Carefree' strain and this was soon followed by the 'Sprinter' range. In the 1970s, the traditional 'Gustav Emich' variety that had always graced the flower beds outside Buckingham Palace was replaced by a scarlet 'Sprinter variety'. The more modern seed strains now used are the multi-bloom types.

Problems raised by hybrid seeds

Seed-raised plants became the norm for local authority parks departments, as they were more cost effective than plants raised from cuttings. But they also had drawbacks. More time was needed to grow the plants than with cutting material, and in order to get them into bloom for early summer they had to be given high temperatures early in their life. For instance, from a January sowing it would be necessary to give a night temperature of 64°F for several weeks. Also the blooms tended to shatter and petal drop was a problem. This was overcome by spraying the blooms with silver thiosulphate. Another problem was the fact that because they came from seed, they tended to form seed heads rapidly – sometimes before the whole bloom had opened up. This gave unsightly beds of half-formed flowers. However, breeding work is still continuing in this field and improvements are being made all the time. The plants raised from seed are single zonal varieties (as doubles do not readily form seed), and they are basically green leaved varieties, though many do have an attractive dark zones.

Cuttings and big-business companies

Now-a-days many plants are raised for the commercial market from cuttings taken from mother stock. There are a number of large wholesale nurseries situated on the Continent, such as Fischer's and Elsner's in Germany, Enthoven and van Veen in the Netherlands, and Philomel in France, that are supplying the garden centres of Europe. They concentrate on breeding varieties that will perform well by being short jointed and bushy and very floriferous, which suits the needs of the vast majority of garden centre customers. Ideally a plant is approximately 2 ft high, 2 ft across and full of flower at the end of May. Fortunately for the horticulture trade, many people will discard the plants at the end of the season and expect to buy more the following year. The balconies of Bavaria, Austria and Switzerland will be filled with cascading ivy-leaf geraniums all summer long, but most amateurs make very little attempt to preserve the plants for use in the following year.

In order to keep costs down, many stock plants will be grown in a climate more equitable to the needs of the plants. The Canary Islands and Kenya are two places chosen for the growing of stock plants. The Canaries are particularly suitable because they are isolated from possible viruses and disease. The wholesale nurserymen take great care to see that the mother stock is free from all viruses, and they have laboratories where micro-propagation takes place. Cuttings are only taken one generation down from this 'cleaned up' material. The wholesale nurseries are run like factories, with every square metre contributing to production, and timetables are strictly adhered to. Production is labour intensive, and costs are kept to a minimum.

Many unrooted cuttings from wholesale nurseries are sold to nurserymen all over Europe, being sent in insulated boxes by air. So the source of most of the plants in garden centres is the same, even though the plants have been grown locally.

Small, family-run nurseries

In a different ball game are the small nurseries, mostly family concerns, that are preserving the many old varieties which are still in existence. They cannot possibly compete with the wholesalers, so instead of growing a vast number of a few varieties, they do the opposite and grow a few of a vast number of varieties. These nurseries are aiming their products at the amateur growers who are interested in finding plants that are not readily available.

Different types of pelargoniums on offer

The types of pelargonium became known under the following sections: zonals, regals, ivy-leaf and scenteds.

Zonals are the ones commonly called 'geraniums' and are largely used as bedding plants and for urns and tubs to decorate buildings.

Ivy-leaf – these are *Pelargonium peltatum* derivatives and are used mainly in hanging baskets, urns and tubs, but rarely for bedding. Recently there has been a distinction made between the 'cascade' (or windowsill) varieties and the normal types. The cascades are multi-flowering single varieties in a limited number of colours, and some have variegated leaves.

Regal are very different from zonals, and known commonly as 'pelargoniums', having large open blooms and only blooming in their season (when there is 14 h of daylight). In the USA, they are known as 'Martha Washingtons'.

Zonals can be subdivided into basic, dwarf and miniature according to their growing habit. In each subdivision, they can have flowers that are either single, semi-double or double. The singles have florets made up of exactly five petals, the semi-doubles have between six and nine petals, and the doubles have ten or more petals to each floret. They can also be sub-divided according to their leaf colour or flower formation. There are golden, bicolours and tricolours in the leaf sub-divisions and cactus, tulip and rosebud in the flower sub-divisions. There are also some varieties which are a cross between the zonal and ivy-leaf varieties, and these are known as 'hybrid ivies'.

The species and primary hybrid plants are not produced in any great numbers commercially, except when they have scented leaves. There is also a wide range of scented leaf cultivars, with odours ranging from apple to camphoric. These are becoming increasingly popular and are the subject of many articles in gardening magazines.

The small nurseries are in some ways altruistic in that they aim to preserve plants that would otherwise be lost for ever, but they also introduce new varieties that have been bred by enthusiastic amateurs. Some dedicated amateurs can produce amazing results with very few facilities at their disposal. Ian Gillam of Canada lives in a flat in Vancouver, but has managed to produce some 'Stellar' varieties which are now renowned throughout the world. The late Reverend Stanley Stringer in Suffolk spent his retirement concentrating mostly on dwarf or miniature growing plants, and left a fine heritage of superb varieties. Bert Pearce of Fareham left a fine series of regal pelargoniums behind when he died recently. Even now Brian West is working in his two greenhouses on the Isle of Wight, taking careful notes, and breeding new plants of distinction. There are many such people working in a small way to increase the family of pelargoniums, to the advantage of us all.

DEVELOPMENT OF PERSONAL NURSERY BUSINESS

Historical background: the Vernon Geranium nursery

On a personal basis, our own nursery developed from a back garden hobby, inspired by the collection of pelargoniums inherited from parents in about 1972 – it would have seemed such a shame to let the plants die, and once hooked on the beauty and diversity of this family of plants, things progressed from there. Grandad was a nurseryman who discovered the joys of the *Pelargonium* family after he retired, and on his death left a considerable collection of varieties. This was the basis of the nursery.

The back garden soon proved to be too small, so the family (including four children still at school) was uprooted and moved to an ancient pub ('The Vernon Arms' – hence the name) that the brewery was selling off for use as a private house. This was situated near to Harefield Hospital in Middlesex, and mother continued her teaching career in the local primary school.

By 1979, the teaching career was abandoned in favour of the geraniums, and by chance father's firm (where he had worked for over 30 years as a civil engineering draughtsman) decided to uproot and move away. At this point care was thrown to the wind and both parents opted for the nursery – with hindsight this was a very foolhardy decision, and not one to be recommended!

In 1983, two sons were fired with enthusiasm and volunteered to join the venture – one was a qualified plumber and the other had been to agriculture college. The site at Harefield was proving to be too small, so the two sons moved to a leasehold nursery at Cheam in Surrey. The drawback to Cheam was that it was to be used for wholesale growing only, so we had hoped to keep Harefield for our main retail outlet.

Initial mail-order business

This was not to be, as the debts mounted, and we had to sell the Harefield nursery. This meant that we had to concentrate all our efforts on our mail-order sales, as the only outlet we had for potted up plants was the wholesale market and shows. The wholesale market was not really 'our scene' as garden centres were only interested in a limited number of varieties for a very few weeks of the year. Shows were more successful, as people appreciated being offered a wide range of plants, and we found that by keeping up a good supply of the more unusual varieties, we could sell plants right on into September.

One morning in 1987, we read that Gambles geraniums of Derbyshire was closing down. This was a nursery famous for introducing some of the best American and European varieties to this country, and for their own breeding of some excellent zonals. We wondered what would happen to the mother plants, so it was decided to approach Mr Ken Gamble as to their fate. We were invited to make him an offer for the complete stock, which was accepted, and the stock was added to our ever increasing collection of geraniums.

Acquisition of retail facility

In 1991, we managed to re-negotiate our lease, so now we are able to sell plants from the nursery, and are open every day from the beginning of March to the end of July. We hold our open weeks during the last week in July, when tours of the nursery are conducted twice a day.

Publicity

We started exhibiting at the Royal Horticultural Society (RHS) summer shows at the Halls in Vincent Square in 1981, and exhibited at the Chelsea flower shows regularly. But in 1989, we decided not to apply for Chelsea any more, as we found that we were a very small fish in a huge sea, and the show came just at the time of year when we were busiest at the nursery. By about 1990, we determined to stop using flower shows as a means of marketing our plants because the amount of time and effort involved was not justified by the monetary return.

Our catalogue is produced in relatively large numbers (100 000 copies) and is now also on the Internet (under www.GeraniumsUK.com), so hopefully, it will reach a much wider public that we are able to by posting out catalogues.

Seed versus cutting-raised plants

The mass sales of pelargoniums (geraniums) has largely gone over to the seed-raised varieties in the last decade, and this is due in no small extent to the marketing power of the seed firms. We have produced our coloured catalogue with the intention of redressing the balance somewhat. The general public do not even seem to be aware that there are such things as miniature, dwarf or golden-leaved geraniums – probably because they rarely read about them, and even more rarely find them offered for sale in their local Garden Centre.

Seed firms market geraniums as an *annual* plant, as it is, of course, in their interest to do so – but the public do seem to be aware that they are a *tender perennial*, and that with a minimum of care in the winter, the plants can live for many years.

THE VERNON GERANIUM NURSERY: SITE AND RUNNING PROTOCOL

Lighting and temperature maintenance

Our site in Surrey that occupies just 1.8 ha, of which some 0.5 ha is under glass or polytunnel. We sell our plants mainly as rooted cuttings via mail-order throughout Europe. Stock plants are kept throughout the year, and are given supplementary lighting during the winter months. The trade is very seasonal, with the vast majority of customers wishing to receive their orders during the first 4 months of the year, which is why we have to spur on the growth with the high pressure sodium lights. An environmental computer measures the lux levels during the day, and brings on the lights in the night to make up the inevitable deficiency during the winter months. By this means we have material growing to be taken as cuttings throughout the year. Coloured leaf varieties, which are lacking in chlorophyll to support growth, are in a special area where there are extra lights, and the temperatures are kept higher than elsewhere. The normal minimum temperatures in the stock house are 10 °C daytime and 7 °C at night. Below 5 °C, the plants will not continue to grow (and at below 0 °C, they will die!). Higher minimum temperatures would speed growth, but the cost would be prohibitive.

Watering and feeding

Watering and feeding of stock is by means of a drip feed system i.e. each pot has a fine tube attached, and the amounts given are regulated by water meters on the taps. The watering is not computerised, as we feel this is best left to personal judgement. The feed is drawn into the hose lines from a central location.

Cuttings

Cuttings are taken in batches according to the time of year and orders pending. For servicing the main mail-order catalogue, batches of 30 or 60 of each variety is taken, so that there is a continuous supply, and orders can be despatched soon after receipt. The orders are entered into a computer, and from this we can ascertain which are the more popular varieties, so that production can be geared accordingly.

We produce a smaller 'collections' catalogue, which is sent out to a larger number of people, using a 'direct marketing' approach. We exchange and buy lists of names and addresses from firms in a similar line of business, so that we know we are targeting people who have gardens, and who have previously bought plants or seeds by mail-order. This means that there is a large number of orders going out during March, April and early May, so cuttings are taken 'en masse' for these, and extra staff are employed at that time of the year to service these orders.

Methodology of taking cuttings

Because we grow from cuttings, production is very labour intensive. Each plant has to be trimmed neatly below a node before insertion into a growing medium. We have found it beneficial to dip each one into a solution of ascorbic acid (Vitamin C) – 1 g/85 ml of cold water. Rooting takes place on heated benches most of the year – from May to August it is not necessary to heat the benches – and the heat is provided by domestic boilers servicing hot pipes running along under the cuttings.

The time taken to form roots can vary with each variety, some are quicker than others, but on the whole it is 2–3 weeks for the zonal types, and 4–5 weeks for the regal types.

We do not use chemical growth inhibitors (like cycocel) as we feel this would be cheating our amateur customers, who do not have access to these products. There are plenty of varieties that will grow in a short jointed, bushy habit without the need for artificial aids. Also there are people who would like to have tall growing plants, and these are no longer available from wholesale growers.

Unrooted cuttings bought from outside

In some instances, it is more economical to buy certain varieties from wholesale nurseries to supplement our own production. This is usually by unrooted material, which we then root ourselves, but occasionally with ready rooted plants. The latter can cause problems because the plants might not be suited to our packaging boxes. When we purchase material from continental wholesalers it is flown in insulated boxes to Heathrow airport, and we collect it from there, so our location in Surrey is a great asset.

Use of insecticides and predators to control insect pests

We use both insecticides and natural predators to control insect pests and fungi. Technical advice is sought from our supplier of these products: Fargro Ltd., Toddington Lane, Littlehampton, West Sussex BN17 7PP. They in turn are suppled with the predators by Novartis BCM (UK) Ltd., Aldham Business Centre, Alham, Colchester, Essex CO6 3PN.

Aphids

Nicotine liquid (95 per cent w/w)
Nemolt 150 g/litre (13.6 per cent w/w) teflubenzuron
Chess 40 g/100 litres (25 per cent w/w) pymetrozine
Pirimor (50 per cent w/w) pirimicarb
Biological control
Encarsia formosa
Macrolophus
Hypoapsis
Aphidius colemani
Mycotel (this contains spores of the entomopathogenic fungus *Verticillium lecanii* which is highly infective to whitefly)

Red spider mite

Dynamec 18 g/litre (1.8 per cent w/w) abamectin – also contains hexan-l-ol

Fungicides

Dorado 200 g/litre (20.8 per cent w/w) pyrifenox
Rovral (50 per cent w/w) iprodione
Plantvax 75 (75 per cent w/w) oxycarboxin

A spreader is used to reduce the amount of chemical needed per application: *Celect* 95 per cent rapeseed oil. Feed is used regularly and is applied via a 'Dosatron' system.

Fertilisers

Early in the year, after the stock plants have been cut once, a higher nitrogen feed will be used to boost growth (along with a minimum temperature of 7 °C and supplementary lights):
'Solufeed', NPK 28:7:14 with magnesium and trace elements.

This is followed by a more balanced formula:
'Sangral' NPK 21:7:24 + 2MgO
The complete feed is:
Total Nitrogen (N) 21 per cent
Ureic Nitrogen (N) 15 per cent
Phosphorus Pentoxide P_2O_5 soluble in neutral ammonium citrate and water 7 per cent (3 per cent P)
Potassium oxide (K_2O) soluble in water 24 per cent (19.9 per cent K)
Magnesium oxide (MgO) soluble in water 2 per cent (1.2 per cent Mg)
Boron (B) soluble in water 0.0022 per cent
Copper (Cu) soluble in water, chelated by EDTA 0.0016 per cent
Iron (Fe) soluble in water, chelated by EDTA 0.007 per cent
Manganese (Mn) soluble in water, chelated by EDTA 0.0042 per cent
Molybdenum (Mo) soluble in water 0.0014 per cent
Zinc (Zn) soluble in water, chelated by EDTA 0.0014 per cent.

Compost

For many years we mixed our own compost, but recently it was decided that it was more economical to buy in a ready prepared mixture. After several not too successful mediums, we have settled for a 'crop-specific compost' from William Sinclair Horticulture Ltd. But the 'specific crop' is poinsettia! We have found it suits our purpose very well, the plants easily send their roots through the open texture, and it retains moisture without becoming waterlogged. The bag states 'a purpose designed compost prepared from a blend of different peat types and a balanced base fertiliser to provide optimum plant growth. Includes an efficient wetting agent. A low initial nutrient charge has been used to allow rapid root establishment and early liquid feeding will be necessary to maintain desired growth rates'.

As commercial nurserymen we have learned by experience how to succeed with this family of plants, but there are technical works on many aspects of their culture. We know that the main conditions that ensure success are a dry atmosphere, moist roots and an abundance of light. Cold, damp conditions will cause pelargoniums to rot, and warm compost is essential to the rooting process.

ALTERNATIVE ROOTING AGENTS

For less perfect conditions, when basal heat cannot be provided and the temperature is low, antioxidants can be used as alternative rooting agents (Lis-Balchin, 1988). Use 1 tablet of Vitamin E and/or the food preservatives, BHT and BHA (butylated hydroxy-toluene and anisole respectively), using a pinch. These are first dissolved in a little alcohol and then diluted with water to 100 ml. Use sparingly for dipping the cut edge of cuttings for a few seconds before rooting.

TOP FIVE PELARGONIUMS

In 1999, the top five pelargoniums from each section sold from the nursery:

Zonals

Single: Mr Wren
Paul Crampel
Simplicity
First Love
Skelly's Pride

Double: Carol Gibbons
Beryl Gibbons
Brenda Kitson
Betty
Regina

Dwarf: Nettlestead
Rosina Read
PAC Millenium Dawn
Jean Beatty
Petite Blache

Miniature: Cotton Tails
Garnet Rosebud
Royal Norfolk
Eileen
Red Black Vesuvius

Rosebud: Appleblossom Rosebud
Red Rambler
Westdale Appleblossom
Happy Appleblossom
Wedding Royale

Fancy leaf: Mrs Pollock
Mrs Henry Cox
Retah's Crystal
Encore
Madame Salleron

Regal: Lord Bute
Springfield Black
Mohawk
Askham Fringed Aztec
Sunset Snow

Angels: The Barle
Swedish Angel
Imperial Butterfly
Black Knight
Velvet Duet

Deacons: Deacon Birthday
Picotee
Lilac Mist
Suntan
Arlon

Scented: Sweet Mimosa
Prince of Orange
Pelargonium dichondraefolium

Crispum variegatum
Clorinda

Ivy-leaf: Jack Gauld
Barbe Bleu
Bonito
Snowdrift
Rio Grande

PELARGONIUMS GROWN ELSEWHERE IN THE WORLD

Commercial growing of the *Pelargonium* family is buoyant throughout the world, and particularly so in the USA, where propagation has increased by almost 60 per cent from 1979 to 1990. There are some 120 million plants grown in the USA and some 70 million in Germany as opposed to some 2 million in the UK. Of course, these figures are meaningless unless compared to the population figures of these countries. But generally, where there are large urban populations, particularly in the western civilisation, there will be a great demand for these plants.

We, ourselves, have a visit at least annually from touring Japanese nurserymen who always show a great interest in our diverse selection of varieties, showing more worldwide appeal.

LITERATURE ON PELARGONIUMS FOR GROWERS

There are considerable number of books and papers written on the cultivation of pelargoniums, with an abundant diversity of opinions on some aspects of their cultivation. Even the scientific works rarely comes to an undisputed conclusion on the best conditions for the propagation and growing of these plants. Possibly the best work to recommend for a detailed study of these plants is *Geraniums: A Grower's Manual* edited by John W. White.

APPENDIX

Addresses of Nurseries – for cuttings of common geranium

Elsner pac Jungpflanzen, Kipsdorfer Straße 146, D-01279 Dresden, Germany.
Tel: 0351 25591-0; Fax: 0351 2517494; e-mail: info@pac-elsner.com.

M van Veen Export BV, PO Box 73, 1430 AB Aalsmeer, Holland.
Tel: 0031 297 326516; Fax: 0031 297 328001.

Philiomel S.A., Establissement Horticole, 66600 Salse-le-Chateau, France.
Tel: 04.68.38.61.28; Fax: 04.68.38.69.93; e-mail: philiomel@smi-telecom.fr; www.littlefrance.tm.fr/philiomel.

M.C.M. Enthoven C.V., Wateringseweg 5, 2685 SP Poeldijk, The Netherlands.
Tel: 0174 245572; Fax: 0174 247977.

REFERENCES

Lis-Balchin, M. (1988) The use of antioxidants as rooting enhancers in the Geraniaceae, *J. Hort. Sci.*, 64, 617–623.

Shellard, A. (1981) *Geraniums for Homes and Garden*, David & Charles, Newton Abbot.

White, J.W. (1993) *Geraniums: A Grower's Manual*. Sponsored by Pennsylvania Flower Growers. 4th Ed. Ball Publishing, Ill. USA. Ed. White, (J.W. published by Ball Publishing, 1 North River Lane, Suite 206 PO Box 532, Geneva, Illinois 60134, USA, ISBN 0–9626796–5–8).

10 Growing pelargoniums in the garden, conservatory and for shows

Maria Lis-Balchin

INTRODUCTION

The main types of Pelargoniums (known by most people as 'geraniums' include: the zonals (*Pelargonium* × *hortorum*), regals (*P.* × *domesticum*), ivy-leaf (*P. peltatum*) and scented-leaf varieties. These can be distinguished by either the leaf shape, flower shape or scent.

The zonals are descended from *P. zonale* and *P. inquinans* and a few other species in the Section *Ciconium*; as the name suggests there is a zone, which is darker than the rest of the leaf and is crescent or horse-shoe shaped. This zone can be very pronounced or almost indistinguishable, depending on the cultivar. There are probably over 2000 named cultivars, some of which may be synonymous with others, but have been given a different name, either in the same country in different nurseries, or in a different country. One of the early varieties is 'Paul Crampel', which used to be the red-flowered 'geranium' growing outside Buckingham Palace.

According to the Muslim legend, geraniums were created when Mohammed went for a walk in an unfrequented place, and in the heat of the sun, took off his shirt, rinsed it in a nearby pool and hung it over some branches, The water from the shirt dripped onto some marshmallows growing underneath and turned them into geraniums!

General cultivation of pelargoniums

Pelargoniums, originating in southern Africa, are more adapted to drier and hotter conditions than most European countries can provide year-round. Cold, wet situations kill them. In winter, they are best kept on the drier side and at a temperature never below 4 °C. They require good soil, either loam-based John Innes or soil-less compost, slightly on the acid-side, preferably with sand, grit, perlite or vermiculite and clay-pots are preferable, as plastic pots retain too much water.

'GERANIUMS' (i.e. PELARGONIUMS) OF DIFFERENT TYPES

Descriptions of different cultivars has been taken from: Clifford (1958); Wood (1966); Witham Fogg (1975); Shellard (1981); Clark (1988) and Taylor (1988). These authors also give instructions on growing, propagating and general tips on the best plants to grow in different areas of the house and garden. For colour photo graphs of 1000 Pelargoniums see Key (2000).

Zonals

The zonals are divided into various groups according to:

1 Size

- Large: including Irenes, which were introduced in 1942 by Charles Behringer in Ohio, USA and were then bred to produce large bushy floriferous plants with huge leaves e.g. 'Irene', medium red flower; 'Sentinel', white; 'Electra', bluish-rose. These are ideal for filling up space in the conservatory or garden.
- Dwarfs: (plants not exceeding 20 cm from the top of the pot to the tip of the foliage) e.g. 'Wendy Read', double pink shading to white; 'Tom Portas', double empire rose.
- Miniatures: Smallest cultivars, known as miniatures (13 cm from pot to tip), are best exhibited in batches of 4–6 in baskets, or individually in conservatories. Miniatures include a wide variety of shades of flowers or leaves and also some scented leaf cultivars e.g. 'Frills', 'Variegated Petit Pierre', 'Silver Kewense'. The original is thought to be a nearly black-leaved 'Red Black Vesuvius'.

2 Colour of the leaf

- Gold leaf or bronze leaf, with/without zone e.g. 'Golden Crest', 'Bridesmaid', 'Morval'.
- Bicolour with white/cream and green leaves e.g. 'Chelsea Gem', 'Frank Headley', 'Freak of Nature', which has white stems and is difficult to take cuttings from.
- Tricolour leaf, with gold, red and green e.g. 'Contrast', 'Dolly Vardon', 'Henry Cox', 'Mrs Strang'.

3 Number of petals

- Single (Five petals), e.g. 'Christopher, Ley', with large orange-red flowers; 'Feurriesse', with velvet-red blooms;
- Double (more than five), e.g. 'Shocking', with bright pink flowers, 'Royal Flush', with purple flowers. The single bloomed geraniums are best for outside, and are easily raised from seed; the double ones are often damaged by rain outside, but are more spectacular.

Appearance of the petals

- 'Spitfire' (quill-shaped), 'Apple Blossom Rosebud', 'Mr Wren' (spotted/striped, white/red), 'Gemini' (stellar/star-like).
- 'Speckled' varieties e.g. 'Gemma' raised by Stringer in 1984; 'Elmsett' raised by Bidwell in 1982.
- Patricia Andrea', a tulip-flowered salmon variety, originating from the USA in 1960.

5 Colour of the flower (Witham Fogg, 1975): New varieties of *Pelargonium* cultivars have appeared over the years, including the 'Deacon' group, derived from miniature zonals × ivy-leaf; these have double flowers and are bushy and floriferous and were introduced by the Rev. Stringer in the 1970s e.g. 'Deacon Lilac Mist'.

The 'Highfield' group introduced by K. Gamble in the 1970s e.g. 'Highfield's Festival', with very bushy but with compact habit and pale lavender-pink blossom. More recently the PAC group (Elsner in Germany) e.g. 'PAC Rosecrystal', which looks like an anemone flower, rose with a white centre was introduced. Sometimes outstanding varieties are on offer for a short time e.g. the yellow-flowered 'Botham's Surprise' raised in Australia and sold through the PAC label. The variegated leaf geraniums were studied by Grieve (1853). His book describes the variegations in other genera, which 'can be due to cells filled with air or gas in immediate contact with the chlorophyll or colouring substance' and apparently this could also be true in geraniums.

Regals

Regals have much larger flowers than the zonals and were therefore called 'pelargoniums' by the general public to distinguish them from the other 'geraniums'. They have been hybridized mainly from *P. cucullatum*, *P. angulosum*, *P. fulgidum* and *P. grandiflorum*. In the USA they are known as 'Martha Washington' or 'Lady Washington' and also 'Show pelargoniums'. They do not do well in the wet summers and have usually a restricted flowering time, often just once a year. They also attract whitefly ferociously. Only a few regals have scented leaves.

The regals are grouped into two sections: the ordinary regals and the Angels or dwarf/miniature regals.

Regals have the most beautifully coloured petals, with a wide range of colours often appearing together e.g.

- 'Aztec' is pale pink with bronze and strawberry pink markings and a fringed margin.
- 'Grand Slam' has rose-red petals with dark markings.
- 'Hazel Cherry' has cherry-red petals with blackish blotches.
- 'Morwenna' is almost totally black.
- 'All My Love', is orchid-mauve and white.
- 'Cezanne' has purple upper petals and the rest is lavender.

Angels

Angels have a distinctive 'pansy-face' and have different foliage to the regals, in general. Many are crosses or back-crosses of *Pelargonium* species with *P. crispum* and *P. grossularioides* (small-leaved scenteds) often implicated. The early Angels were possibly the old 'Angelines', catalogued in the 1820s, derived from *P. dumosum*, which has since disappeared (Taylor, 1988). The founding father of the Angels was Arthur Langley Smith, a school teacher, who first started hybridizing in the early 1900s. *P. crispum* was thought to have been hybridized with 'The Shar' and his first creation was named 'Catford Belle' after his place of residence in London (Catford). The Angels all have mauvish/white or pinkish, cerise/white colourings with different blotches and most have scented leaves. Like the regals, they attract whitefly.

Examples of the best known Angels include:

- 'Beromunster', pale pink with cerise blotches towards edges.
- 'Tip Top Duet', mauve base with two top petals maroon.
- 'Velvet Duet' is dark purplish-maroon.
- 'Wayward Angel' is pale mauve with upper markings of light maroon.

Many new Angel varieties are introduced each year, but have very similar appearances to the well-known originals.

Ivy-leaved

These are very distinctive, with shiny five angular-lobed, ivy-shaped leaves, all derived from *P. peltatum* and are known as trailing, window-box, Swiss/German or balcony varieties etc. The leaves have a characteristic odour, which is vaguely like that of real ivy, *Hedera helix*. Examples include:

- 'Balcon Royal', originating in the 1960s, with bright red petals.
- 'Duke of Edinburgh', with pink flowers and green/white variegated foliage.
- 'Galilee', originating in 1882 in France, with double pink flowers.
- 'L'Elegante', again early 1860s, with white single flowers in profusion.
- 'Red/Pink/mini Cascade', with small leaves and single primitive flowers on short branching stems.
- 'Sugar Baby' has double candy-pink flowers on a dwarf plant.

Hybrid ivies

These can have the appearance of either an ivy-leaf or a zonal as they have resulted from a *P. peltatum* × *P.* × *hortorum* cross.

Zonal-type, the Deacon group (see above) e.g. 'Deacon Lilac Mist', a lilac flowered variety turning almost white; 'Deacon Moonlight', 'Deacon Birthday', 'Deacon Regatta', 'Deacon Regalia'.

Ivy-type, e.g. 'Millfield Gem', amaranth-rose blotches and feathered a rosy red, 'Millfield rose', 'Jack of Hearts', 'Elsi' and 'Auden Ken an Emil Eschbach'.

Other recent innovations (early 1970s) include the 'Harlequin' group, developed through grafting: these have numerous red/white striped varieties e.g. 'Rouletta' known also as 'Mexicarin'. There are also veined leaf varieties e.g. 'Crocodile', 'White Mesh'; these originated through the innoculation with a beneficial virus, passed on by insects (possibly whitefly) and thence they were hybridized.

Scented-leaf varieties

There are numerous scented-leaf varieties which originated from hybridizations between scented-leaf species, e.g. *P. denticulatum*, *P. capitatum*, *P.* × *fragrans*, *P. odoratissimum* etc. The size and shape of the leaf varies greatly from the very large peppermint, furry leaves of *P. tomentosum* to the tiny, pungent *P.* × *fragrans* (resembling Vick's ointment). The scenteds have been placed in various groups by different authors, and it appears that there is an individual aspect to this, based on the person's appreciation of each odour and like/dislike of it. Witham Fogg (1975) puts *P. denticulatum* in the 'rosy' group, but to the present author it is distinctly pungent/camphoric and Vick's-like. Some species and cultivars are however unanimously appraised e.g. *P. odoratissimum* is always apple-like; *P. tomentosum* is pepperminty; *P. capitatum* is rosy; *P. crispum* is citrusy/lemony. There is often a change in odour of the leaves during the year, with most changes occurring during flowering e.g. *P. graveolens* is more minty when not flowering, and more rosy when flowering.

Some common scented cultivars include:

- 'Chocolate peppermint' syn. 'Chocolate tomentosum', derived from *P. tomentosum*, with massive peppermint-smelling green, furry, leaves with a brown blotch centrally
- *P. 'filicifolium'* has a sticky, very finely divided leaf, with a 'Vick's-like' odour; described by Andrews in his Geraniums around 1805. It is probably a derivative of
- *P. denticulatum* with its divided leaves, but there is some theoretical justification to include it under a derivative of *P. glutinosum*, which has the more viscid secretion (Abbott, 1994).
- 'Clorinda', vigorous, pink-flowered, camphoraceous/exotic leaves.
- 'Lady Plymouth', derived from *P. graveolens*, with a minty odour and leaves edged in white. It was described by Andrews in 1805 as *Geranium capitatum variegatum* and then in the current name prior to 1880.
- 'Copthorne', similar to 'Clorinda', but palish pink flowers with bright cerise markings
- 'Village Hill Oak', derived from *P. quercifolium*.
- 'Bolero', one of the 'Hartsook' hybrids produced in California as Uniques from regal crosses, this one is salmon-orange with dark brown to black-purple markings. Others include: 'Carefree', 'Hula', 'Polka' and 'Voodoo'. All have huge spectacular flowers.
- 'Purple Unique', one of the original Uniques introduced in the 1880s, has huge purple flowers. The first Uniques were however in existence before 1820, as they were mentioned by Sweet in his 'Geraniaceae' volume 1 (Abbott, 1994). One of the parents for this group is *P. fulgidum*, but other species were used later. The *P. fulgidum* flower markings have been carried across, even where the red colour has been replaced by mauve etc. as in 'Rollinson's Unique' which has the distinctive markings in the upper petals. Others in the group include: 'Unique Aurore', 'Crimson Unique'.
- 'Jessel's Unique', 'Madam Nonin', 'Paton's Unique', 'Scarlet Unique' and 'Shrubland Pet'. Many of these have very distinctive, camphoraceous, fruity-exotic leaf odours. 'Sweet Mimosa' was introduced in USA in the late 1970s, and has similar leaves to that of *P. graveolens*, but with a rosy odour, although its origin is unknown.

Seedlings

The first *Pelargonium* seeds sold were all F1 hybrids which were all zonals and were initially very expensive e.g. 'Carefree', which appeared in different flower colours; there were also F2 hybrids, which were much cheaper but less dependable and of lesser quality as these had been 'open-pollinated'. The plantlets, sold in garden centres, are all raised from seed and although the variety has increased over the years, there is still a lack of the majority of different shapes, sizes and scents available from cuttings. These seedlings are ideal for gardens.

Species

Many species can be grown in the garden, and include a number of the scented-leaf species, which form huge bushes e.g. *P. tomentosum*, *P. denticulatum* and *P. quercifolium* hybrids in particular. Most species are however best kept in the conservatory or greenhouse and include the more succulent or cactus-like species e.g. *P. carnosum*, *P. hirtum* and the geophytes e.g. *P. rapaceum*, *P. longifolium*, which have a short life above ground.

Show plants

Pelargonium and *Geranium* shows have categories in each of the groups mentioned and they are judged specifically for each category. There are easier and harder plants to grow and for shows, it is generally the most easy which are usually exhibited. These are also in strict order of popularity and according to Shellard (1981) this as follows:

1. 'Regina' (zonal), introduced in 1964, with pinkish-salmon flowers and is bushy, self-branching and short jointed.
2. 'Henry Cox' (zonal), a tricolor, first introduced prior to 1879 with red zone over a pale green centre and yellow margin to the leaf and salmon-pink flowers.
3. 'Aztec' (regal), introduced in 1962, compact and self-branching, white-base with red/maroon markings in the centre of the petals, shading to strawberry towards the edge. It flowers for months.
4. 'Jane Eyre' (zonal), a miniature introduced by Stringer in 1970, with deep lavender flowers and glossy dark-green, zoned leaves.
5. 'Deacon Lilac Mist' (zonal), a dwarf introduced by Stringer in 1970, with double pale lilac petals.
6. 'Hazel Cherry' (regal), introduced in 1971, with cherry-red flowers with brown/black feathering on all petals; it has a similar habit to 'Aztec'.
7. 'Highfield's Festival' (zonal), introduced in 1974, with pale rose-pink flowers with a white eye on the upper two petals and zoned deep green leaves.
8. 'L'Elegante' (ivy), old variety from 1860s with white flowers.
9. 'Mabel Grey' (scented), with a very citrusy odour.
10. 'Burgenland Girl' (regal), introduced in 1964, red-pink double florets, and dark green leaves.

Latest favourites listed from the 1999 Show organised by the British Pelargonium and Geranium Society include:

- 'Claydon', 'Tuddenham', 'Orion' – miniature zonal
- 'Elmsett', 'Brackenwood', Clatterbridge, Bold Carmine' – dwarf zonal
- 'Burghi', 'Aztec', 'Georgi', 'Ginny Reeves' – regal
- 'Dolly vardon', Mrs. Henry Cox, 'Silver Delight' – bi- and tri-colour zonal
- 'Phyllis', Jessel's Unique – unique
- 'Voodoo' – hybrid unique
- 'Blooming Gem', 'Lakeland' – Ivy-leaf
- 'The Lowman', 'The Culm', 'The Mole', 'Spanish Angel', 'Quantock Rory' – Angel

REFERENCES

Abbott, P. (1994) *A Guide to Scented Geraniaceae.* Hill Publicity Services, West Sussex.
Clark, D. (1988) *Pelargoniums.* Collingridge Books, The Hamlyn Publishing Group Ltd, London.
Clifford, D. (1958) *Pelargoniums including the popular 'Geranium'.* Blandford Press, Ltd, London.
Key, H. (2000) 1001 Pelargoniums, Batsford, B.T. Ltd.
Shellard, A. (1981) *Geraniums for Home and Garden.* David & Charles, Newton Abbot.
Taylor, J. (1988) *Geraniums and Pelargoniums.* The Crowood Press, Wiltshire.
Witham Fogg, H.G. (1975) *Geraniuims and Pelargoniums.* John Gifford Ltd, London.
Wood, H.J. (1966) *Pelargoniums: A Complete Guide to their cultivation.* A. Wheaton & Co. Ltd, Exeter.

11 Phytochemistry of the genus *Pelargonium*

Christine A. Williams and Jeffrey B. Harborne

INTRODUCTION

Species of *Pelargonium* have long been in cultivation because of their attractive scents and bright colours. Geranium oil, which is derived from several *Pelargonium* species, is an important plant crop. It is among the top 20 of plant volatile oils and the annual world production is worth nearly 7 million pounds. It is not surprising therefore that the chemical study of *Pelargonium* has concentrated on the essential oils present. Over 120 volatile constituents have been detected. Work has also been devoted to producing these volatiles in shoot-organ culture of *Pelargonium* material.

The essential oils of *Pelargonium* leaves are generally located in glandular hairs on both leaf surfaces. Other classes of chemical have been recorded in these leaf hairs or trichomes. These include both flavonoid methyl ethers and salicylic acid derivatives. Protection against herbivory by two-spotted spider mites has already been demonstrated in the case of the salicylic acid derivatives. However, it is likely that all the various glandular hair components have some role in protecting *Pelargonium* plants from herbivory or microbial attack in the natural environment. An interesting chemical ecology is already beginning to develop in the case of these plants and their natural enemies.

TERPENOIDS

Essential oils

Many of the *Pelargonium* species and the cultivars derived from them by artificial hybridisation have scented leaves. Members of the Sections *Pelargonium*, *Polyactium* and *Cortusina* are particularly rich in essential oils (Webb, 1984). The 'Geranium oil' of commerce is in fact obtained from steam distillation of the leaves of several *Pelargonium* species, namely: *P. graveolens*, *P. capitatum*, and *P. radula*. It is a complex mixture of over 120 monoterpenes and sesquiterpenes and other low molecular weight aromatic compounds (Vernin *et al.*, 1983). The oil varies in composition depending on the country of origin. However, the major components, comprising *ca.* 60–70 per cent of the oil, are citronellol (1), geraniol (2) and linalol (3), either free or in ester combination. Other terpenoids present in oils from all localities are: isomenthone (4), menthone (5), nerol (6), *cis*-rose oxides (7) and *trans*-rose oxides (8), α-terpineol (9), α-pinene (10), myrcene (11) and β-phellandrene (12) (Figure 11.1).

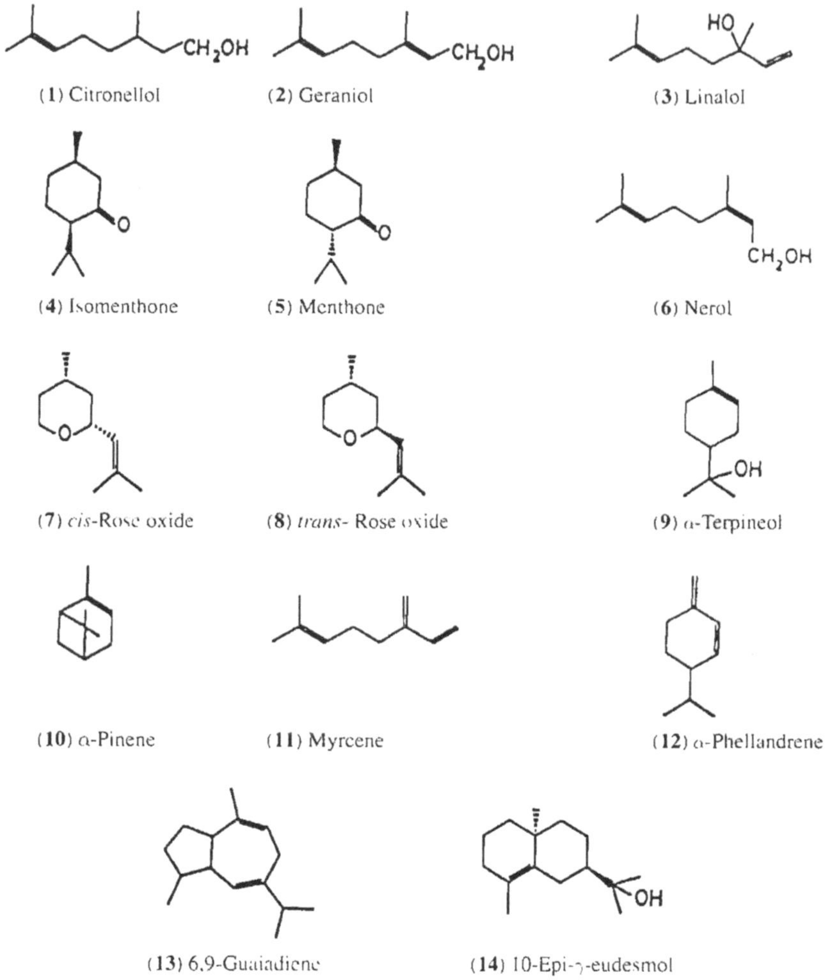

Figure 11.1 Some simple terpenoids found in *Pelargonium* species.

Egyptian oils and Bourbon oil from Réunion differ from the Chinese oil in having much higher levels of geraniol, geranyl esters and linalol and smaller amounts of citronellol and its ester and the absence of the sesquiterpene hydrocarbon, 6,9-guiadiene (13) Figure 11.1. African oil can be distinguished by the unique presence of 10-epi-γ-eudesmol (14) Figure 11.1 (Teisseire, 1987). The Geranium oil from Réunion also contains some unusual tetrahydropyrans (15–17) Figure 11.2 (Naves *et al*., 1961, 1963). The oils of a number of other scented *Pelargonium* species, which are not used commercially have been analysed (Charlwood and Moustou, 1988; Charlwood and Charlwood, 1991). They also contain complex mixtures of terpenoids. Thus, *P.* × *fragrans* produces many different classes of lower isoprenoids including camphene (18), 1,8-cineole (19), *p*-cymene (20), farnesene (21), fenchone (22), limonene (23) and sabinene (24) Figure 11.2. By contrast, the oil of *P. tomentosum* is made up almost

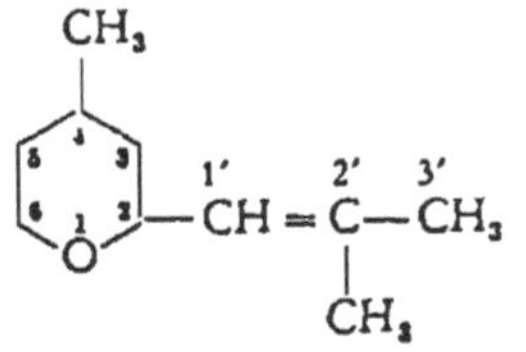

(**15**) 2-(1′,2′-Butenyl)-4′-methyl tetrahydropyran

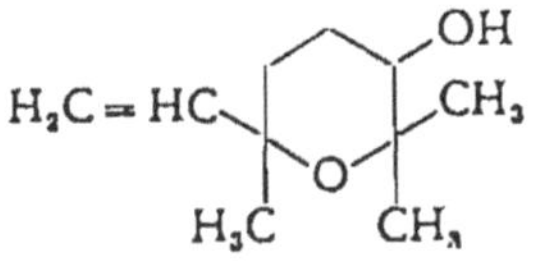

(**16**) 3-Hydroxy-2,2,6-trimethyl-6-vinyltetrahydropyran

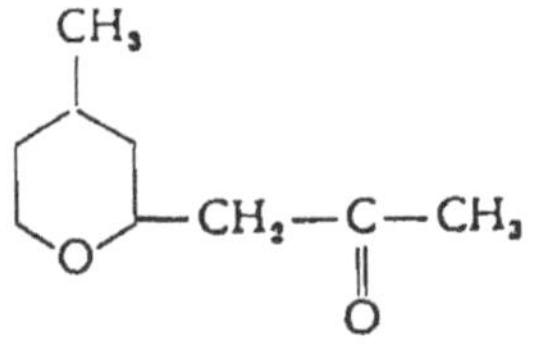

(**17**) 2-Acetonyl-4-methyl-tetrahydropyran

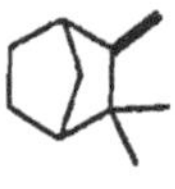

(**18**) Camphene

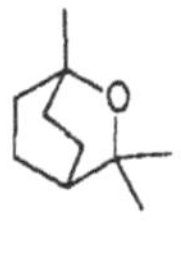

(**19**) 1,8-Cineole

(**20**) *p*-Cymene

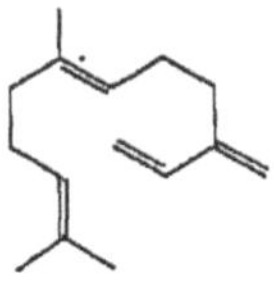

(**21**) Farnescene

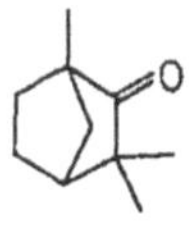

(**22**) Fenchone

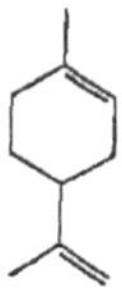

(**22**) Limonene

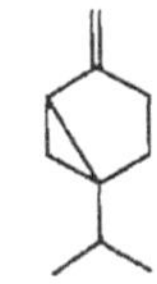

(**24**) Sabinene

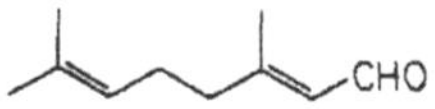

(**25**) Geranial

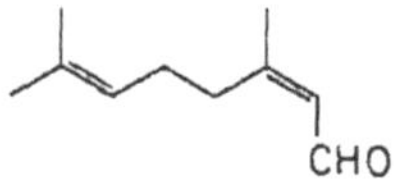

(**26**) Neral

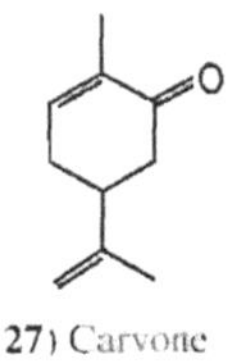

(**27**) Carvone

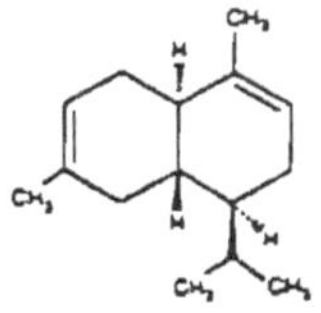

(**28**) Cadinene

Figure 11.2 Further simple terpenoids found in *Pelargonium* species.

entirely of isomenthone (96 per cent) and menthone, while that of the cultivar *Pelargonium* 'Mabel Grey' contains 88 per cent citral isomers, geranial (25) and neral (26) Figure 11.2. The former is utilised in the kitchen to provide peppermint flavour, and the latter to give a lemon flavour to food. The camphoraceous vapour from steamed leaves of *P. betulinum* has been used to relieve coughs and chest complaints and the powdered leaves of *P. inquinans* to cure headaches and relieve the symptoms of the common cold and as a body deodorant by native tribesmen (van der Walt, 1977). Three *Pelargonium* species, *P.* × *fragrans*, *P. tomentosum* and *P. graveolens* have been grown in tissue culture by Charlwood and Moustou (1988). The callus tissue in all three taxa failed to form the oil of the parent plant but shoot–organ cultures produced 50 per cent of a qualitatively similar oil to the original plants. Thus, shoot–organ cultures of

P. × *fragrans* accumulated pinene, sabinene, farnesene, carvone (27) and cadinene (28) (Charlwood and Moustou, 1988) (Figure 11.2), while *P. graveolens* shoot cultures produced geraniol, citronellol and citronellyl formate (Katagi *et al.*, 1986). As with the parent plant *P. tomentosum* accumulated equal amounts of menthone and isomenthone (Charlwood and Moustou, 1988; Charlwood *et al.*, 1990). However, shoot–organ cultures of *P.* × *fragrans* produced much less fenchone and accumulated more α- and β-pinene and sabinene and that of *P. graveolens* significantly less citronellol than the parent plant. However, oil production in shoot–organ culture is unlikely at present to compete successfully with the yields obtained by growing the whole plant.

Sterols and triterpenoids

The well known sterols: cholesterol, campesterol, and sitosterol, have been reported both in free form and in esterified form from *Pelargonium hortorum*, while stigmasterol was found only in the free form. The major triterpenoids in this plant are α- and β-amyrin, which also occur both free and esterifed, and a partially identified compound named, isomultifluorenol. Traces of free cycloeucalenol and its acetate, obtusifoliol acetate, cycloartenol acetate and 24-methylene cycloartanol acetate were also detected (Axel *et al.*, 1972).

MONOMERIC FLAVONOIDS

Vacuolar leaf flavonoids

There have been a number of flavonoid aglycone surveys of *Pelargonium* species but no records of flavonoid glycosides. In an early study, Bate-Smith (1973) found wide variation in the flavonoid constituents between the ten taxas he examined with a preponderance of myricetin (29) and quercetin (30) Figure 11.3 in species from Sections *Hoarea* and *Pelargonium*. In later chemotaxonomic studies of four Sections: *Glaucophyllum*, *Cortusina*, *Chorisma* and *Jenkinsonia*, quercetin was found to be universally present. Myricetin and kaempferol (31) Figure 11.3 were detected in 71 per cent of the taxa in section *Glaucophyllum* (van der Walt *et al.*, 1990) and myricetin was found also in all the species examined in section *Cortusina* except *P. dichondrifolium* (Dreyer *et al.*, 1992) but was absent from sections *Chorisma* (Albers *et al.*, 1995) and *Jenkinsonia* (van der Walt *et al.*, 1997). The major components of the section *Chorisma* were flavonols but two flavones, luteolin (32) and apigenin (33) Figure 11.3 were additionally detected in two taxas – *P. mollicomum* and *P. tetragonum*. In a more definitive survey of 58 *Pelargonium* species from 19 sections Williams *et al.* (2000) confirmed that flavonols are the major leaf vacuolar flavonoid constituents in the genus. Quercetin was present in all species except *P. spinosum*, where luteolin was the major component accompanied by flavone *C*-glycosides. However, luteolin was more usually present as a trace constituent in 49 per cent of taxa and flavone *C*-glycosides as the major flavonoids in 36 per cent of the species. The regular occurrence of other flavonols, myricetin in 38 per cent and kaempferol in 50 per cent of taxa, respectively, confirms the previous findings. Additionally both quercetin 3-methyl ether (34) and isorhamnetin (35) (Figure 11.3) were detected in 10 per cent of the sample. However, apigenin was not detected in any taxon examined in this study. The results are presented in more detail in Table 11.1.

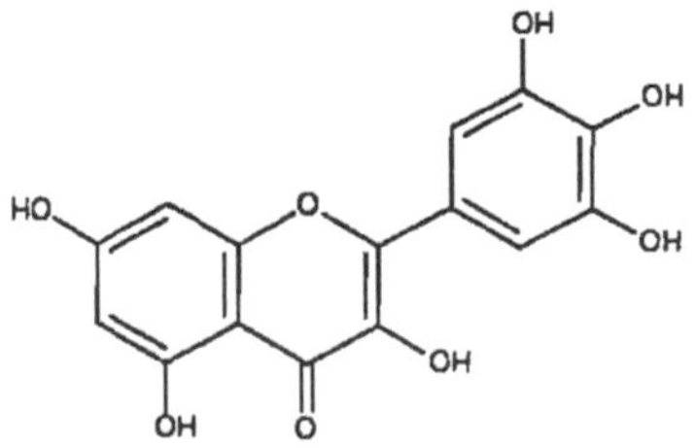

(**29**) Myricetin

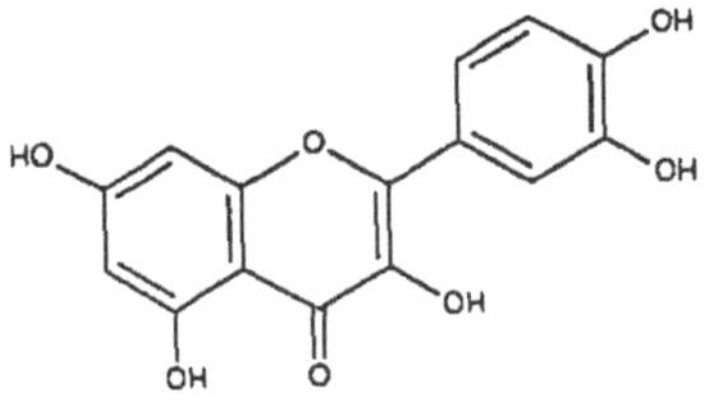

(**30**) Quercetin

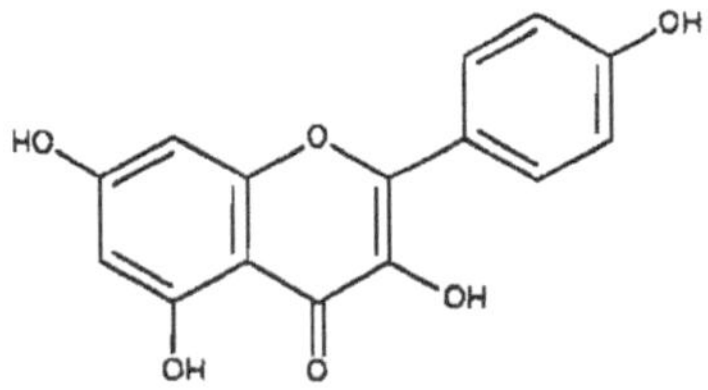

(**31**) Kaempferol

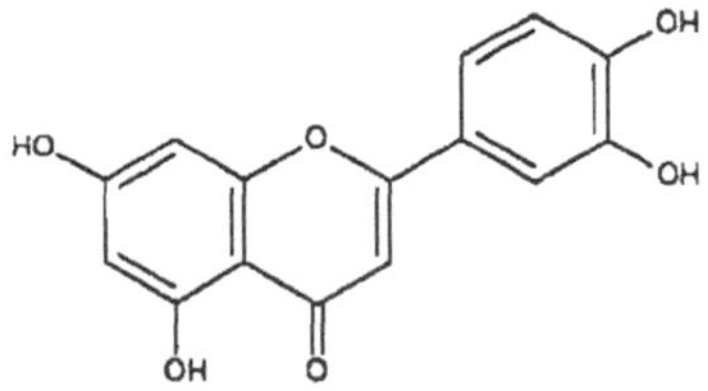

(**32**) Luteolin

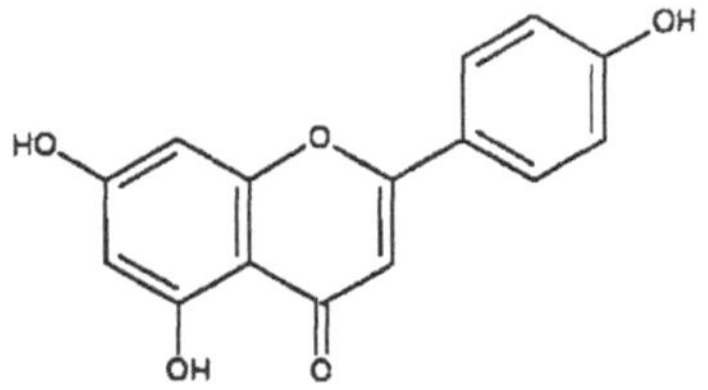

(**33**) Apigenin

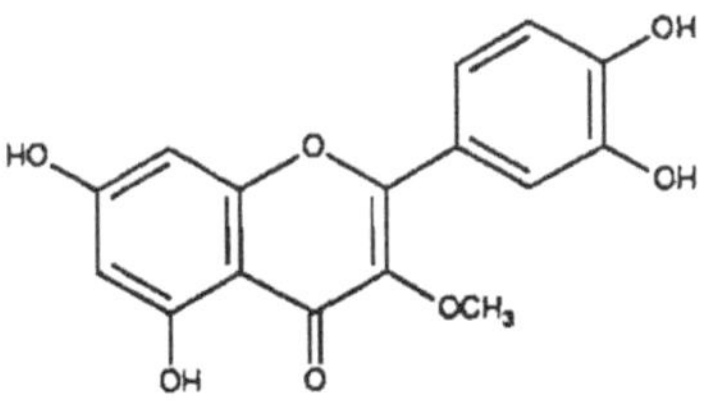

(**34**) Quercetin 3-methyl ether

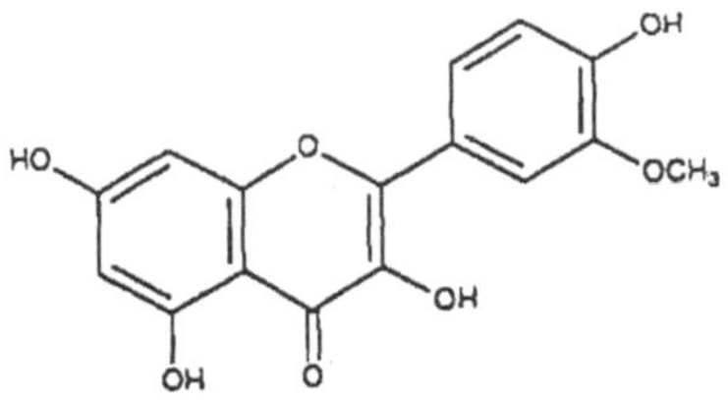

(**35**) Isorhamnetin

Figure 11.3 Leaf flavonoids from *Pelargonium* species.

Myricetin and flavone *C*-glycosides are the most useful flavonoid taxonomic markers in the genus but the co-occurrence of flavonoids with various other phenolic constituents i.e. ellagitannins and proanthocyanidins, is more valuable (see below). In summary, the leaf flavonoid pattern is broadly similar to that encountered in *Geranium* (see Chapter 4).

Floral flavonoids

The occurrence of pelargonin (pelargonidin 3,5-diglucoside) (**36**) Figure 11.4 in flowers of *Pelargonium* cultivars was established in the 1930s, following the synthesis of this pigment. Robinson and Robinson (1932) found that the salmon pink petals of zonal

Table 11.1 The distribution of vacuolar leaf flavonoids* in *Pelargonium* species†

Section/Species/Authority		*Flavonols*					*Flavones*	
		Myricetin (6)	*Quercetin (7)*	*Kaempferol (8)*	*Isorh (12)*	*Qu3ME (11)*	*Lu (9)*	*C-glycs*
PELARGONIUM								
P. quercifolium	(L.f) L'Hér.	+	+	(+)	–	+	(+)	–
P. graveolens	L'Hér.	++	+++	+	–	–	+	–
P. tomentosum	Jacq.	++	+++	(+)	(+)	+	(+)	–
P. cucullatum	(L.) L'Hér.	+	++	(+)	–	–	–	–
P. crispum 'Whiteknights'	(Berg.) L'Hér.	+	+++	(+)	(+)	–	–	–
c.v. Lady Plymouth		++	+++	+	–	–	(+)	–
CICONIUM								
P. transvaalense	Knuth	–	+	–	–	–	–	–
P. acraeum	R.A. Dyer	–	+	+	+	–	–	–
P. mutans	P. Vorster	–	++	(+)	–	+	(+)	+
P. multibracteatum	Hochst.	–	+	–	–	–	(+)	–
P. alchemilloides	(L.f.) L'Hér.	–	++	++	–	–	–	–
P. ranunculophyllum	(Eckl. and Zeyh.) Bak.	–	(+)	++	–	–	–	–
OTIDIA								
P. laxum	(Sweet) G. Don	+++	++	(+)	–	(+)	–	–
P. ceratophyllum	L'Hér.	++	++	–	–	–	–	–
P. alternans	Wendl.	(+)	++	–	–	(+)	–	–
P. carnosum	(L.) L'Hér.	+++	+	–	–	–	–	–
LIGULARIA								
P. fulgidum	(L.) L'Hér.	–	(+)	–	–	–	++	++
P. oreophilum	Schltr.	+	++	–	–	–	–	–
P. pulchellum	Sims	+	++	(+)	–	–	–	–
SUBSUCCULENTIA								
P. karooicum	Compton	–	(+)	–	–	–	(+)	+++
P. otaviense	Knuth	–	(+)	(+)	–	–	(+)	+++
P. spinosum	Willd.	–	–	–	–	–	++	+
CORTUSINA								
P. echinatum	Curt.	–	+	++	–	–	–	–

P. magenteum	J.J.A. v.d. Walt	+++	++	(+)	–	–	–	–
P. crassicaule	L'Hér.	+	++	–	–	–	–	–
P. cortusifolium	L'Hér.	+	++	–	–	–	–	–
CHORISMA								
P. worcesterae	Knuth	–	(+)	–	–	–	(+)	+
P. tetragonum	(L.f.) L.Hér.	–	+	+	–	–	(+)	+
P. mollicomum	Fourcade	–	+	–	–	–	+	++
PERISTERA								
P. drummondii	Turcz.	–	(+)	–	–	–	+	++
ISOPETALUM								
P. cotyledonis	(L.) L'Hér.	–	+	(+)	–	–	(+)	++
P. cotyledonis 2	(L.) L'Hér.	–	+++	–	–	+	+	–
P. cotyledonis 3	(L.) L'Hér.	–	+	–	–	–	(+)	–
POLYACTIUM								
Caulescentia								
P. gibbosum hybrid	(L.) L'Hér.	+	++	–	–	–	–	+
POLYACTIUM								
P. triste	(L.) L'Hér.	++	+++	(+)	–	–	–	–
MAGNISTIPULACEA								
P. luridum	(Andr.) Sweet	–	(+)	–	–	–	+	++
SCHIZOPETALA								
P. amatymbicum	(Eckl. & Zeyh.) Harv.	–	+++	–	–	–	–	–
P. bowkeri	Harv.	–	+	+	–	–	(+)	–
GLAUCOPHYLLUM								
P. grandiflorum	(Andr.) Willd.	–	+++	(+)	–	–	–	–
CAMPYLIA								
P. tricolor	Curt.	–	+++	(+)	–	–	–	–
P. ovale ssp. *hyalinum*	(Burm.f.) L'Hér.	+	+++	(+)	–	–	–	–
RENIFORMIA								
P. abrotanifolium	(L.f.) Jacq.	+	+	–	–	–	+	–
P. reniforme	Curt.	–	++	+	+	–	(+)	++
P. exstipulatum	(Cav.) L'Hér.	–	+	(+)	(+)	–	(+)	++
P. album	J.J.A. v.d. Walt	–	+	++	(+)	–	(+)	+++

Table 11.1 (Continued)

Section/Species/Authority		*Flavonols*					*Flavones*	
		Myricetin (6)	*Quercetin (7)*	*Kaempferol (8)*	*Isorh (12)*	*Qu3ME (11)*	*Lu (9)*	*C-glycs*
P. sidoides	DC.	–	+	–	–	–	+	+
P. odoratissimum	(L.) L'Hér.	+	++	+	–	–	(+)	–
HOAREA								
P. bubonifolium	(Andr.) Pers.	–	(+)	–	–	–	–	–
P. appendiculatum	(L.f.) Willd.	+	++	(+)	–	–	–	–
P. incrassatum	(Andr.) Sims	+	+++	+	–	–	–	–
MYRRHIDIUM								
P. surburbanum	Clifford ex Boucher	–	+	–	–	–	–	–
P. myrrhifolium	(L.) L'Hér.	–	+	+	–	–	(+)	–
JENKINSONIA								
P. antidysentericum	(Eckl. & Zeyh.) Kostel	–	(+)	–	–	–	(+)	+
P. trifidum	Jacq.	–	+	–	–	(+)	+	++
P. praemorsum	(Andr.) Dietr.	–	+++	–	–		–	–
P. griseum	Knuth	–	+	–	–		(+)	+
P. senecioides	L'Hér.	–	(+)	–	–	–	+	++
UNKNOWN								
P. rodneyanum	Mitch.	–	(+)	–	–	–	+	+
P. quercetorum	Agnew	–	++	++	–	–	–	++

Source: † Data from Williams *et al.* (2000).

Notes

* Flavonoid aglycones identified after acid hydrolysis.

Isorh = isorhamnetin; Lu = luteolin; Qu3ME = quercetin 3-methyl ether; C-glycs = flavone C-glycosides.

Pelargonium were pigmented by pure pelargonium 3,5-diglucoside, while the cultivar 'Henry Jacoby' contained the same pigment together with malvidin 3,5-diglucoside (37) Figure 11.4. A survey of the three wild species: *P. veitchianum*, *P. bertiana* and *P. cucullatum* showed that malvidin 3,5-diglucoside was the major pigment in each.

Later investigation of *Pelargonium* cultivar petals showed that pelargonidin and malvidin 3,5-diglucoside was accompanied by small amounts of the relatively rare peonidin 3,5-diglucoside (**38**) (Harborne, 1961) (Figure 11.4). A further survey of petal pigments in *P.* × *hortorum* showed that all six common anthocyanidins (including delphinidin, petunidin and cyanidin) could be identified as the 3,5-diglucosides in different colour forms (Asen and Griesbach, 1983). The 3,5-diglucosides are invariably accompanied by the corresponding acetates, where an acetyl residue is substituted at the 6-position of the glucose of C-5 (Mitchell *et al.*, 1998). It may be noted that the acetate of the malvidin pigment has also been detected in Geranium (see Chapter 4). A range of colourless flavonol

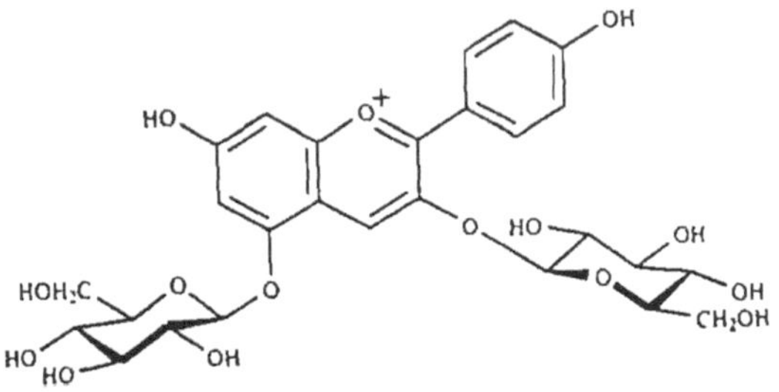

(**36**) Pelargonidin 3,5-diglucoside

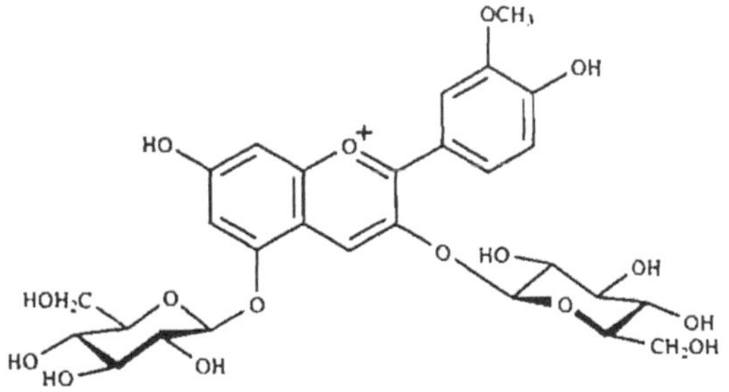

(**37**) Malvidin 3,5-diglucoside

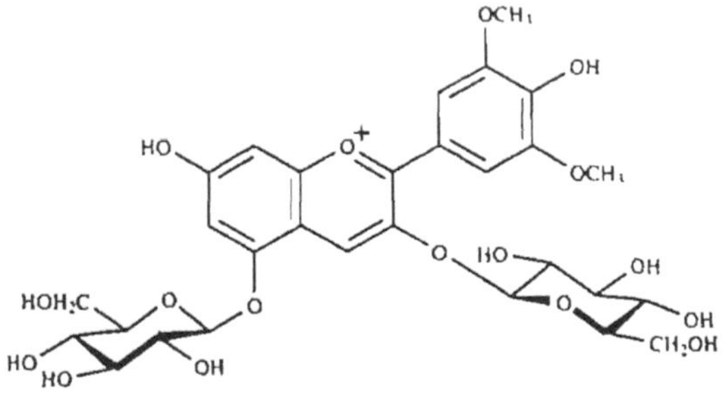

(**38**) Peonidin 3,5-diglucoside

Figure 11.4 Anthocyanins found in flowers of *Pelargonium* species.

glycosides based on kaempferol and quercetin occur with the above anthocyanins in *Pelargonium* × *hortorum* petals. Besides several monoglycosides, the 3-rutinosides and 3-rhamnosylgalactosides of these two flavonols are present (Asen and Griesbach, 1983).

Exudate flavonoids

Many *Pelargonium* species have glandular hairs on the leaf surface. Besides producing terpenoid constituents (see under terpenoids), these structures may also yield lipophilic flavonoids. Such compounds were detected in 35 per cent of the *Pelargonium* taxa surveyed by Williams *et al.* (1997), but mostly only in trace amounts. However, in *P. crispum* the simple flavone, chrysin (5,7-dihydroxyflavone) (39) Figure 11.5 is the major leaf exudate constituent, accompanied by a *C*-methyl derivative (either 6- or 8-) of the corresponding 5,7-dihydroxyflavanone. Two unidentified flavanones were detected in one other species, *P. drummondii*. Exudate flavones were further found in four taxa: chrysoeriol (40) Figure 11.5 and acacetin (41) (luteolin 3′- and apigenin 4′-methyl ethers, respectively) in *P. album* and some unidentified apigenin-based flavones in *P. abrotanifolium*, *P. exstipulatum* and *P. fulgidum*. A variety of flavonol (quercetin and kaempferol) methyl ethers (see Table 11.2) were found in *P. fulgidum*, *P. quercifolium* and

Table 11.2 Exudate flavonoids detected in *Pelargonium* species

Species	*Leaf exudate flavonoids*		
	Flavones	*Flavanones*	*Flavonols*
P. abrotanifolium (L.f.) Jacq.	Apigenin (33)-based, possibly acylated	–	–
P. album J.J.A. v.d. Walt	Chrysoeriol (40) Acacetin (41)	–	–
P. crispum (Berg.) L'Hér.	Chrysin (39)	6-or-8-*C*-methyl 5,7-dihydroxyflavanone	–
P. drummondii Turcz.	–	Two unidentified flavanones	–
P. exstipulatum (Cav.) L'Hér.	Apigenin-based	–	–
P. fulgidum (L.) L'Hér.	Apigenin-based	–	Qu 3ME (34) Km MME Qu 3,7-diME (43) Km 3,7-diME (42) Km triME Qu tetraME
P. quercifolium (L.f.) L'Hér.	–	–	Qu 3ME Km 3ME Qu 3,7-diME Qu triME
P. tomentosum Jacq.	–	–	Km triME Qu tetraME

Notes
Qu = quercetin; Km = kaempferol; ME = methyl ether.

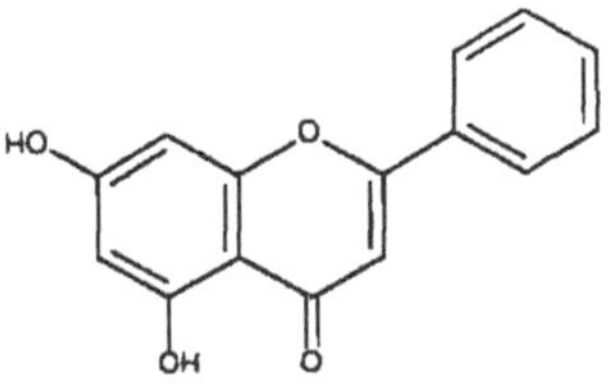

(**39**) Chrysin

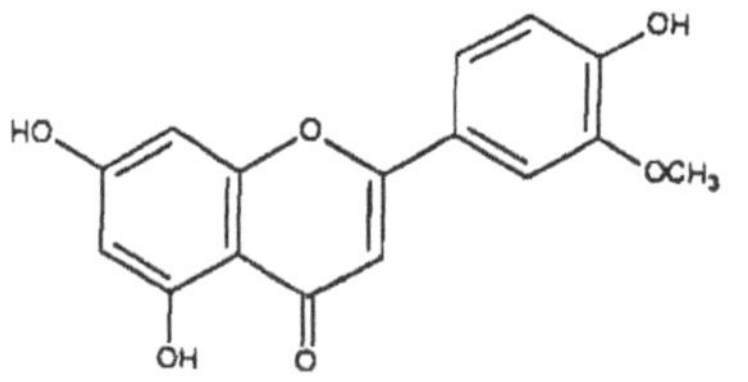

(**40**) Chrysoeriol

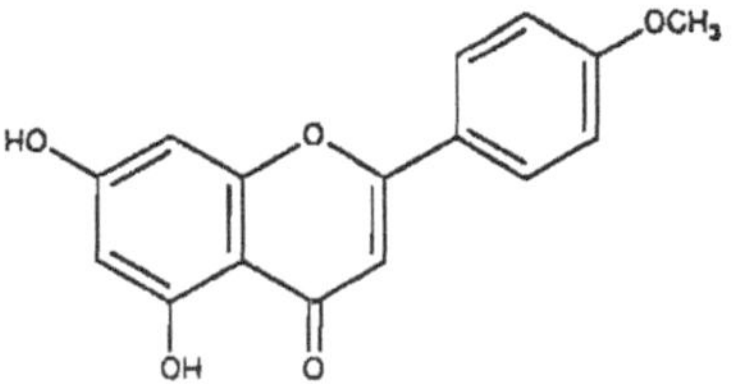

(**41**) Acacetin

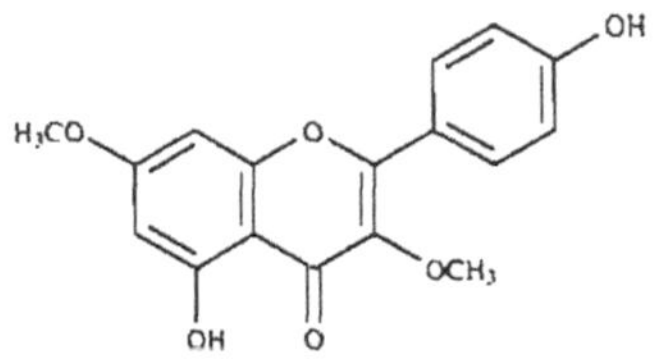

(**42**) Kaempferol 3,7-dimethyl ether

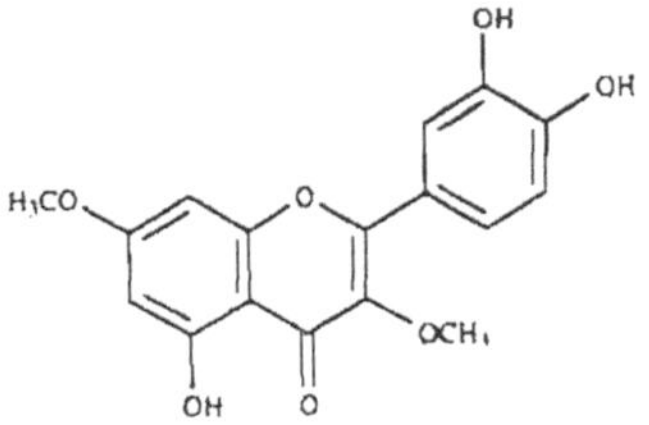

(**43**) Quercetin 3,7-dimethyl ether

Figure 11.5 Leaf exudate flavonoids from *Pelargonium* species.

P. tomentosum. Two typical structures are kaempferol and quercetin 3,7-dimethyl ethers (42 and 43) (Figure 11.5). Some partially characterised higher methyl ethers of both these flavonols were also present.

Tannins

The genus *Pelargonium*, like *Geranium* is unusual in synthesising both hydrolysable (ellagitannins) and non-hydrolysable (proanthocyanidins) tannins in abundance in many of its species. The two types of tannin are produced by different pathways. The proanthocyanidins are formed by condensation of single catechin units (e.g. 44) to form dimers, trimers, tetramers and higher oligomers (45) (Figure 11.6). The ellagitannins originate from gallic acid (46), which condenses with glucose to give a digalloyl ester, which then forms the hexahydroxydiphenoyl ester (47) (Figure 11.6) as a key intermediate. The proanthocyanidins can be detected by the production of some of the parent anthocyanidin on treatment with hot acid and the ellagitannins by the release of ellagic acid (48) (Figure 11.6) on acid hydrolysis. However, some *Pelargonium* species produce free ellagic acid in the absence of ellagitannins. In the survey of Williams *et al.* (2000) prodelphinidin was found in 53 per cent of the sample, procyanidin in 29 per cent, ellagitannins in 53 per cent and free ellagic acid in 50 per cent of the taxa. Gallic acid was also recorded from 62 per cent of the species and protocatechuic acid

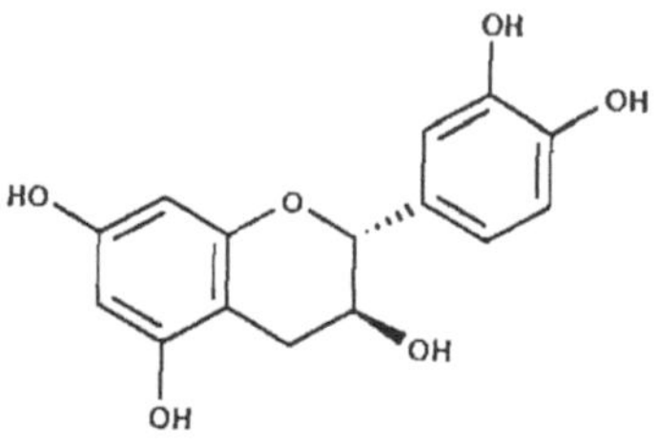

(**44**) (+)-Catechin

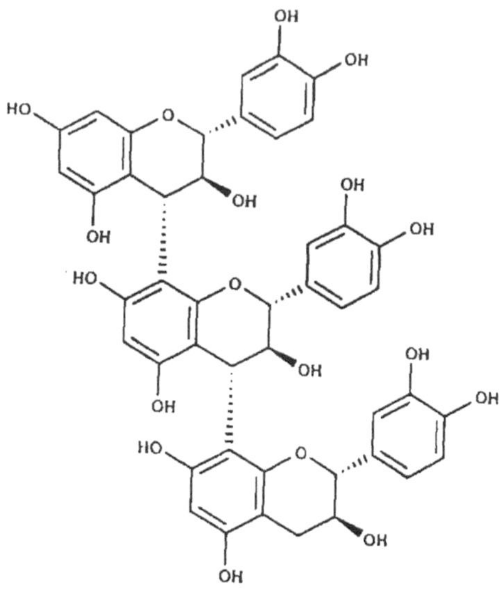

(**45**) Procyanidin C_2

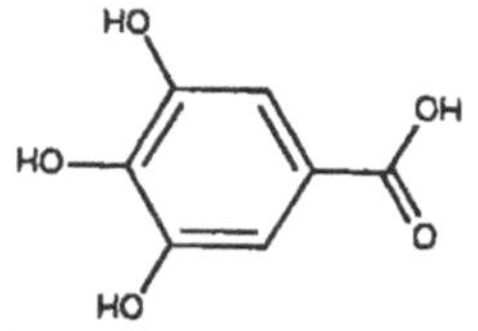

(**46**) Gallic acid

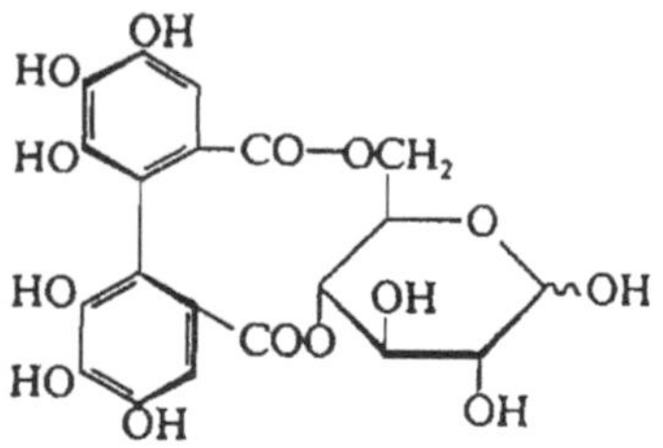

(**47**) Hexahydroxydiphenoyl ester

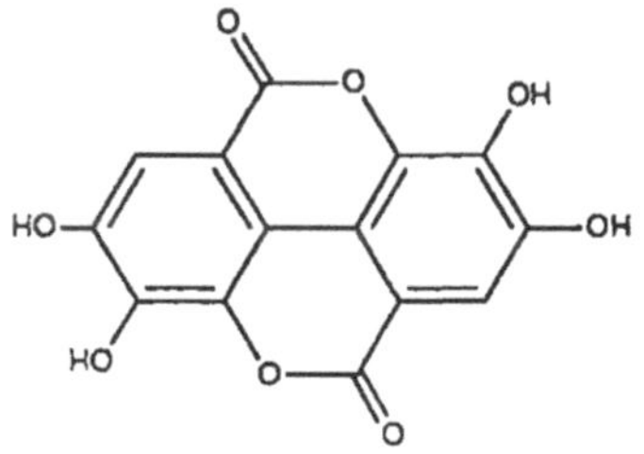

(**48**) Ellagic acid

Figure 11.6 Tannins and their precursors in *Pelargonium*.

from 38 per cent of the sample. In *Pelargonium* both types of tannin may occur in the same plant, separately, or both may be absent. Their co-occurrence with flavonoids has some taxonomic and evolutionary significance. Thus, the presence of myricetin is correlated with the presence of proanthocyanidins (but not *vice versa*) and the absence of ellagitannins. Conversely, the absence of myricetin shows a correlation with the absence of ellagitannins. On this basis, the various species may be divided into six chemical groups: i.e. those containing (1) myricetin + proanthocyanidins; (2) myricetin + proanthocyanidins + free ellagic acid; (3) myricetin + proanthocyanidins + ellagitannins + free ellagic acid; (4) ellagitannins + free ellagic acid but no myricetin; (5) ellagitannins but no free ellagic acid or myricetin; (6) no tannins or myricetin. There also seems to be a correlation between flavone *C*-glycosides and the presence of luteolin in some species. Bate-Smith (1962) suggested that both ellagitannins and proanthocyanidins are indicators of primitiveness in the dicotyledons and that they are both associated with woodiness. He also suggested that there have been two lines of evolution in the dicots and that in one line ellagitannins were lost before the proanthocyanidins and in the other the proanthocyanidins were lost first. This was supported by a statistical analysis of the Bate-Smith data by Sporne (1975), which also indicated that the procyanidins were lost more quickly than the ellagitannins. These evolutionary trends can be clearly seen in *Pelargonium* and are illustrated in Figure 11.7, where the species analysed by Williams *et al.* (2000) are grouped in sections according to their tannin constituents and overlayed with data for the presence of myricetin and glycoflavones.

Detailed chemical investigations of the *Pelargonium* tannins remain for the future. All that is known is that the ellagitannin geraniin, which is so widespread in *Geranium* species, is apparently not present in the several *Pelargonium* taxa so far surveyed (Okuda *et al.*, 1980), which included *P. graveolens* and *P. tomentosum*.

MINOR PHENOLIC COMPOUNDS

Coumarins

Four coumarins: the common scopoletin (49), the rare 7-hydroxy-5,6-dimethoxy-coumarin (50) and its 7-methyl ether (51) and its 7-glucoside (52) (Figure 11.8) were identified in roots of *Pelargonium reniforme* and detected in roots of 11 other species: *P. betulinum*, *P. capitatum*, *P. cucullatum*, *P. hirtum*, *P. luridum*, *P. moreanum*, *P. myrrifolium*, *P. radula*, *P. reniforme*, *P. salmoneum*, *P. sidaefolium*, *P. triste* and *P. zonale* (Wagner and Bladt, 1975).

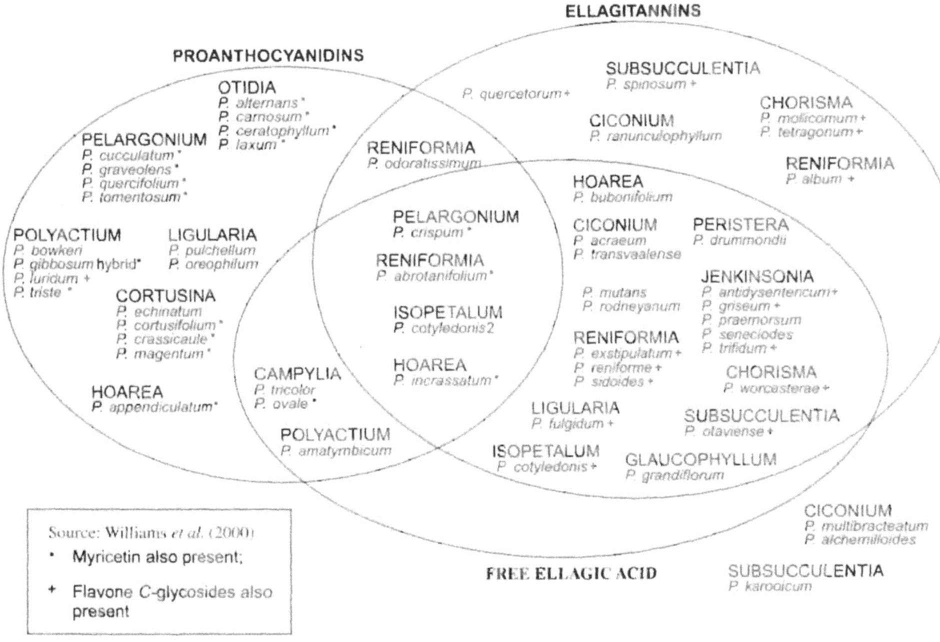

Figure 11.7 *Pelargonium* species grouped in sections according to their tannin constituents.

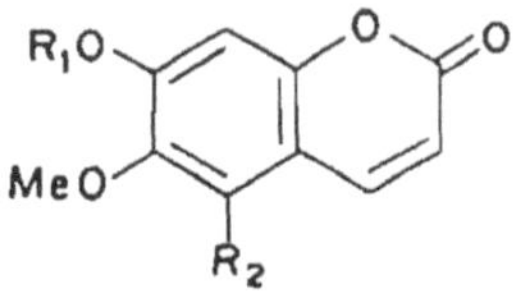

(**49**) Scopoletin $R_1=R_2=H$
(**50**) 7-hydroxy-5,6-dimethoxycoumarin $R_1=H$, $R_2=OMe$
(**51**) 5,6,7-trimethoxycoumarin $R_1=Me$, $R_2=OMe$
(**52**) 7-hydroxy-5,6-dimethoxycoumarin 7-glucoside $R_1=Glc$, $R_2=OMe$

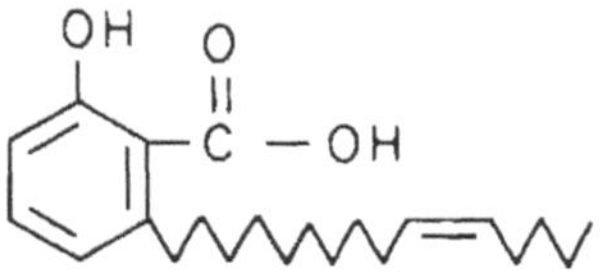

(**53**) 6-[(Z)-10′-Pentadecenyl]salicylic acid

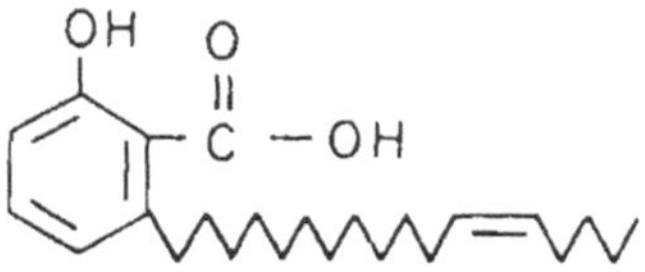

(**54**) 6-[(Z)-12′-Heptadecenyl]salicylic acid

Figure 11.8 Coumarins and salicylic acid derivatives from *Pelargonium* species.

Salicylic acid derivatives

Plants of *Pelargonium* × *hortorum*, (the so-called geraniums sold in garden centres for summer bedding), were found to be largely resistant to attack by the two-spotted spider mite, *Tetranychus urticae* Koch. This resistance was tracked down to the presence of glandular trichomes on both upper and lower surfaces of the leaves, which produce a toxic, sticky exudate (Craig *et al.*, 1986). Morphologically similar trichomes occur in equal number on susceptible plants but they do not produce any exudate. The two major toxins of the exudate were identified as 6-[(*Z*)-10′-pentadecenyl]salicylic acid (53) and 6-[(*Z*)-12′-heptadecenyl]salicylic acid (54) (Walters *et al.*, 1988) (Figure 11.8). In a study of the relative resistance to the two-spotted spider mite of seven *Pelargonium* species, four namely: *P. capitatum*, *P. australe*, *P. tomentosum* and *P. rapaceum* were found to be more resistant than the other three: *P. fulgidum*, *P. odoratissimum* and *P. graveolens* (Snetsinger *et al.*, 1966). These workers also evaluated the susceptibility of six accessions of *P.* × *hortorum* and found that the three commercial cultivars (two diploids and one tetraploid) were resistant or moderately resistant to attack and of the other cultivars 'King Midas' was susceptible and G71 and their hybrid were resistant. The source of the resistance was not investigated. Variation among cultivars of *P. peltatum*, the ivy-leaved *Geranium*, in their resistance to this spider mite has been reported (Potter and Anderson, 1982) and similar variation has been observed among the tetraploid cultivars of *P.* × *hortense* of European origin (Craig *et al.*, 1986).

Organic acids

Tartaric acid (55) (Figure 11.9) is a characteristic constituent of the genus *Pelargonium*. It represents more than 1.5 per cent of the dry weight of the aerial parts of 23 species and hybrids, 0.75–1.5 per cent in six other species and hybrids and less than 0.75 per cent in three further species (Stafford, 1961). Oxalic acid (56)

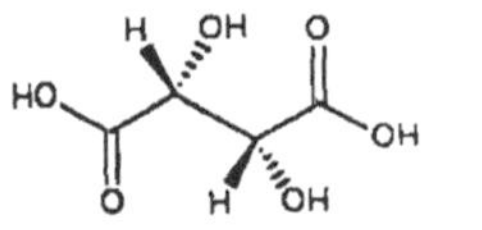

(**55**) Tartaric acid

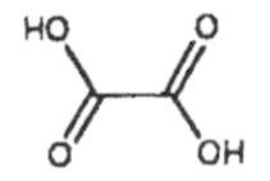

(**56**) Oxalic acid

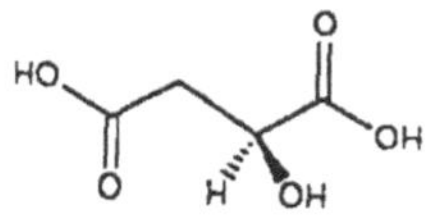

(**57**) Malic acid

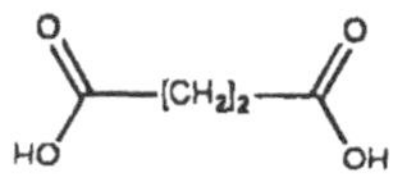

(**58**) Succinic acid

Figure 11.9 Organic acids from *Pelargonium* species.

(Figure 11.9) has been isolated from *P. peltatum*. This plant is potentially poisonous to livestock since oxalic acid is present in a water-soluble form; instead of the more usual insoluble calcium salt. It occurs together with malic (57), tartaric and succinic (58) acids in this species (Figure 11.9). The presence of malic acid is expected because *P. peltatum* is a Crassulacean acid metabolism CAM plant, and thus accumulates malic acid in the dark, which disappears during the day.

Alkaloids

Screening studies on *Pelargonium* species had revealed the presence of alkaloids, especially in 'zonal' cultivars (Lis-Balchin, 1996, 1997), which had previously only been found in the *Erodium* genus of the Geraniaceae (Lis-Balchin and Guittoneau, 1995). The identity of the alkaloids detected was not determined until studies by Lis-Balchin *et al.* (1996) on the cultivar 'Appleblossom'. The simple amines tyrosine 1 and tryptamine 2 were first detected by thin-layer chromatography (TLC) comparison with authentic samples but of more interest was the isolation of three other, more complex, indole alkaloids. The structures of two of these were determined by spectroscopic methods and they were shown to be the two isomers elaeocarpidine 3 and isoelaeocarpidine. These alkaloids were found in all the different types of zonal *Pelargonium* cultivars, but not in the pure ivy-leaf or regals; the hybrid ivy-leaf pelargoniums, with the appearance of either ivy-leaf or zonal *Pelargonium* also inherited the alkaloids (Lis-Balchin, 1996). These alkaloids only appeared in half of the section *Ciconium*, which were originally allocated to the Section and some of whose genera gave rise to the modern zonals (Lis-Balchin, 1996). The indole alkaloids appear to have an insect repellent activity against whitefly (Woldemariam *et al.*, 1997) and concentrate in the darker zonal area of the leaves.

CONCLUSION

The essential oil of *Pelargonium* (often termed Geranium oil) has already been intensively investigated. However, in view of its possible medicinal and other useful properties, it probably deserves continuing studies. There is still much to be learnt about the volatiles present in those *Pelargonium* species yet to be cultivated.

In terms of the polyphenolic pattern, the genus *Pelargonium* is broadly similar in the range and type of constituent to the genus *Geranium*. And yet there are significant differences. Thus, the characteristic ellagitannin of *Geranium* is not produced by *Pelargonium* species. We know that ellagitannins are abundant in the two genera. What is formed in *Pelargonium* to replace the geraniin of *Geranium*?

REFERENCES

Albers, F., van der Walt, J.J.A., Gibby, M., Marschewski, D.E., Price, R.A. and du Preez, G. (1995) A biosystematic study of *Pelargonium* section *Chorisma*. *S. Afr. J. Bot.*, 61, 339–346.

Asen, S. and Griesbach, R. (1983) High pressure liquid chromatographic analysis of flavonols in *Geranium* florets as an adjunct for cultivar identification. *J. Am. Soc. Hortic. Sci.*, 108, 845–850.

Axel, R.T., Ramsey, R.B. and Nicholas, H.J. (1972) Sterols and triterpenes of *Pelargonium hortorum*. *Phytochemistry*, 11, 2353–2354.

Bate-Smith, E.C. (1962) The phenolic constituents of plants and their taxonomic significance. 1. Dicotyledons. *J. Linn. Soc. (Bot.)*, 58, 39–54.

Bate-Smith, E.C. (1973) Chemotaxonomy of *Geranium*. *Bot. J. Linn. Soc.*, 67, 347–359.

Charlwood, B.V. and Charlwood, K.A. and Molina-Torres, J. (1990) Accumulation of secondary compounds by organised plant cultures. In: B.V. Charlwood and M.J.C. Rhodes (eds), Secondary Products from Plant Tissue Culture, Oxford Science Publications, Clarendon Press, Oxford, pp. 187–189.

Charlwood, B.V. and Charlwood, K.A. (1991) *Pelargonium* spp. (*Geranium*): *In vitro* culture and the production of aromatic compounds. In: Y.P.S. Bajaj (ed.), *Biotechnology in Agriculture and Forestry. vol. 15. Medicinal and Aromatic Plants III*, Springer-Verlag, Berlin and Heidelberg, pp. 339–352.

Charlwood, B.V. and Moustou, C. (1988) Essential oil accumulation in shoot-proliferation cultures of *Pelargonium* species: In: R.J. Robins and M.J.C. Rhodes (eds), *Manipulating Secondary Metabolism in Culture*, Cambridge University Press, Cambridge, pp. 187–194.

Craig, R., Mumma, R.O., Gerhold, D.L., Winner, B.L. and Snetsinger, R. (1986) Genetic control of a biochemical mechanism for mite resistance in geraniums. In: M.B. Green and P.A. Hedin (eds), *Natural Resistance of Plants to Pests*, American Chemical Society, Washington, DC, USA, pp. 168–176.

Dreyer, L.L., Albers, F., van der Walt, J.J.A. and Marschewski, D.E. (1992) Subdivision of *Pelargonium* sect. *Cortusina* (Geraniaceae). *Pl. Syst. Evol.*, 181, 83–97.

Harborne, J.B. (1961) The anthocyanins of roses. Occurrence of peonin. *Experientia*, 17, 72–73.

Katagi, H., Takahashi, E., Nakao, K. and Inui, M. (1986) Shoot-forming cultures of *Pelargonium graveolens* by jar fermentation. *Nippon Nogeikagku Kaishi*, 60, 15–17.

Lis-Balchin, M. (1996) A chemotaxonomic reappraisal of the Section Ciconium *Pelargonium* (Geraniaceae). *S. Afr. J. Bot.*, 62, 277–279.

Lis-Balchin, M. (1997) A chemotaxonomic study of the *Pelargonium* (Geraniaceae) species and their modern cultivars. *J. Hort. Sci.*, 72, 791–795.

Lis-Balchin, M. and Guittonneau, G.-G. (1995) Preliminary investigations on the presence of alkaloids in the genus *Erodium* L'Herit. (*Geraniaceae*). *Acta Bot. Gallica*, 141, 31–35.

Lis-Balchin, M., Houghton, P. and Woldermariam, T. (1996) Elaeocarpidine alkaloids from *Pelargonium* (Geraniaceae). *Nat. Products Lett.*, 8, 105–112.

Mitchell, K.A., Markham, K.R. and Boase, M.R. (1998) Pigment chemistry and colour of *Pelargonium* flowers. *Phytochemistry*, 47, 355–361.

Naves, J.-R., Lamparsky, D. and Ochsner, P. (1961) Études sur les matières végétales volatiles CLXXIV (1). Présence de tétrahydropyrannes dans l'huile essentielle de géranium. *Bull. Soc. Chim. France*, 645–647.

Naves, J.-R., Ochser, P., Thomas, A.F. and Lamparsky, D. (1963) Études sur les matières végétales volatiles CLXXXVI (1). Présence d'acétonyl-2-méthyl-4-tétrahydropyranne dans l'huile essentielle de géranium. *Bull. Soc. Chim. France*, 1608–1611.

Okuda, O., Mori, K. and Hatano, T. (1980) The distribution of geraniin and mallotusinic acid in the order Geraniales. *Phytochemistry*, 19, 547–551.

Potter, D.A. and Anderson, R.G. (1982) Resistance of ivy geraniums to the two-spotted spider mite. *J. Am. Soc. Hort. Sci.*, 107, 1089–1092.

Robinson, M. and Robinson, R. (1932) A survey of anthocyanins. Part II. *Biochem. J.*, 26, 1647–1664.

Snetsinger, R., Balderston, C.P. and Craig, R. (1966) Resistance to the two-spotted spider mite in *Pelargonium*. *J. Econ. Ent.*, 59, 76–78.

Sporne, K.R. (1975) A note on ellagitannins as indicators of evolutionary status in dicotyledons. *New Phytol.*, 75, 613–618.

Stafford, H.A. (1961) Distribution of tartaric acid in the Geraniaceae. *Am. J. Botany*, 48, 699–701.

Teisseire, P. (1987) Industrial quality control of essential oils by capillary GC. In: P. Sandra and C. Bicchi (eds), *Capillary Gas Chromatography in Essential Oil Analysis*, Huethig, Heidelberg, pp. 215–258.

Vernin, G., Metzger, J., Fraisse, D. and Scarf, C. (1983) Etude des huiles essentielles par GC-SM-banque specma: essences de geranium. *Parf. Cosm. Arom.*, 52, 51–61.

van der Walt, J.J.A. (1977) In *Pelargoniums of Southern Africa*, vol. 1., Purnell, Cape Town, pp. 5, 23.

van der Walt, J.J.A., Albers, F. and Gibby, M. (1990) Delimitation of *Pelargonium* sect. *Glaucophyllum* (Geraniaceae). *Pl. Syst. Evol.*, 171, 15–26.

van der Walt, J.J.A., Albers, F., Gibby, M., Marschewski, D.E., Hellbrugge, D., Price, R.A. and der Merwe, A.M. (1997) A biosystematic study of *Pelargonium* section *Ligularia*, 3, Reappraisal of section *Jenkinsonia*, *S. Afr. J. Bot.*, 63, 4–21.

Wagner, H. and Bladt, S. (1975) Coumarine aus südafrikanischen *Pelargonium*-arten. *Phytochemistry*, 14, 2061–2064.

Walters, D.S., Minard, R., Craig, R. and Mumma, R.O. (1988) Geranium defensive agents. III. Structural determination and biosynthetic considerations of anacardic acids of geranium. *J. Chem. Ecol.*, 14, 743–751.

Webb, W.J. (1984) The *Pelargonium* family, Croom Helm, London.

Williams, C.A., Harborne, J.B., Newman, M., Greenham, J. and Eagles, J. (1997) Chrysin and other leaf exudate flavonoids in the genus *Pelargonium*. *Phytochemistry*, 46, 1349–1353.

Williams, C.A., Newman, M. and Gibby, M. (2000) The application of leaf phenolic evidence for systematic studies within the genus *Pelargonium* (Geraniaceae). *Biochem. Syst. Ecol.*, 28, 119–132.

Woldemariam, T.Z., Houghton, P.J., Lis-Balchin, M. and Simmonds, S.J. (1997) Alkaloid and tannin distribution in the leaves of Pelargonium zonale with reference to insect behaviours. The 135th British Pharmaceutical Conference, Scarborough. Sept. 15–18.

12 Pharmacology of Pelargonium essential oils and extracts *in vitro* and *in vivo*

Stephen Hart and Maria Lis-Balchin

INTRODUCTION

Pelargonium species originated in southern Africa where a large number were used as folk medicines with mainly antidysenteric properties (Watt and Breyer-Brandwijk, 1962). A few species were brought over to Europe in the 17th Century and hybridised over the centuries resulting in thousands of modern cultivars. Commercial 'Geranium oil' is obtained from the scented leaves of a number of *Pelargonium* cultivars grown mainly in Reunion, China, Egypt and Morocco. There are, however, large numbers of scented species, hybrids and cultivars which are at present unexploited, but exhibit some potential as odourants for the perfumery and food industry, antimicrobial agents and insecticides (Lis-Balchin, 1988, 1997; Lis-Balchin *et al.*, 1996b).

FOLK-MEDICINAL USAGE OF *PELARGONIUM* SPECIES

The *Pelargonium* species used in South Africa by the local population were also used by the Boers mainly for antidysenteric purposes e.g. root of *P. transvaalense* and *P. triste* and the leaves of *P. bowkeri* and *P. sidaefolium*. Some *Pelargonium* species were also used to treat specific maladies e.g. *P. cucullatum* for nephritis, *P. tragacanthoides* for neuralgia, *P. luridum* and *P. transvaalense* root for fever; *P. minimum*, *P. reniform* and *P. grossularioides* for menstrual flow (Pappe, 1868; Watt and Brandwijk, 1962). The last was also used as an emmanogogue and abortifacient by both Zulus and Boers and has recently been studied further (Lis-Balchin and Hart, 1994) and shown to have spasmogenic properties on the uterus and smooth muscle preparations *in vitro*.

All the folk-medicinal attributes have been due to the water-soluble or ethanolic extracts, as they involved teas and infusions and not volatile essential oils themselves. This differs to the practice of aromatherapy, where essential oils are used, and furthermore the plant extracts were not massaged into the skin but mainly used internally. The *Pelargonium* species used were often non-scented and the scented ones used were different *Pelargonium* species to those utilised in the production of commercial Geranium oil for aromatherapy.

STUDY OF *IN VITRO* PHARMACOLOGICAL ACTIVITY

The preparation chosen to assess the *in vitro* pharmacological activity of an essential oil or its components on smooth muscle must respond to spasmogenic agents i.e. those causing a contraction and spasmolytic agents which will relax smooth muscle. Preparations of intestinal smooth muscle are robust and, although spasmogenic and spasmolytic activity can be identified on those from rabbit and the large intestine of the guinea-pig, the field stimulated guinea-pig ileum allows quantitative experiments to be performed readily. The field stimulated guinea-pig ileum also enables a neurogenic response to be distinguished from a myogenic response and the receptors involved to be elucidated. When the reported use of the essential oil indicates a targeted organ such as the uterus or lungs then appropriate preparations of these smooth muscles can be studied. For a more complete study of the *in vitro* pharmacology it would be necessary to include vascular and cardiac tissues and to examine activity on skeletal muscle such as the rat phrenic nerve hemi-diaphragm preparation. It must of course be remembered that essential oils contain many components and the observed pharmacological action of the oil may be due to several active principles acting by different mechanisms or, indeed, by opposing mechanisms which may give rise to biphasic responses. It is also important to verify that the solvent for an essential oil or its components, in the concentration that will be in contact with the tissue, does not itself have a spasmogenic or spasmolytic action. Methanol is satisfactory for most smooth muscle preparations at a dilution of not less than 1:125 but it has been reported that hexane extracts may give misleading results (Lis-Balchin *et al.*, 1997). It is questionable whether aqueous emulsions are appropriate but some workers prefer a vehicle such as Arlatone 285 (Reiter and Brandt, 1985) or dimethylsulphoxide, DMSO (Aqel, 1991).

STUDY OF *IN VIVO* PHARMACOLOGICAL ACTIVITY

There are three appropriate approaches to the study of the effects of *Pelargonium* in the whole animal, which differ in the route of administration. The classical approach is the parenteral administration into the conscious or anaesthetised animal. Results from such experiments will demonstrate the pharmacology of the compound but may be difficult to relate to their normal use. Exposure of animals to the aroma of an essential oil, or the application of the diluted oil to depilated skin, obviously mimics in animals the normal use of the compounds in humans. The ideal investigation will use human volunteers and two approaches are possible. In the first, the purpose of the experiment is to determine the pharmacology of the essential oil and a dose-related effect would be expected whilst in the second, one is interested in whether, at the concentrations normally used, the oil has any significant effects. The latter experiments require careful design and this will differ depending on whether the aroma alone, or aroma plus massage plus absorption across skin, is being investigated.

INVESTIGATIONS OF PELARGONIUM ESSENTIAL OILS

The bioactivity of essential oils has been studied for many years with much of the work originating from Germany but very little has been published on *Pelargonium*. Treibs (1956) and Sticher (1977) have reviewed early work which includes research on some of the constituents of *Pelargonium*. Gunn (1921) studied the actions of several volatile oils (in the form of emulsions) on the activity of intestinal muscle from rabbit, rat and cat *in vitro* and *in vivo* and observed spasmolysis, which appeared to be a direct action on the muscle, but did not include *Pelargonium* (or *Geranium*). A year later Stross (quoted by Sticher, 1977) reported the spasmolytic activity of several naturally occurring compounds including geranyl acetate. Imaseki and Kitabatake (1962) report an antispasmodic action of citronellol, geraniol and linalool on the mouse small intestine. Buchbauer *et al.* (1993) have reviewed the biology of essential oils and Tisserand and Balacs (1995) have described their general activity and safety.

EXPERIMENTS ON *IN VITRO* GUINEA-PIG ILEUM PREPARATIONS

Method

Segments of ileum (2 cm) from a freshly killed guinea-pig are mounted in an organ bath (25 ml) containing Krebs solution (composition, mM: NaCl 118.1, KCl 4.7, $NaHCO_3$ 25, Glucose 11.1, $MgSO_4$ 1, KH_2PO_4 1 and $CaCl_2 \cdot 2H_2O$ 2.5) maintained at 37 °C and gassed continually with 95 per cent oxygen in carbon dioxide. The preparation is placed under a tension of 1 g and contractions recorded through an isometric force transducer connected to a pen recorder or Mac-Lab. The method of Paton (1954) is used to stimulate nerves within the wall of the intestine. Two platinum electrodes, attached to a stimulator, are placed on either side of the intestine and a square wave (width, 0.5 msec) delivered every 10 sec (0.1 Hz) at an appropriate voltage (about 50 v) produces a regular and reproducible contraction of the intestine. This contraction is not affected by the presence of a ganglion blocking agent but is inhibited by atropine indicating that it is due to the stimulation of post ganglionic parasympathetic nerves leading to the release of acetylcholine which acts on muscarinic receptors on smooth muscle. The addition of an extract with spasmogenic activity during field stimulation is recognised as a rise in the baseline and/or an increase in the size of the electrically induced contraction. A reduction in the size of the contraction indicates spasmolytic activity which can be further analysed as described below.

General action on intestinal smooth muscle

Most Pelargonium essential oils and commercial Geranium oils are spasmolytic on intestinal smooth muscle and reduce the response to electrical stimulation (Figure 12.1).

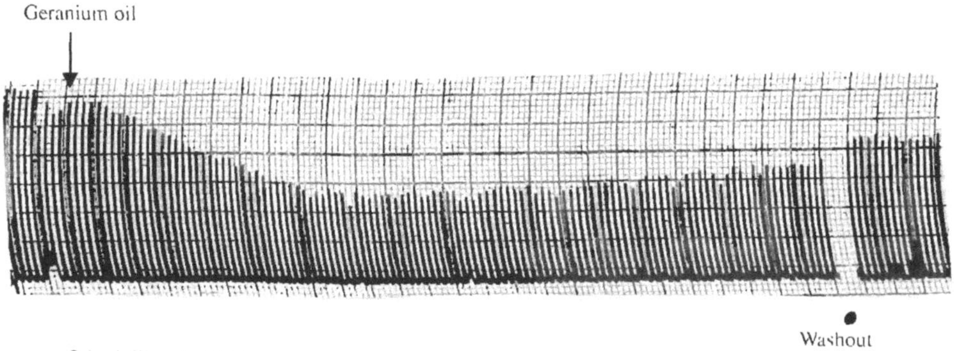

Figure 12.1 The spasmolytic effect of Geranium oil on the electrically-stimulated guinea-pig ileum *in vitro*.

Reiter and Brandt (1985) and Lis-Balchin and Hart (1997) have shown that citronellol, citronellal and geraniol have a spasmolytic action on guinea-pig ileum. One exception is the oil from *P. grossularioides* which contracts the tissue but on washout, when tone has returned to normal, inhibits briefly the electrically induced contractions (Lis-Balchin and Hart, 1994). One of the components of Pelargonium oil, geranyl formate, contracts guinea-pig ileum (Lis-Balchin *et al.*, 1996b). Alpha- and β-pinene are present in very small amounts and are interesting components because, although they are pure compounds, they initially contract intestinal smooth muscle and then produce a spasmolytic effect (Lis-Balchin *et al.*, 1999). The species from which essential oils have been studied, and the components that have been investigated, by the authors are shown in Tables 12.1 and 12.2. The concentrations of essential oils which inhibit the electrically induced contraction of the guinea-pig ileum by 50 per cent range from $6{\cdot}10^{-7}$ to $4.8{\cdot}10^{-5}$ (w/v). In all cases the effect of the essential oil lasts for as long as it is in contact with the ileum but the rate of recovery of the tissue after washout varies between different species of *Pelargonium*.

Table 12.1 Summary of investigations of spasmolytic activity of various *Pelargonium* species and cultivars on guinea-pig ileum *in vitro*

Pelargonium	*Atropine-like*	*Adrenoceptor-involvement*	*cAMP-involvement*		*cGMP-involvement*
			ea	*edp*	
Attar of Roses	No		No		No
Brunswick				No	
Chocolate Peppermint	No			Yes	No
Clorinda	No		Yes	No	No
Copthorne				No	
Geranium oil –					
Bourbon (rose)	No	No	Yes	Yes	No
Geranium oil – China	No	No			
Geranium oil – Egypt	No	No			
Lady Plymouth			Yes		No
Madam Nonin			Yes		No
P. citronellum J.J.J. v.d.Walt	No			No	No
P. cucullatum (L.) L'Her.	No				
P. filicifolium		No	Yes	No	No
P. glutinosum			Yes		No
P. graveolens L'Her.	No				
P. grossularioides					
P. odoratissimum L'Her.	No			No	No
P. quercifolium				No	No
P. tomentosum Jacq.	No				
P. viscossimum Sweet	No				
P. vitifolium (L.) L'Her.	No	No		No	No
P. × fragrans	No			No	
Paton's Unique			Yes	No	No
Purple Unique	No				
Radula	No				
Royal Oak	No				
Sweet Mimosa	No	No	Yes	No	No
Sweet Rosina	No	No	No	Yes	No

Sources: Data from Lis-Balchin *et al.* (1997, 1998) and Lis-Balchin and Hart (2000).

Table 12.2 Summary of investigations of spasmolytic activity of components of Pelargonium essential oils on guinea-pig ileum *in vitro*

Components	*Atropine-like*	*Adrenoceptor-involvement*	*cAMP-involvement*		*cGMP-involvement*
			trq	*cdp*	
Citronellal	No	No			
Citronellol			Yes		
Citronellyl formate					
Damascenone			No		
Geraniol			Yes		
Geranyl acetate					
Geranyl formate					
Guaiadiene					
Linalool			Yes		
Linalyl acetate					
Linalyl butyate					
Menthone					
α-pinene	No	No			
β-pinene	No	No			
Rose oxide			No		

Sources: Data from Lis-Balchin *et al.* (1998, 1999) and Lis-Balchin and Hart (2000).

As well as the essential oils obtained from scented *Pelargonium* leaves, non-scented *pelargoniums* were also studied (Table 12.3). These include numerous zonals, regals and ivy-leaved Pelargoniums, the leaves or roots of which were tested both as methanolic and water-soluble extracts and all but the *P. grossularioides* water-soluble and ethanolic extracts (Figure 12.2) were found to be spasmolytic on guinea-pig ileum.

The alkaloid extracts obtained from the zonals (Lis-Balchin *et al.*, 1996c; Lis-Balchin, 1997) were also all spasmolytic. The methanolic and water-soluble extracts, and alkaloid fractions of *P. luridum* (root) and the leaves of *P. inquinans* and *P. cucullatum* were also found to be spasmolytic.

Recently several hydrophilic extracts of numerous *Pelargonium* species and cultivars were tested on guinea-pig ileum. Three different hydrophilic extracts were made from each plant: a cold methanolic extract; a tea, made with boiling water and a hydrosol i.e. the water remaining after steam/water distillation. And in each case the activity was compared against the steam/water-distilled essential oils.

Table 12.3 Examples of non-scented leaved *Pelargonium* cultivars showing a spasmolytic effect on guinea-pig ileum *in vitro*: methanolic and water-soluble extracts

'Penny'	'Golden Harry Hieover'
'Appleblossom'	'Princess Alexandra'
'Frank Headley'	'Daydream'
'Magda'	'Fire Dragon'
'Els'	'Lass O'Gowrie'
'Mr Wren'	'Occold Shield'
'Mr Henry Cox'	'Appleblossom'
'Redondo'	'Golden Staph'

Note
These samples were stored for many months in the refrigerator prior to pharmacological assays.

Table 12.4 Evidence for contraction, C and or relaxation, R for extracts and essential oils in guinea-pig ileum *in vitro (electrically stimulated)*

Plant	*Extract*	*Tea*	*Hydrosol*	*EO*
Lady Plymouth	R	C/R	C/R	R
Clorinda	C/R	C/R	C/R	R
Sweet Mimosa	C/R	CR	C/R	R
Attar of Roses	C/R	*C/R	*C/R	R
P. tomentosum	C/R	C/R	C/R	R
Chocolate Tomentosum	R	R	R	R
P. × fragrans	C/R	*C/R	*C/R	R
Madam Nonin	C/R**	C/R	C/R	R
P. odoratissimum	R (occasional C)	***C/R	C/R (high) R (low)	R
P. graveolens	R/C	*C/R	C/R C (low)	R
Orsett	C/R	*C/R	C/R	R
Ardwick Cinnamon	C/R	C/R	C/R	R
Atomic Snowflake	*C/R	*C/R	C/R	R
Paton's Unique	*C/R	C/R	*C/R	R
Lady Plymouth	C/R	C/R	C/R (only R, low)	R
Appleblossom	C/R	C/R	N/T	No EO
Geranium robertianum	R	R	R	N/A

Notes
Contraction means an increase in tone, unless stated.
* contraction stronger than relaxation; ** relaxation greater than contraction; *** contraction seems to be very constant and not dose-related.

The extracts were tested on guinea-pig ileum, with and without electical stimulation *in vitro*, using the same conditions as described previously All Pelargonium essential oils studied to date have again shown a relaxant effect on field-stimulated guinea-pig ileum as previously shown (Lis-Balchin *et al.*, 1997; Lis-Balchin and Hart, 1998). More hydrophilic extracts corresponding to most of these scented Pelargoniums, however, showed a strong initial contractile effect followed in most cases by a relaxation at low concentrations for the teas and hydrosols as well as the methanolic extracts and showed a very strong contraction at high concentrations, except for 'Chocolate Peppermint' extract (Table 12.4).

In contrast, the three extracts of *Geranium robertianum* all showed a relaxant effect.

Neurogenic or myogenic

A concentration of Pelargonium essential oil which produces a significant inhibition of the contraction of the field stimulated guinea-pig ileum *in vitro*, also reduces the size of the contraction of the smooth muscle induced by exogenous acetylcholine. This shows that the oil is not affecting nerve conduction (local anaesthetic action) or the release of acetylcholine (as seen with opioids such as morphine) but is acting directly on the smooth muscle. All samples studied have a myogenic action.

Atropine-like action

The concentration of Pelargonium oil, which reduces the size of the contraction of the smooth muscle to acetylcholine, has a similar effect on the contraction due to exogenous histamine. If the essential oil contains an atropine-like compound the response to histamine would be unaffected by a concentration of the oil which reduces the response to acetylcholine: as this did not occur it can be concluded that the oil does not contain an atropine-like compound and that it is relaxing the smooth muscle by some other mechanism. None of the extracts studied have been reported to possess atropine-like activity (Tables 12.1 and 12.2).

Adrenoceptor mediated response

The presence of an adrenoceptor agonist in the essential oil would explain the relaxation of the intestinal smooth muscle and such activity would be sensitive to inhibition by a combination of α- and β-adrenoceptor antagonists. A concentration of phentolamine and propranolol, which inhibits the spasmolytic action of exogenous noradrenaline, has no effect on the inhibition of the contraction of the field-stimulated guinea-pig ileum produced by the Pelargonium oils. The Pelargonium essential oils do not, therefore, appear to contain a substance which stimulates adrenoceptors (Tables 12.1 and 12.2).

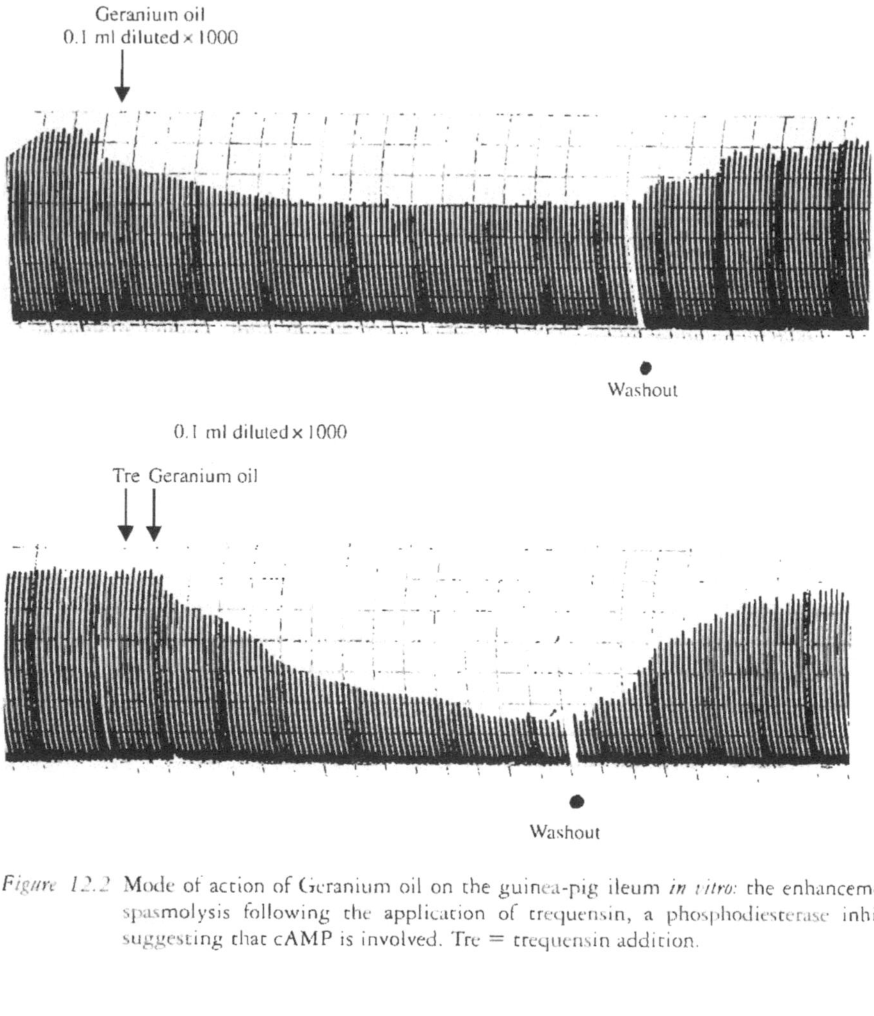

Figure 12.2 Mode of action of Geranium oil on the guinea-pig ileum *in vitro*: the enhancement of spasmolysis following the application of trequensin, a phosphodiesterase inhibitor, suggesting that cAMP is involved. Tre = trequensin addition.

Involvement of cyclic adenosine monophosphate (cAMP)

The relaxation of intestinal smooth muscle, which occurs as a result of adrenoceptor activation, is mediated by an increase in the level of cAMP as a result of the stimulation of adenylate cyclase. The duration of action of the cAMP so produced is limited by its metabolism by phosphodiesterase. It is thus possible that the essential oil contains a substance which is capable of stimulating adenylate cyclase in which case the spasmolytic action would be expected to be potentiated by a phosphodiesterase inhibitor.

Several phosphodiesterase inhibitors are available and they may be classified by their selectivity in inhibiting the seven or so different isoenzymes of phosphodiesterase which have been identified. Theophylline and trequinsin are non-selective inhibitors whilst rolipram (Banner *et al.*, 1995) and CDP 840 (Hughes *et al.*, 1996) inhibit the type IV isoenzyme.

Many of the Pelargonium oils and some of their components appear to mediate a spasmolytic action on intestinal smooth muscle via a rise in cAMP (Tables 12.1 and 12.2, Lis-Balchin *et al.*, 1998). Thus a concentration of trequinsin, which potentiates the response of isoprenaline from 22.6 to 41.8 per cent (per cent inhibition of twitch response to electrical stimulation), potentiates the spasmolytic activity of Geranium oil from 14.1 to 30.4 (Figure 12.3). Some, such as rose oxide and damascenone, were potentiated by the phosphodiesterase inhibitor but the potentiation did not reach statistical significance.

Involvement of cyclic guanosine monophosphate (cGMP)

Sodium nitroprusside, and other compounds which lead to the local production of nitric oxide, produce a spasmolytic action which is mediated by an increase in cGMP as a result of the stimulation of guanylate cyclase by nitric oxide. A further possible mechanism of action for the Pelargonium oils is therefore via the stimulation of guanylate cyclase, a mechanism which is sensitive to inhibition by ODQ (1H-(1,2,4) oxadia-zolo (4,3-α) quinoxalin-1-one) (Garthwaite *et al.*, 1995). A concentration of ODQ which blocks the spasmolytic action of sodium nitroprusside has no effect on the activity of Pelargonium essential oils or their components.

Calcium channel blockade

Intestinal smooth muscle contraction involves the influx of calcium ions through L-type channels and calcium channel blockers, such as verapamil and nifedipine, which are used therapeutically for their ability to block calcium channels in vascular and cardiac muscle, have a spasmolytic effect on intestinal smooth muscle. It is therefore possible that the essential oils contain components capable of blocking calcium channels.

There are several methods by which possible calcium channel blockade can be identified and in each method one uses an accepted calcium channel blocker as a positive control. Such screening methods have been reviewed by Neuhaus-Carlisle *et al.* (1997) and Vuorela *et al.* (1997) but neither group mention Pelargonium oils.

When normal Krebs solution is replaced by calcium-free Krebs solution the contraction due to field stimulation is lost but returns gradually when the tissue is returned to normal Krebs solution. The rate of recovery of the twitch is delayed by the presence of a calcium channel blocking agent, thus in the presence of nifedipine

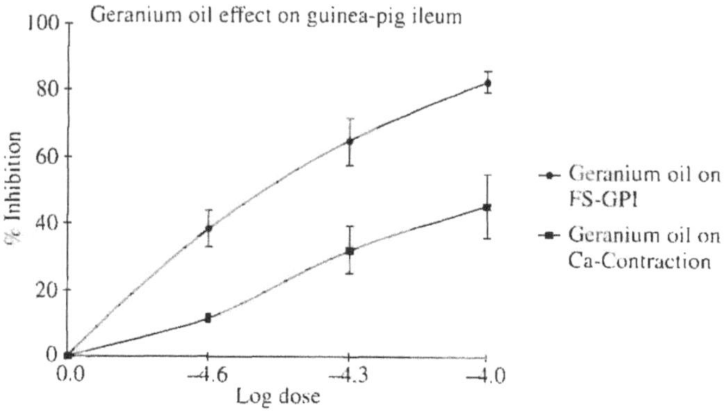

Figure 12.3 The effect of Geranium oil on field-stimulated guinea-pig ileum (FS-GPI) under normal physiological conditions and on calcium induced contractions in depolarising-Krebs solution. The graph shows that the spasmolytic action of Geranium oil is only partly explained by calcium channel blockade.

(10^{-6} M) recovery of the twitch is delayed from 8.6 ± 1.9 min to 30 ± 5.3 min. Pelargonium oils, at concentrations which inhibit the twitch and are potentiated by a phosphodiesterase inhibitor, do not delay the recovery (Lis-Balchin *et al.*, 1998).

When guinea-pig ileum is bathed with a depolarising-Krebs solution the addition of calcium chloride solution produces dose-related contractions that are sensitive to inhibition by calcium channel blockers. Nifedipine ($4{\cdot}10^{-5}$ M) inhibits the calcium

Table 12.5a Evidence for use of Ca^{++} channel blocking

Plant	*Extract*
'Ardwick Cinnamon'	EO
P. odoratissimum	MeOH extract
P. graveolens	MeOH extract
G. robertianum	Hydrosol

Table 12.5b No evidence for use of Ca^{++} channel blocking at normal concentrations

Plant	*Extract*
'Sweet Mimosa'	MeOH extract and hydrosol and EO
'Attar of Roses'	MeOH extract and hydrosol and EO
'Lady Plymouth'	MeOH extract and EO
P. graveolens	Hydrosol and EO
P. odoratissimum	Hydrosol and EO
'Ardwick Cinnamon'	Hydrosol and EO

contraction by 84 per cent whilst Geranium oil, at a concentration which produces a significant inhibition of the twitch response (8.10^{-6} w/v), has no effect. However, at higher concentrations of Geranium oil there is evidence of calcium block with 2.10^{-5} giving 56 per cent inhibition (Figure 12.3) with a complete block occurring at 2.10^{-4} (Lis-Balchin and Hart, 2000).

Recent preliminary studies using methanolic and water extracts of various *Pelargonium* species and cultivars have shown that there is some evidence for calcium channel blockage by *P. odoratissimum* and *P. graveolens* methanolic extracts as well as the 'Ardwick cinnamon' essential oil and methanolic extract, whilst the latter's hydrosol did not exhibit this action (Table 12.5a, b). The different extracts tested of 'Sweet Mimosa', 'Attar of Roses', 'Lady Plymouth', and the hydrosols of *P. odoratissimum* and *P. graveolens* did not show this effect either (Table 12.5b). The essential oils of the last mentioned pelargoniums had previously all shown that they mediate their activity via cAMP (Lis-Balchin *et al.*, 1998).

Potassium channel opening

Some drugs which are used therapeutically to relax vascular smooth muscle act by opening potassium channels and thereby hyperpolarise the cell. Cromokalim is such a compound and it has a spasmolytic action on intestinal smooth muscle thus emphasising another possible mode of action for the spasmolytic action of Pelargonium essential oils.

In the presence of a potassium channel opener, guinea-pig ileum will not respond to a high concentration of potassium (60 mM) with a contraction. If tissue is contracted with a high concentration of potassium then a relaxation can be obtained with a calcium channel blocker but not with a potassium channel opener. There is no experimental evidence to suggest that Pelargonium oils contain components capable of opening potassium channels (Lis-Balchin *et al.*, 1998).

SPASMOGENIC ACTION ON GUINEA-PIG ILEUM

The oil of *P. grossularioides* and some components of Pelargonium oils, namely citronellyl formate, α- and β-pinene are spasmogenic on guinea-pig ileum. Provisional results show that α-pinene is not acting via muscarinic cholinoceptors or histamine receptors and that the two enantiomers of α-pinene do not have identical pharmacological activity (Lis-Balchin *et al.*, 1999).

Recent studies on freshly-prepared hydrophilic extracts and methanolic extracts have shown that many scented *Pelargonium* species and cultivars have a spasmogenic effect initially, followed by a relaxation (Table 12.4). This suggests that some of the many phenolic components may produce this spasmogenic effect. It also appears that there could be some residual essential oil components present in all these extracts, but even the unscented 'Appleblossom' cultivar, which in previous experiments had shown only a spasmolytic effect (Table 12.3) now showed this dual effect; this was probably due to two reasons, one being that the samples had previously been stored for many months in a refrigerator and personal experience has indicated that spasmolytic components disappear quite rapidly under these conditions and secondly the samples were studied at very low concentrations, where the contraction does not show up.

EXPERIMENTS ON *IN VITRO* UTERINE PREPARATIONS

Method

Uterus from a freshly killed rat is mounted in an organ bath containing Krebs solution maintained at 37 °C and gassed continually with 95 per cent oxygen in carbon dioxide. Activity of the tissue is monitored with an isometric force transducer connected to a pen recorder. Uterine tissue may be quiescent but usually exhibits regular contractions and relaxations, after a short period of equilibration, depending upon the oestrus cycle of the rat. The ability of an essential oil to increase or decrease the overall activity of the uterus is readily demonstrated but the elucidation of the mode of action is not as straightforward as with intestinal tissue.

General action on uterus

The essential oil of *P. grossularioides* stimulates contraction and relaxation of quiescent uterus for as long as it is in contact with the tissue. When added to tissue which is contracting and relaxing regularly, the oil produces a maintained contraction followed by an increase in pendular activity after washout (Lis-Balchin and Hart, 1994).

Geranium oil and geraniol both reduce uterine activity at concentrations which are spasmolytic on intestinal muscle (Lis-Balchin and Hart, 1997) but the mechanism of action has not been studied (Figure 12.4).

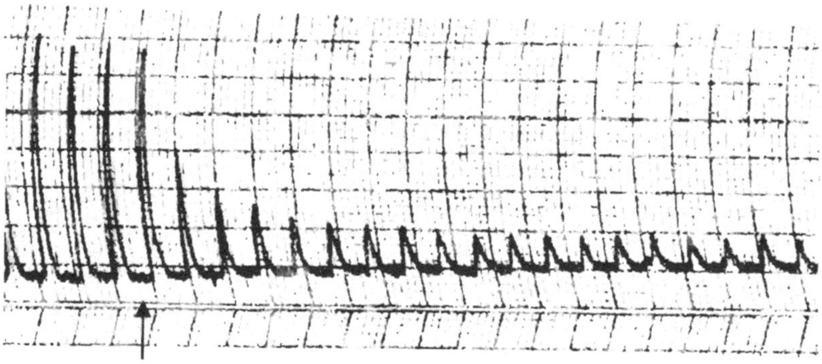

Figure 12.4 The effect of Geranium oil on the spontaneously contracting uterus of the rat *in vitro*, showing the inhibition of contractions, which at higher Geranium oil concentrations actually ceased altogether.

EXPERIMENTS ON *IN VITRO* BRONCHIAL PREPARATIONS

Method

Activity on bronchial muscle is usually studied on guinea-pig tracheal muscle which is cut transversally to produce rings which are then tied together and mounted in an organ bath. An alternative preparation, which reduces the involvement of cartilage, is favoured by Reiter and Brandt (1985).

General actions on bronchial muscle

Reiter and Brandt (1985) reported the actions of several essential oils and their components on tracheal and intestinal smooth muscle including some components of Pelargonium oils but not the oil itself. Citronellal, citronellol, geraniol and linalool are all spasmolytic with intestinal muscle being more sensitive than tracheal. Linalool is one of many components of essential oils studied by Brandt (1988) and was found to be spasmolytic on guinea-pig ileum and trachea. α- and β-pinene also relax guinea-pig tracheal muscle (Lis-Balchin and Hart, 2000).

EXPERIMENTS ON *IN VITRO* CARDIAC PREPARATIONS

Method

A heart from a freshly killed rat or rabbit can be perfused with Krebs solution through the aortic arch by the method of Langendorf (1895) and the overall activity recorded by attaching a thread from the ventricles to a transducer. The perfused heart remains viable for several hours and the effect of essential oils on rate and force can be assessed on addition to the perfusion fluid and the mode of action investigated by the use of standard antagonists.

Alternatively, the auricles can be dissected from the heart and set up in an organ bath in a manner similar to that used for intestinal tissue.

General action on cardiac tissue

The essential oil of *P. grossularioides* has been reported to reduce the force of contraction of the spontaneously beating perfused isolated heart (Lis-Balchin and Hart, 1994).

EXPERIMENTS ON *IN VITRO* SKELETAL MUSCLE PREPARATIONS

Method

It is more difficult to study skeletal muscle than smooth muscle *in vitro* because preparations of the former are usually thicker and do not remain viable in the organ bath. Two preparations which are successful are the chick biventer cervices and the rat phrenic nerve hemi-diaphragm.

The diaphragm with attached phrenic nerves is dissected from a freshly killed rat and the diaphragm cut in half to give two preparations that are mounted in a 50 ml organ bath on a support which keeps the diaphragm secure and enables the stimulation of the phrenic nerve (Bulbring, 1946). Stimulation of the phrenic nerve (0.1 Hz, 5 ms, 20 V) causes regular and reproducible contractions of the diaphragm which are recorded via a transducer and pen recorder. The introduction of an essential oil into the organ bath during stimulation can raise the tone of the tissue and/or affect the size of the nerve induced contraction. It is also possible to stimulate the muscle directly which allows differentiation between compounds acting via the nerve or directly on the muscle.

General action on skeletal muscle

The essential oil of *P. grossularioides* increases the tone of the rat diaphragm and reduces the size of the contraction in response to stimulation of the phrenic nerve and also when the muscle is stimulated directly (Lis-Balchin and Hart, 1994). Thus the action is myogenic but the mechanism has not been studied. The commercial Geranium oil, on the other hand, decreased the tone and did not alter the size of contraction (Figures 12.5a,b).

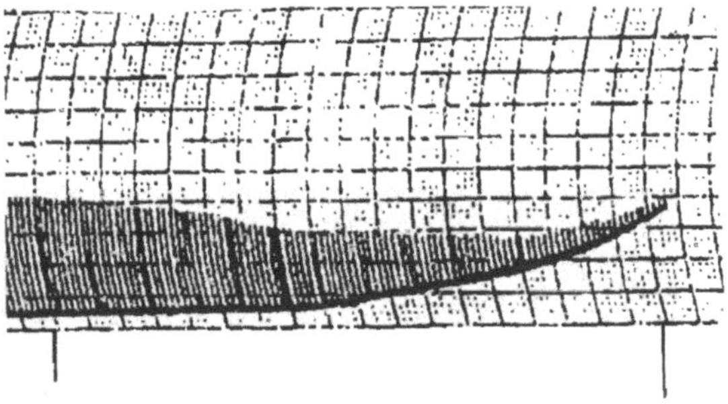

Geranium oil
0.2 ml × 25 dilution

Figure 12.5a The action of Geranium oil on skeletal muscle using the chick biventer muscle preparation *in vitro*. This shows an inhibition of electrically-stimulated contractions with a rise in the baseline.

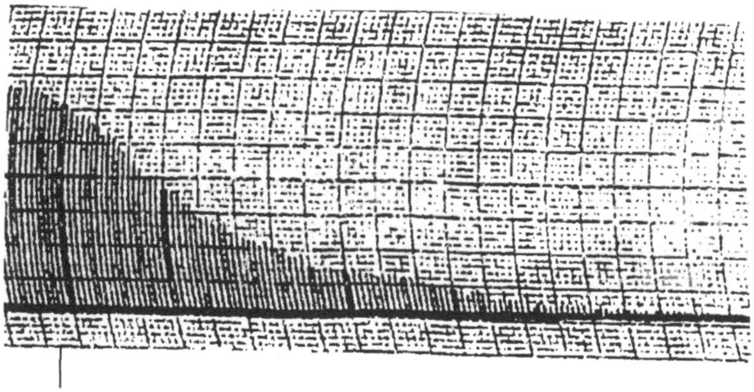

Geraniol
0.2 ml × 25 dilution

Figure 12.5b The action of geraniol on the skeletal muscle showing a spasmolytic action with no rise in the baseline at the same concentration as that of Geranium oil.

RECEPTOR BINDING STUDIES

A precise method of determining whether a compound interacts with a receptor is by ligand binding studies in which the ability of the compound to displace a radio-labelled ligand from the receptor is measured. The receptors are present on membranes which are often prepared from a specific brain region. Such experiments could be performed with the individual components of an essential oil but not on the oil itself.

Elisabetsky *et al.* (1995) report that linalool produces a dose-dependent inhibition of the binding of glutamate (an excitatory neurotransmitter in the brain) to its receptors on membranes prepared from the cerebral cortex of the rat. The authors propose that this action is consistent with their observation that linalool is sedative in animals.

EXPERIMENTS *IN VIVO*

Cardiovascular system

Geranium oil and the components geraniol, linalool and citronellol produce a fall in blood pressure in experimental animals (Tisserand and Balacs, 1995) but reliable observations from human experiments are not available.

Central nervous system

Inhalation of vapour from essential oils affects overall activity of mice and although Pelargonium oils do not appear to have been studied, geraniol, linalyl acetate and citronellal are sedative (Buchbauer *et al.*, 1993).

In humans, exposure to the vapour of Pelargonium oils affects certain brain waves (contingent negative variation, CNV) but both depression and stimulation have been reported (Torii *et al.*, 1988; Manley, 1993).

General activity

Pelargonium oils have not been the subject of well controlled clinical trials for either the inhalation of the vapour or application through massage.

CONCLUSION

Results from experiments on isolated guinea-pig ileum demonstrate that the majority of Pelargonium oils, and their components, produce a relaxation of smooth muscle through a mechanism that involves the enzyme adenylate cyclase and a rise in the concentration of the second messenger, cAMP. There is some evidence of calcium channel blockade but only at concentrations higher than those required to produce a significant spasmolytic effect. This is in contrast to other essential oils, such as peppermint oil, for which there is good evidence of calcium channel blockade (Hills and Aaronson, 1991). The pharmacology of the central actions of *Pelargonium* has not been studied sufficiently to reach any conclusions. Preliminary results using more hydrophilic extracts of *Pelargonium* species and cultivars (using methanol and water) indicate that most have a contractile effect initially which is followed by a relaxation. There is also some evidence that a few methanolic extracts use calcium channels at normal concentrations, but many have shown no such activity, in keeping with the mode of action of their equivalent essential oils.

REFERENCES

Aqel, M.B. (1991) Relaxant effect of the volatile oil of Rosmarinus officinalis on tracheal smooth muscle. *J. Ethnopharmacol.*, **33**, 57–62.

Banner, K.H., Marchini, F., Buschi, A., Moriggi, E., Semeraro, C. and Page, C.P. (1995) The effect of selective phosphodiesterase inhibitors in comparison with other anti-asthma drugs on allergen-induced eosinophilia in guinea-pig airways. *Pulm. Pharmacol.*, **8**, 37–42.

Brandt, W. (1988) Spasmolytische wirkung atherischer Ole. *Zeit. Phytother.*, **9**, 33–39.

Buchbauer, G., Jager, W., Jirovetz, I., Ilmberger, J. and Dietrich, H. (1993) Therapeutic properties of essential oils and fragrances. In: R. Teramishu, R.G. Buttery and H. Sugisaca (eds), Bioactive Volatile compounds from plants, ACS symposium series 525. Washington, DC, American chemical society, 159–165.

Bulbring, E. (1946) Observations on the isolated phrenic nerve diaphragm preparation of the rat. *Br. J. Pharmacol.*, **1**, 38–61.

Elisabetsky, E., Marschner, J. and Souza, D.O. (1995) Effects of linalool on glutamatergic system in the rat cerebral cortex. *Neurochemical Res.*, **20**, 461–465.

Garthwaite, J., Southam, E., Boulton, C.L., Nielsen, E.B., Schmidt, K. and Mayer, B. (1995) Potent and selective inhibition of nitric oxide-sensitive guanylyl cyclase by 1H-[1,2,4] Oxadiazolo[4,3-a]quinoxalin-1-one. *Mol. Pharmacol.*, **48**, 184–188.

Gunn, J.W.C. (1921) The carminative action of volatile oils. *J. Pharmacol.*, **16**, 39–47.

Hills, J.M. and Aaronson, P.I. (1991) The mechanism of action of peppermint oil on gastrointestinal smooth muscle. An analysis using patch clamp eletrophysiology and isolated tissue pharmacology in rabbit and guinea pig. *Gastroenterol.*, **101**, 55–65.

Hughes, B., Howat, D., Lisle, H., Holbrook, M., James, T., Gozzard, N., Blease, K., Highes, P., Kingaby, R., Warrellow, G., Alexander, R., Head, J., Boyd, E., Eaton, M., Perry, M., Wales, M., Smith, B., Owens, R., Catterall, C., Lumb, S., Russell, A., Allen, R., Merriman, M., Bloxham, D. and Higgs, G. (1996) The inhibition of antigen-induced eosinophilia and bronchoconstriction by CDP840, a novel stereo-selective inhibitor of phosphodiesterase type 4. *Br. J. Pharmacol.*, **118**, 1183–1191.

Imaseki, I. and Kitabatake, Y. (1962) Studies on effect of essential oils and their components on the isolated intestines of mice. *J. Pharmaceut. Soc. Jpn.*, 82, 1326–1328.

Kitchen, I. (1984) Textbook of *in vitro* practical pharmacology. Blackwell Scientific Publications. Oxford.

Langendorff (1895) *Pflugers Arch ges Physiol.*, **190**, 280.

Lis-Balchin, M. (1988) The commercial usefulness of the Geraniaceae, including their potential in the Perfumery, Food manufacture and Pharmacological industries. *Proc. Int. Geraniaceae Symp.*, Stellenbosch, South Africa, 24–26 Sept., pp. 269–277.

Lis-Balchin, M. (1997) A chemotaxonomic study of *Pelargonium* (Geraniaceae) species and their modern cultivars, *J. Hort. Sci.*, **72**, 791–795.

Lis-Balchin, M. and Hart, S. (1994) A pharmacological appraisal of the folk medicinal usage of *Pelargonium grossularioides* and *Erodium cicutarium* (Geraniaceae). *J. Herbs. Spices Med. Plants*, **2**, 41–48.

Lis-Balchin, M. and Hart, S. (1997) Pharmacological effect of essential oils on the uterus compared to that on other tissue types. *Proc. 27th Int. Symp. Ess. Oils*, Vienna, Austria, Ch. Franz, A. Mathé and G. Buchbauer, (eds) Allured Pub. Corp., Carol Stream, Ill. 29–32.

Lis-Balchin, M. and Hart, S. (1998) Studies on the mode of action of essential oils of scented-leaf *Pelargonium* (Geraniaceae). *Phytother. Res.*, **12**, 215–217.

Lis-Balchin, M. and Hart, S. (2000) Unpublished observations.

Lis-Balchin, M., Deans, S.G. and Hart, S. (1996a) Bioactivity of Geranium oils from different commercial sources. *J. Essent. Oil Res.*, **8**, 281–290.

Lis-Balchin, M., Hart, S., Deans, S.D. and Eaglesham, E. (1996b) Comparison of the pharmacological and antimicrobial action of commercial plant essential oils. *J. Herbs. Spices Med.Plants*, 4, 69–86.

Lis-Balchin, M., Houghton, P. and Woldernarian, T. (1996c) Elaeocarpidine alkaloids from *Pelargonium* species (Geraniaceae). *LAM*, 23, 205–207.

Lis-Balchin, M., Hart, S. and Roth, G. (1997) The spasmolytic activity of the essential oils of scented *Pelargoniums* (Geraniaceae). *Phytother. Res.*, 11, 583–584.

Lis-Balchin, M., Patel, J. and Hart, S. (1998) Studies on the mode of action of essential oils of scented-leaf *Pelargonium* (Geraniaceae). *Phytother. Res.*, 12, 215–217.

Lis-Balchin, M., Ochocka, R.J., Deans, S.G., Asztemborska, M. and Hart, S. (1999) Differences in bioactivity between the enantiomers of α-pinene. *J. Essent. Oil Res.*, 11, 393–397.

Manley, C.H. (1993) Psychophysiological effect of odor. *Crit. Rev. Food Sci. Nutr.*, **33**, 57–62.

Neuhaus-Carlisle, K., Vierling, W. and Wagner, H. (1997) Screening of plant extracts and plant constituents for calcium-channel blocking activity. *Phytomedicine*, 4, 67–69.

Paton, W.D.M. (1954) The response of the guinea-pig ileum to electrical stimulation by coaxial electrodes. *J. Physiol. (Lond.)*, 127, pp. 40–41.

Pappe, L. (1868) *Florae Capensis Medicae*, Prodromus, 3rd ed., Cape Town.

Reiter, M. and Brandt, W. (1985) Relaxant effects on tracheal and ileal smooth muscles of the guinea-pig. *Arzneim. Forsch/Drug Res.*, 35, 408–414.

Sticher, O. (1977) Plant mono-, di- and sesquiterpenoids with pharmacological or therapeutical activity. In: H. Wagner and P. Wolf (eds), New natural products and plant drugs with pharmacological, biological or therapeutic activity. Springer-Verlag, Berlin.

Tisserand, R. and Balacs, T. (1995) *Essential oil safety*. Churchill Livingstone, London.

Torii. S., Fukuda, H., Kanemoto, H., Miyanchio, R., Hamauzu, Y. and Kawasaki, M. (1988) Contingent negative variation and the psychological effects of odor. In S. Toller and G.H. Dodds (eds), Perfumery: The Psychology and Biology of Fragrance, Chapman & Hall, New York.

Triebs, W. (1956) *Die Atherischen Ole*, Akademie-Verlag, Berlin.

Vuorela, H., Vuorela, P., Tornquist, K. and Alaranta, S. (1997) Calcium channel blocking activity: screening methods for plant derived compounds. *Phytomedicine*, 4, 167–181.

Watt, J.M. and Breyer-Brandwijk (1962) *The Medicinal Plants of Southern Africa*, Livingstone Ltd, Edinburgh.

13 Antimicrobial properties of *Pelargonium* extracts contrasted with that of *Geranium*

Stanley G. Deans

INTRODUCTION

The preservative properties of aromatic and medicinal plant volatile (essential) oils and extracts have been recognised since Biblical times, while attempts to characterise these properties in the laboratory date back to the 1900s (Martindale, 1910; Hoffman and Evans, 1911).

Plant volatile oils

Plant volatile oils are generally isolated from non-woody plant material by steam or hydrodistillation, and are variable mixtures of principally terpenoids, specifically monoterpenes (C_{10}) and sesquiterpenes (C_{15}) although diterpenes (C_{20}) may also be present, and a variety of low molecular weight aliphatic hydrocarbons (linear, ramified, saturated and unsaturated), acids, alcohols, aldehydes, acyclic esters or lactones and exceptionally nitrogen- and sulphur-containing compounds, coumarins and homologues of phenylpropanoids.

Terpenes are amongst the chemicals responsible for the culinary, medicinal and fragrant uses of aromatic and medicinal plants. Most terpenes are derived from the condensation of branched five-carbon isoprene units and are categorised according to the number of these units in the carbon skeleton (Dorman, 1999). Traditionally, plants and their extracts, including geranium (*Pelargonium* sp.) (Geraniaceae) have been used to extend the shelf life of foods, beverages and pharmaceutical/cosmetic products through their antimicrobial and antioxidant properties (Baratta *et al.*, 1998a,b; Cai and Wu, 1996; Gallardo *et al.*, 1987; Janssen *et al.*, 1988; Jay and Rivers, 1984; Pélissier *et al.*, 1994; Shapiro *et al.*, 1994; Shelef, 1984; Ueda *et al.*, 1982; Youdim *et al.*, 1999).

More recently, attempts have been made to identify the component(s) responsible for such bioactivities (Daferera *et al.*, 2000; Deans and Ritchie, 1987; Deans *et al.*, 1994a, b; Dorman and Deans, 2000; Jeanfils *et al.*, 1991; Lis-Balchin, 1997; Lis-Balchin and Deans, 1997; Lis-Balchin *et al.*, 1998a; Vokou *et al.*, 1993). The antimicrobial properties of volatile oils and their constituents from a wide variety of plants have been assessed and reviewed (Carson *et al.*, 1995, 1996; Garg and Dengre, 1986; Inouye *et al.*, 1983; Jain and Kar, 1971; Janssen *et al.*, 1987; Larrondo *et al.*, 1995; Nenoff *et al.*, 1996; Pattnaik *et al.*, 1995, 1996; Rios *et al.*, 1987, 1988; Sherif *et al.*, 1987). Investigations into the antimicrobial activities, mode of action and potential uses of plant volatile oils have regained momentum and there appears to be a revival in the use of traditional approaches to livestock welfare as well as food preservation. The activity of the oils would be expected to relate and reflect the respective composition of the volatile oils, the structural configuration of the constituent components and their functional groups along with potential synergistic interactions between components. A correlation of the antimicrobial activity of compounds under test in the study by Dorman and Deans (2000) and their relative percentage composition in the oils, with their chemical structure, functional groups and configuration suggested a number of observations on the structure–function relationship.

Components with phenolic structures, such as carvacrol and thymol, were highly active against test bacteria, despite their low capacity to dissolve in water. The importance of the hydroxyl group in the phenolic ring was confirmed in terms of activity when carvacrol was compared to its methyl ester. The high activity of the phenolic components may be further explained in terms of the alkyl substitution into the phenolic nucleus, which is known to enhance the antimicrobial activity of phenols (Pelczar *et al.*, 1988). The introduction of alkylation has been proposed to alter the distribution ratio between the aqueous and non-aqueous phases, including bacterial phases, by reducing the surface tension or altering the species selectivity. Alkyl-substituted phenolic compounds form phenoxyl radicals which interact with isomeric alkyl substituents (Pauli and Knobloch, 1987). This does not occur with etherified/esterified isomeric molecules, possibly explaining their relative lack of activity.

The presence of an acetate moiety in the structure appears to increase the activity of the parent compound as in the case of geraniol where geranyl acetate demonstrated an increase in activity. Aldehydes, notably formaldehyde and glutaraldehyde, are known to possess powerful antimicrobial activity. It has been suggested that an aldehyde group conjugated to a C=C is a highly electronegative arrangement which may explain their activity (Moleyar and Narasimham, 1986), suggesting an increase in electronegativity increases the antibacterial activity (Kurita *et al.*, 1979, 1981). Such electronegative compounds may interfere in biological processes involving electron transfer and react with vital nitrogen components such as proteins and nucleic acids, and therefore inhibit growth of the microorganisms. The aldehydes *cis* and *trans* citral displayed moderate activity against test bacteria while citronellal was less active. Alcohols are known to possess bactericidal rather than bacteriostatic activity against vegetative bacterial cells. The alcohol terpenoids studied did show some activity against test bacteria, acting as protein denaturing agents, solvents or dehydrating agents.

A number of oil components are ketones, wherein the presence of an oxygen function in the framework increases the antimicrobial properties of the terpenoids (Naigre *et al.*, 1996). From this study, and by using the contact method, the bacteriostatic and fungistatic action of terpenoids was increased when carbonylated. Menthone was shown to have modest activity with *Clostridium sporogenes* and *Staphylococcus aureus* being the most significantly affected (Dorman and Deans, 2000). An increase in activity dependant upon the type of alkyl substituent incorporated into a non-phenolic ring structure appeared to occur in this study. An alkenyl substituent [1-methylethenyl] resulted in increased antibacterial activity, as seen in limonene [1-methyl-4-(1-methylethenyl)-cyclohexene], compared to an alkyl [1-methylethyl] substituent as in *p*-cymene [1-methyl-4-(1-methylethyl)-benzene]. The inclusion of a double bond increased the activity of limonene relative to *p*-cymene, which demonstrated no activity against the test bacteria. In addition, the susceptible organisms were principally

Gram-negative, which suggests alkylation influences Gram reaction sensitivity of the bacteria. The importance of the antimicrobial activity of alkylated phenols in relation to phenol has been previously reported (Pelczar *et al.*, 1988). Their data suggest that an allylic side chain seems to enhance the inhibitory effects of a component and chiefly against Gram-negative organisms.

Furthermore, the stereochemistry has an influence on bioactivity. It was observed that α-isomers are inactive relative to β-isomers, for example, α-pinene; *cis*-isomers are inactive contrary to *trans*-isomers, for example geraniol and nerol; compounds with methyl-isopropyl cyclohexane rings are most active; or unsaturation of the cyclohexane ring further increases the antibacterial activity, for example terpinolene, terpineol and terpineolene (Hinou *et al.*, 1989).

Investigations into the effects of terpenoids upon isolated bacterial membranes suggest that their activity is a function of the lipophilic properties of the constituent terpenes, (Knobloch *et al.*, 1986), the potency of their functional groups and their aqueous solubility (Knobloch *et al.*, 1988). Their site of action appeared to be at the phospholipid bilayer, caused by biochemical mechanisms catalysed by these phosholipid bilayers of the cell. These processes include the inhibition of electron transport, protein translocation, phosphorylation steps and other enzyme-dependant reactions (Knobloch *et al.*, 1986). Their activity in whole cells appears more complex (Knobloch *et al.*, 1988). Although a similar water-soluble tendency is observed, specific statements on the action of single terpenoids *in vivo* have to be assessed singularly, taking into account not only the structure of the terpenoid, but also the chemical structure of the cell wall (Knobloch *et al.*, 1988). The plant extracts clearly demonstrate antibacterial properties, although the mechanistic processes are poorly understood.

Chemotherapeutic agents, used orally or systemically for the treatment of microbial infections of humans and animals, possess varying degrees of selective toxicity. Although the principle of selective toxicity is used in agriculture, pharmacology and diagnostic microbiology, its most dramatic application is the systemic chemotherapy of infectious diseases. The tested plant products appear to be effective against a wide spectrum of microorganisms, both pathogenic and non-pathogenic. Administered orally, these compounds may be able to control a wide range of microbes, but there is also the possibility that they may cause an imbalance in the gut microflora, allowing opportunist pathogenic bacteria, such as coliforms, to become established in the gastrointestinal tract with resultant deleterious effects. Further studies on therapeutic applications of volatile oils should be undertaken to investigate these issues, especially when consideration is made of the substantial number of analytical/bioactivity studies carried out on these natural products.

ANTIMICROBIAL ACTIVITY

Real *Geranium* species in contrast to *Pelargonium* species

There is a great difference in the chemical composition and bioactivity between the true *Geranium* and *Pelargonium* species. The commercial 'Geranium oil' is in fact obtained from a *Pelargonium* cultivar and there is, therefore, great confusion in the literature regarding the actual species studied (see General Introduction).

In the past only *Geranium* species e.g. *G. robertianum* not *Pelargonium* have been studied and it was reported in various herbals that these *Geranium* species were useful as an astringent and tonic, in the treatment of diarrhoea, cholera and chronic dysentery (Grieve, 1994). The volatile oil of *Geranium* species has been attributed a number of biological properties including: antibacterial (Ivancheva *et al.*, 1992; Pattnaik *et al.*, 1996), antifungal (Pattnaik *et al.*, 1996) and antiviral (Zgorniak-Nowosielska *et al.*, 1989; Ivancheva *et al.*, 1992; Serkedjieva, 1995) and also some other pharmacological properties (Petkov *et al.*, 1974; Manolov *et al.*, 1977). Recent studies have shown that root extracts from *Geranium pratense* exhibited antimicrobial activity against *Streptomyces scabies* (Ushiki *et al.*, 1996) and these authors later reported on the isolation of an antimicrobial substance, geraniin, at approximately 15 per cent root dry weight based on quantitative analysis by high-performance liquid chromatography (HPLC), and its antimicrobial activity corresponded to 1.25 per cent of that of streptomycin, based on the paper-disc assay. These findings suggest that *G. pratense* has potential agronomic applications as an organic amendment or companion crop for the control of common scab of potato (Ushiki *et al.*, 1998).

Polyphenolic complex (PC)

Studies on PC have now shown activity against herpes simplex, vaccinia and HIV-I in cell cultures. PC has also been shown to inhibit the *in vitro* growth of *Staphylococcus aureus* and *Candida albicans* (Serkedjieva, 1997). High concentrations of PC (>200 μg/mL) exhibited a strong virucidal effect. Although the action was directed against an early stage of infection (within 3 h of infection), the process affected was not identified. The selectivity of antiviral action was confirmed by the variation in sensitivity of different influenza viruses to PC and the selection of variants with reduced drug sensitivity (Serkedjieva and Hay, 1998). The aqueous extract from aerial roots was the least toxic for cell cultures, yet significantly inhibited the replication of herpes simplex virus (Serkedjieva and Ivancheva, 1999). Okuda *et al.* (1992) report on antitumour as well as antiviral activities of numerous polyphenolic compounds isolated from a number of Asian plants including *Geranium thunbergii*.

Finally, the antiviral properties of the polyphenol complex (PC) isolated from *Geranium sanguineum* have received considerable attention from the Bulgarian Academy of Science group. Ivancheva *et al.*, 1987 and Serkedjieva and Manolova, 1988 determined the chromatographic characteristics of the complex shown to possess antiviral activity. A later study was undertaken to investigate the effect of PC on virus-specific protein synthesis in influenza virus-infected cells, where it was found that the expression of viral glycoproteins on the surface of chick embryo fibroblasts infected with such viruses was suppressed (Serkedjieva, 1995). The inhibitory effect was dose-dependant and more pronounced when PC was applied post-viral infection. The studies have been extended to human and equine influenza viruses where it was shown that the inhibitory effect was strain-related, and consistent with a selective antiviral action (Serkedjieva, 1996).

Pelargonium species

In one of the larger studies in this field, Deans and Ritchie (1987) studied 50 commercial volatile oils at four concentrations against a range of 25 bacterial genera. In the case of 'Geranium oil', it was found to be most effective against the dairy products

Table 13.1 Antibacterial activity of geranium volatile oil against 25 test bacteria. Inhibition zone diameter in mm, including well diameter of 4 mm. Each value is the mean of three replicates

Organism	*Source**	*Inhibition zone*
Acinetobacter calcoacetica	NCIB 8250	11.5
Aeromonas hydrophila	NCTC 8049	9.0
Alcaligenes faecalis	NCIB 8156	6.0
Bacillus subtilis	NCIB 3610	4.0
Beneckea natriegens	ATCC 14048	6.5
Brevibacterium linens	NCIB 8456	12.0
Brocothrix thermosphacta	Sausage meat	5.5
Citrobacter freundii	NCIB 11490	9.0
Clostridium sporogenes	NCIB 10696	4.0
Enterobacter aerogenes	NCTC 10006	11.0
Enterococcus faecalis	NCTC 775	4.0
Erwinia carotovora	NCPPB 312	8.5
Escherichia coli	NCIB 8879	✧
Flavobacterium suaveolens	NCIB 8992	9.5
Klebsiella pneumoniae	NCIB 418	✧
Lactobacillus plantarum	NCDO 343	7.5
Leuconostoc cremoris	NCDO 543	4.0
Micrococcus luteus	NCIB 8165	8.0
Moraxella sp.	NCIB 10762	10.5
Proteus vulgaris	NCIB 4175	5.0
Pseudomonas aeruginosa	NCIB 950	5.5
Salmonella pullorum	NCTC 10704	6.0
Serratia marcescens	NCIB 1377	10.0
Staphylococcus aureus	NCIB 6571	5.0
Yersinia enterocolitica	NCTC 10460	13.0

Source: Deans and Ritchie, 1987.

Notes

* NCIB – National collection of industrial bacteria; NCTC – National collection of type cultures; ATCC – American type culture collection; NCPPB – National collection of plant pathogenic bacteria; NCDO – National collection of dairy organisms.

✧ Enhancement of growth.

organism *Brevibacterium linens* and the toxin-producing *Yersinia enterocolitica*. With *Klebsiella pneumoniae* and *Escherichia coli*, in contrast, its presence resulted in enhancement of growth (Table 13.1).

Pattnaik *et al.* (1996) tested Geranium oil for antibacterial activity against 22 bacteria (Gram-positive cocci and rods, Gram-negative rods) and 12 fungi (three yeast-like, nine filamentous) by disc diffusion. Only 12 bacterial strains were inhibited by the Geranium oil, but all the fungi were inhibited.

Differences between samples of 'Geranium oil'

There were significant differences in the chemical composition of a number of Geranium samples of various geographical origins used as antibacterial agents (Lis-Balchin *et al.*, 1996c). There was a wide range of citronellol and geraniol levels in each source category, Reunion, China, Egypt or Morocco, with the ratio of citronellol:-geraniol also being very variable, ranging from 1.7 to 10.9. The antibacterial activity against 25 different bacteria also varied between samples of oil, ranging from 8 to 19 inhibited. All samples actively inhibited *Clostridium sporogenes*, *Bacillus subtilis* and *Brevibacterium linens*, and all but one inhibited *Acinetobacter calcoacetica* and *Staphylococcus aureus*. There were however, differences in the inhibition of all the other organisms which could not be correlated with the chemical composition of the samples.

The differences in antimicrobial activity reflect those shown for different *Pelargonium* species and cultivars. The anti-*Listeria* action of the Geranium oils was again very variable, the number of strains affected ranging from 3 to 16 (Lis-Balchin *et al.*, 1996c). This study highlights the wide variability of biological activity between commercial samples, which is not directly correlated with the chemical composition. This can be explained by the following considerations: a high degree of blending, if not adulteration or the use of synthetics in the commercial oil industry; some of the samples were diluted when sold; minor components play a substantial part in biological activities. The study by Lis-Balchin and Deans (1997) indicated considerable variation in the activity of the 'same' volatile oil, such as that of Geranium, against different strains of *L. monocytogenes*, this being in agreement with the findings of Aureli *et al.* (1992) and Lis-Balchin *et al.* (1996a). In the latter study, 16 Geranium samples were active against 8–18 of the test bacteria and the number of active oils against *Listeria monocytogenes* ranged from 3 to 16 out of a set of 20 bacteria. In the antifungal studies, the Geranium samples varied from 0 to 94 per cent inhibition of *Aspergillus niger*, 12–95 per cent against *A. ochraceus* and 40–86 per cent against *Fusarium culmorum*. This highlights the wide variation within the group of samples when considering that the samples should be very similar. The adulteration of volatile oils with synthetic components may give rise to a different proportion of enantiomers for a large number of components than in a pure botanical sample: this can greatly influence the bioactivity of the subsequent material (Lis-Balchin *et al.*, 1996b; 1998a).

These conclusions were in agreement with the results of an earlier study (Lis-Balchin *et al.*, 1995) wherein 24 cultivars were tested for antimicrobial activity against 25 test bacteria and *Aspergillus niger*. Of these test bacteria, the most sensitive bacteria were *Alcaligenes faecalis*, *Bacillus subtilis*, *Brevibacterium linens*, *Brocothrix thermosphactum*, *Clostridium sporogenes*, *Flavobacterium suaveolens* and *Staphylococcus aureus*, a group of predominantly Gram-positive organisms, whilst the least sensitive organisms were *Aeromonas hydrophila*, *Beneckea natriegens*, *Escherichia coli*, *Klebsiella pneumoniae*, *Leuconostoc cremoris* and *Pseudomonas aeruginosa*, a group of predominantly Gram-negative organisms. The inhibition against *A. niger* ranged from 3 to 98 per cent with two extracts causing growth enhancement. The activity in some cultivars was attributed to the level of monoterpenes present.

Synergistic antibacterial agents

In a study into the potential usage of mixtures of plant volatile oils as synergistic antibacterial agents in foods, Lis-Balchin and Deans (1998) included 'Geranium oil' in a mixture with nutmeg and bergamot oils. In the case of *Bacillus subtilis*, the mixture was more active than Geranium on its own whilst *Enterobacter aerogenes* was less affected by the mixture compared with Geranium alone. The activity of the mixture was marginally improved in the case of seven bacteria and lessened in five bacteria with *Serratia marcescens* being unchanged (Table 13.2). These results indicate that mixing

Table 13.2 Antibacterial activity of geranium volatile oil alone and in combination with nutmeg and bergamot volatile oils against test bacteria. Inhibition zone diameter in mm, including well diameter of 4 mm. Each value is the mean of three replicates. Bacterial sources as in Table 13.1

Organism	*Geranium*	*Nutmeg*	*Bergamot*	*Mean*	*Mixture*
Aeromonas hydrophila	7.0	8.6	9.2	8.3	7.6
Bacillus subtilis	7.3	6.3	11.4	8.3	10.5
Brocothrix thermosphacta	8.8	8.7	4.0	7.2	7.1
Enterobacter aerogenes	5.4	5.7	4.0	5.0	4.0
Enterococcus faecalis	5.6	4.0	17.0	8.9	7.5
Escherichia coli	8.5	6.0	14.1	9.5	5.8
Lactobacillus plantarum	5.5	6.3	8.2	6.7	6.8
Leuconostoc cremoris	6.2	4.0	5.2	5.1	8.4
Micrococcus luteus	5.6	4.0	8.2	5.9	5.8
Salmonella pullorum	4.0	4.0	6.4	4.8	5.6
Serratia marcescens	4.0	6.3	6.0	5.4	4.0
Staphylococcus aureus	9.0	4.0	16.4	9.8	7.4
Yersinia enterocolitica	6.7	4.0	9.1	6.6	5.7

Source: Lis-Balchin and Deans, 1998.

three oils did not have any real synergistic effect. The chemical components of the individual oils were diluted by a third in the 1:1:1 mixture and the most active oil therefore became less potent compared with its solo use. Although there was clearly no real synergistic antibacterial activity, it may still be reasonable to use mixtures of volatile oils to lower the individual odour potential when used in processed foodstuffs as at least some of the mixtures referred to showed a similar net effect compared with the mean of the individual effects.

Antimicrobial activity of *Pelargonium* oil components

A number of individual oil components were tested for antibacterial activity (Dorman and Deans, 2000). In the case of 'Geranium oil' constituents, the ranking order of activity was linalool (mean inhibition zone diameter of 12.5 mm) > geranyl acetate (9.4 mm) > nerol (8 mm) > geraniol (7 mm) > menthone (6.8 mm) > β-pinene (6.3 mm) > limonene (6.1 mm) > α-pinene (5.8 mm). Compared with more phenolic compounds, these activities are relatively modest. The bacteria showing the greatest level of inhibition were *Clostridium sporogenes* (11.5 mm) > *Lactobacillus plantarum* (10.5 mm) > *Citrobacter freundii* (9.9 mm) > *Escherichia coli* (9.6 mm) > *Flavobacterium suaveolens* (9 mm): all remaining bacteria had mean inhibition zone diameters of <9 mm (Table 13.3).

Antifungal properties of different Pelargonium species and cultivars

In addition to antibacterial activities, Geranium oil also possesses antifungal properties. *Pelargonium × hortorum* leaves were reported as being most active against *Candida albicans*, *Trichophyton rubrum* and *Streptococcus mutans*, organisms causing common dermal, mucosal or oral infections in humans (Heisey and Gorham, 1992). The action of 16 samples of 'Geranium oil' against three filamentous fungi was investigated and revealed a wide range of activities. In the presence of 1 μl of Geranium oil per ml of

Table 13.3 Antibacterial activity of geranium methanolic extracts against 25 test bacteria. Inhibition zone diameter in mm, including well diameter of 4 mm. Each value is the mean of three replicates. Bacterial sources as in Table 13.1

Organism	*A**	*B*	*C*	*D*	*E*	*F*	*G*	*H*
Acinetobacter calcoacetica	12	11	6	17	11	12	12	10
Aeromonas hydrophila	19	17	4	13	11	10	8	4
Alcaligenes faecalis	17	15	6	15	15	15	14	11
Bacillus subtilis	19	17	6	20	16	16	10	7
Beneckea natriegens	10	6	4	4	4	4	6	9
Brevibacterium linens	18	18	7	23	17	17	14	8
Brocothrix thermosphacta	9	13	6	14	10	11	12	8
Citrobacter freundii	11	10	5	13	7	17	4	4
Clostridium sporogenes	24	24	5	24	20	18	4	18
Enterobacter aerogenes	11	10	6	4	4	4	4	4
Enterococcus faecalis	15	12	6	18	16	14	13	13
Erwinia carotovora	16	13	6	14	15	16	13	4
Escherichia coli	12	9	6	11	8	8	4	4
Flavobacterium suaveolens	20	15	5	15	15	16	15	15
Klebsiella pneumoniae	16	15	i	11	14	15	13	9
Lactobacillus plantarum	16	14	5	17	13	12	11	5
Leuconostoc cremoris	17	14	6	14	12	13	11	11
Micrococcus luteus	18	17	6	21	17	17	15	10
Moraxella sp.	12	10	6	4	4	4	12	8
Proteus vulgaris	16	13	6	26	15	16	8	6
Pseudomonas aeruginosa	16	16	5	15	15	15	13	9
Salmonella pullorum	11	10	6	11	12	9	10	4
Serratia marcescens	20	16	7	20	17	17	16	10
Staphylococcus aureus	9	8	6	12	8	9	4	4
Yersinia enterocolitica	23	20	6	22	20	18	17	9

Source: Lis-Balchin and Deans, 1996.

Notes
* Key to *Pelargonium*: A – *P. zonale*; B – *P. inquinans*; C – *P. capitatum*; D – *P. acraem*; E – *White Boar*; F – *P. scandens*; G – *P. hybridum*; H – *P. cucullatum*.

YES broth, *Aspergillus niger* was inhibited by 0–94 per cent, *A. ochraceus* by 12–95 per cent and *Fusarium culmorum* by 40–86 per cent. Correlation between chemical composition and antimycotic activity is not always clear. The test organisms, *A. niger*, *A. ochraceus* and *Fusarium culmorum* reacted differently to the oils, with the plant pathogenic *F. culmorum* being less affected, with two exceptions, than the two *Aspergilli*. The spoilage organism *A. niger* was more inhibited in its growth, again with two exceptions, than the mycotoxigenic *A. ochraceus* whilst antifungal action was poor with <40 per cent inhibition. The antioxidant properties of the *Pelargonium* oils is also given for these eight methanolic extracts (Table 13.4).

Antioxidant properties of Pelargonium *species*

Pelargonium species, including the commercial 'Geranium oil' have been shown to have antioxidative properties (Dorman *et al.*, 2000; Fukaya *et al.*, 1988; Youdim *et al.*, 1999), though these properties had very variable activities in different commercial samples of 'Geranium oil' (Lis-Balchin *et al.*, 1996c). Various pharmacological properties have also

Table 13.4 Antibacterial activity of *Pelargonium* volatile oils, petroleum ether and methanol extracts against four bacteria. Inhibition zone diameter in mm, including well of 4 mm. Each value is the mean of three replicates

Pelargonium type	*Extract type*	*A**	*B**	*C**	*D**
Attar of Roses	Steam distill	17	11	12	8
	Pet ether	10	12	11	10
	Methanol	4	8	9	13
Lady Plymouth	Steam distill	ND	11	7	4
	Pet ether	7	4	4	4
	Methanol	11	12	12	11
Pink Little Gem	Steam distill	ND	11	8	4
	Pet ether	13	11	8	4
	Methanol	13	16	14	15
Radula	Steam distill	10	7	8	8
	Pet ether	4	4	6	4
	Methanol	13	13	13	18
Rober's Lemon Rose	Steam distill	7	4	4	8
	Pet ether	7	4	7	4
	Methanol	17	20	18	20
Sweet Mimosa	Steam distill	ND	10	7	4
	Pet ether	11	10	8	4
	Methanol	12	17	14	18
P. tomentosum	Steam distill	ND	4	6	4
	Pet ether	8	10	12	11
	Methanol	9	10	10	10
Chocolate Tomentosum	Steam distill	ND	8	8	8
	Pet ether	13	11	10	8
	Methanol	9	12	8	10
Lemon Fancy	Steam distill	18	13	9	12
	Pet ether	8	8	19	8
	Methanol	10	11	12	8
Crispum variegatum	Steam distill	8	4	4	7
	Pet ether	7	9	9	8
	Methanol	7	10	14	4
Clorinda	Steam distill	ND	10	11	4
	Pet ether	16	13	16	11
	Methanol	11	11	9	9
Copthorne	Steam distill	7	4	4	10
	Pet ether	8	8	8	9
	Methanol	17	16	16	18
Oak cultivar	Steam distill	ND	8	11	4
	Pet ether	11	9	9	6
	Methanol	ND	ND	ND	ND
Village Hill Oak	Steam distill	9	8	8	8
	Pet ether	4	8	8	8
	Methanol	11	11	9	12
P. denticulatum	Steam distill	8	7	8	7
	Pet ether	4	7	8	7
	Methanol	4	8	8	7
P. × fragrans	Steam distill	ND	9	10	8
	Pet ether	19	10	17	14
	Methanol	16	23	28	19

P. odoratissimum	Steam distill	8	8	8	7
	Pet ether	ND	ND	ND	ND
	Methanol	ND	ND	ND	ND
Orsett	Steam distill	ND	10	9	4
	Pet ether	12	9	9	8
	Methanol	12	15	11	13

Source: Lis-Balchin *et al.*, 1998b.

Notes
* Key to bacteria and their source: A – *Bacillus cereus* NCIMB 6349; B – *Proteus vulgaris* ATCC 13315; C – *Staphylococcus aureus* ATCC 9144; D – *Staphylococcus epidermidis* ATCC 12228.

been determined for Pelargonium essential oils (Lis-Balchin *et al.*, 1995), which were not found to be related to the antimicrobial activity to any great extent.

METHANOLIC EXTRACTS OF *PELARGONIUM*

Bioactivity of methanolic extracts of *Pelargonium* against 25 species of bacteria

Methanolic extracts of representative species and cultivars of *Pelargonium* were assessed for bioactivity against the set of 25 test bacteria (listed in Table 13.1). The antibacterial of all the extracts was very pronounced (Lis-Balchin and Deans, 1996). The ranking order of the greatest inhibition of growth was *Clostridium sporogenes* (mean inhibition zone diameter of 17.1 mm) > *Yersinia enterocolitica* (16.9 mm) > *Serratia marcescens* (15.4 mm) > *Micrococcus luteus* (15.1 mm) > *Flavobacterium suaveolens* (14.5 mm) > *Bacillus subtilis* (13.9 mm) > *Alcaligenes faecalis* (13.5 mm) > *Enterococcus faecalis* (13.3 mm) ~ *Proteus vulgaris* (13.3 mm) > *Pseudomonas aeruginosa* (13 mm): all remaining bacteria had mean inhibition zone diameters of < 13 mm. The individual extracts were ranked by activity as *P. zonale* (15.5 mm) > *P. acraem* (15.2 mm) > *P. inquinans* (13.7 mm) > *P. scandens* (12.9 mm) > 'White Boar' (12.6 mm) > *P. hybridum* (10.5 mm) > *P. cucullatum* (8.2 mm) > *P. capitatum* (5.6 mm) (Table 13.5).

Comparison of antibacterial activity between different extractives of the same *Pelargonium*

In a similar study, the volatile oil, petroleum spirits and methanol extracts from 18 *Pelargonium* cultivars were assayed against four bacteria, *Bacillus cereus*, *Proteus vulgaris*, *Staphylococcus aureus* and *Staphylococcus epidermidis*. There was considerable variation in the activity both between the cultivars themselves and the different extraction techniques (Table 13.6) (Lis Balchin *et al.*, 1998b).

In summary, the genus *Pelargonium* is a rich source of bioactive compounds having applications as antibacterial, antifungal, and antioxidant agents in addition to their contribution to the food/beverage, cosmetic and pharmaceutical sectors (Lis-Balchin *et al.*, 1995; Dorman and Deans, 2000; Pattnaik *et al.*, 1996), whilst *Geranium* species are apparently more active as antiviral agents (Serkedjieva, 1996).

Table 13.5 Antibacterial activity of Geranium oil constituents against 25 test bacteria. Inhibition zone diameter in mm, including well diameter of 4 mm. Each value is the mean of three replicates. Bacterial sources as in Table 13.1

Organism	*A**	*B*	*C*	*D*	*E*	*F*	*G*	*H*
Acinetobacter calcoacetica	6.1	10.3	4.0	9.3	9.7	11.4	4.0	11.2
Aeromonas hydrophila	6.4	9.0	4.0	11.5	7.0	7.7	4.0	7.1
Alcaligenes faecalis	7.0	10.5	4.0	12.1	6.2	7.1	4.0	7.8
Bacillus subtilis	6.4	10.8	4.0	14.0	7.1	12.4	4.0	4.0
Beneckea natriegens	6.2	10.8	4.0	11.4	5.9	11.3	4.0	6.5
Brevibacterium linens	7.3	12.5	4.0	12.5	4.0	11.7	4.0	4.0
Brocothrix thermosphacta	7.4	9.2	4.0	8.1	6.8	9.0	4.0	5.9
Citrobacter freundii	9.2	6.8	7.8	27.5	7.8	7.8	6.0	5.9
Clostridium sporogenes	12.9	20.4	10.3	20.3	10.7	4.0	5.7	7.5
Enterobacter aerogenes	6.4	7.6	7.1	9.7	6.3	7.2	4.0	4.0
Enterococcus faecalis	12.9	7.5	4.0	16.7	4.0	4.0	9.2	7.8
Erwinia carotovora	8.1	8.7	7.4	12.3	6.5	7.7	8.7	4.0
Escherichia coli	9.7	11.0	11.2	13.8	6.6	7.6	8.9	7.8
Flavobacterium suaveolens	7.0	11.0	10.6	15.7	5.8	7.0	6.5	8.4
Klebsiella pneumoniae	4.0	7.8	7.0	12.6	5.9	4.0	8.1	7.9
Lactobacillus plantarum	6.2	12.9	4.0	25.3	8.8	19.1	4.0	4.0
Leuconostoc cremoris	4.0	4.0	4.0	4.0	4.0	4.0	4.0	4.0
Micrococcus luteus	6.1	8.0	4.0	13.4	7.1	7.4	7.6	6.3
Moraxella sp.	6.1	9.0	7.9	10.3	6.9	4.0	6.2	4.8
Proteus vulgaris	5.6	9.8	7.4	12.2	6.2	4.0	7.5	6.6
Pseudomonas aeruginosa	5.7	6.5	4.0	4.0	4.0	13.6	4.0	6.5
Salmonella pullorum	6.3	8.7	11.2	7.5	6.2	4.0	7.9	6.0
Serratia marcescens	5.7	6.8	6.5	8.8	7.1	8.5	4.0	5.4
Staphylococcus aureus	5.2	6.6	4.0	9.0	10.2	9.4	8.3	7.4
Yersinia enterocolitica	8.0	8.2	7.1	9.5	8.0	7.1	6.6	5.8

Source: Dorman and Deans, 2000.

Notes

* Key to compounds: A – geraniol; B – geranyl acetate; C – limonene; D – linalool; E – menthone; F – nerol; α – -pinene; β – -pinene.

Table 13.6 Bioactivity of geranium methanolic extracts: antimycotic action against *Aspergillus niger* at 10 L mL^{-1} YES broth; antioxidant activity expressed as zone of colour retention in mm, including well diameter of 4 mm, followed by indication of intensity of colour retention on scale + mild, ++ moderate, +++ intense

Pelargonium	*Antimycotic inhibition (%*)*	*Antioxidant activity/intensity*	
P. zonale	31	16.6	+++
P. inquinans	28	19.5	+++
P. capitatum	36	12.6	+++
P. acraem	27	18.6	+++
White Boar	25	18.3	+++
P. scandens	38	18.5	+++
P. hybridum	35	21.4	+++
P. cucullatum	20	15.4	+++

Source: Lis-Balchin and Deans, 1996.

Note

* Fungal Inhibition Index calculated as (C-T)/C × 100 where C is the weight of mycelium from control flasks and T is the weight of mycelium from test flasks.

ACKNOWLEDGEMENTS

SAC received financial support from the Scottish Executive and Executive Rural Affairs Department.

REFERENCES

Aureli, P., Costantini, A. and Zolea, S. (1992) Antimicrobial activity of some plant essential oils against *Listeria*. *J. Food Prot.*, 55, 344–348.

Baratta, M.T., Dorman, H.J.D., Deans, S.G., Biondi, D.M. and Ruberto, G. (1998a) Chemical composition, antibacterial and antioxidative activity of laurel, sage, rosemary, oregano and coriander essential oils. *J. Essent. Oil Res.*, 10, 618–627.

Baratta, M.T., Dorman, H.J.D., Deans, S.G., Figueiredo, A.C., Barroso, J.G. and Ruberto, G. (1998b) Antimicrobial and antioxidant properties of some commercial essential oils. *Flav. Frag. J.*, 13, 235–244.

Cai, L. and Wu, C.D. (1996) Compounds from *Syzygium aromaticum* possessing growth inhibitory activity against oral pathogens. *J. Nat. Prod.*, 59, 987–990.

Carson, C.F., Cookson, B.D., Farrelly, H.D. and Riley, T.V. (1995) Susceptibility of methicillin-resistant *Staphylococcus aureus* to the essential oil of *Melaleuca alternifolia*. *J. Antimicrob. Chemother.*, 35, 421–424.

Carson, C.F., Hammer, K.A. and Riley, T.V. (1996) *In vitro* activity of the essential oil of *Melaleuca alternifolia* against *Streptococcus* spp. *J. Antimicrob. Chemother.*, 37, 1177–1181.

Daferera, D.J., Ziogas, B.N. and Polissiou, M.G. (2000) GC-MS analysis of essential oils from some Greek aromatic plants and their fungitoxicity on *Penicillium digitatum*. *J. Agric. Food Chem.*, 48, 2576–2581.

Deans, S.G. and Ritchie, G. (1987) Antibacterial properties of plant essential oils. *Int. J. Food Microbiol.*, 5, 165–180.

Deans, S.G., Kennedy, A.I., Gundidza, M.G., Mavi, S., Waterman, P.G. and Gray, A.I. (1994a) Antimicrobial activities of the volatile oil of *Heteromorpha trifoliata* (Wendl.) Eckl. and Zeyh. (Apiaceae). *Flav. Frag. J.*, 9, 245–248.

Deans, S.G., Hiltunen, R., Wuryani, W., Noble, R.C. and Penzes, L.G. (1994b) Antimicrobial and antioxidant properties of *Syzygium aromaticum* (L.) Merr. and Perry: Impact upon bacteria, fungi and fatty acid levels in ageing mice. *Flav. Frag. J.*, 10, 323–328.

Dorman, H.J.D. (1999) Phytochemistry and bioactive properties of plant volatile oils: antibacterial, antifungal and antioxidant activities. PhD Thesis, Strathclyde Institute of Biomedical Sciences, University of Strathclyde, Glasgow, Scotland.

Dorman, H.J.D. and Deans, S.G. (2000) Antimicrobial agents from plants: antibacterial activity of plant volatile oils. *J. Appl. Microbiol.*, 88, 308–316.

Dorman, H.J.D., Surai, P. and Deans, S.G. (2000) *In vitro* antioxidant activity of a number of plant essential oils and phytoconstituents. *J. Essent. Oil Res.*, 12, 241–248.

Fukaya, Y., Nakazawa, K., Okuda, T. and Iwata, S. (1988) Effect of tannin on oxidative damage of ocular lens. *Jpn. J. Ophthalmol.*, 32, 166–175.

Gallardo, P.P.R., Salinas, R.J. and Villar, L.M.P. (1987) The antimicrobial activity of some spices on microorganisms of great interest to health. IV: Seeds, leaves and others. *Microbiologie aliments Nutr.*, 5, 77–82.

Garg, S.C. and Dengre, S.L. (1986) Antibacterial activity of essential oil of *Tagetes erecta* Linn. *Hindustan Antibiotic Bull.*, 28, 27–29.

Grieve, M. (1994) *A Modern Herbal: The Medicinal, Culinary, Cosmetic and Economic Properties, Cultivation and Folklore of Herbs, Grasses, Fungi, Shrubs and Trees with All their Modern Scientific Uses*, pp. 1–912. Tiger Books International, London.

Heisey, R.M. and Gorham, B.K. (1992) Antimicrobial effects of plant extracts on *Streptococcus mutans*, *Candida albicans*, *Trichophyton rubrum* and other microorganisms. *Lett. Appl. Microbiol.*, 14, 136–139.

Hinou, J.B., Harvala, C.E. and Hinou, E.B. (1989) Antimicrobial activity screening of 32 common constituents of essential oils. *Pharmazie*, 44, H4.

Hoffman, C. and Evans, A.C. (1911) The use of spices as preservatives. *J. Indian Eng. Chem.*, 3, 835–838.

Inouye, S., Goi, H., Miyouchi, K., Muraki, S., Ogihara, M. and Iwanami, I. (1983) Inhibitory effect of volatile components of plants on the proliferation of bacteria. *Bokin Bobai*, 11, 609–615.

Ivancheva, J., Maximova, V.A., Manolova, N.H., Serkedjieva, Y.P. and Gegova, G.A. (1987) Chromatographic characteristics of a polyphenolic complex with antiviral activity isolated from *Geranium sanguineum* L. *Dokladi na Bolgarskata Akademiya na Naukite*, 40, 95–97.

Ivancheva, S., Manolova, N., Serkedjieva, J., Dimov, V. and Ivanovska, N. (1992) Polyphenols from Bulgarian medicinal plants with anti-infectious activity. *Basic Life Sci.*, 59, 717–728.

Jain, S.R. and Kar, A. (1971) The antibacterial activity of some essential oils and their combinations. *Planta Medica*, 20, 118–123.

Janssen, M.A., Scheffer, J.J.C., Parhan Van Atten, A.W. and Baerheim-Svendsen, A. (1988) Screening of some essential oils for their activities on dermatophytes. *Pharmaceutische Weekblad (Scientific Edition)*, 10, 277–280.

Jay, J.M. and Rivers, G.M. (1984) Antimicrobial activity of some food flavouring compounds. *J. Food Saf.*, 6, 129–139.

Jeanfils, J., Burlion, N. and Andrien, F. (1991) Antimicrobial activities of essential oils from different plant species. *Landbouwtijdschrift-Revue de l'Agriculture*, 44, 1013–1019.

Knobloch, K., Weigand, N., Weis, H.M. and Vigenschow, H. (1986) *Progress in Essential Oil Research*, E.J. Brunke (ed.), p. 429. Walter de Gruyther, Berlin.

Knobloch, K., Pauli, A., Iberl, N., Weis, H.M. and Weigand, N. (1988) Modes of action of essential oil components on whole cells of bacteria and fungi in plate tests. In: P. Schreier (ed.), *Bioflavour*, pp. 287–299. Walter de Gruyther, Berlin.

Kurita, N., Miyaji, M., Kurane, R., Takahara, Y. and Ichimura, K. (1979) Antifungal activity of molecular orbital energies of aldehyde compounds from oils of higher plants. *Agric. Biol. Chem.*, 43, 2365–2371.

Kurita, N., Miyaji, M., Kurane, R., Takahara, Y. and Ichimura, K. (1981) Antifungal activity of components of essential oils. *Agric. Biol. Chem.*, 45, 945–952.

Larrondo, J.V., Agut, M. and Calvo-Torras, M.A. (1995) Antimicrobial activity of essences from Labiatae. *Microbios*, 82, 171–172.

Lis-Balchin, M. (1997) Essential oils and aromatherapy: their modern role in healing. *J. R. Soc. Health*, 117, 324–329.

Lis-Balchin, M. and Deans, S.G. (1996) Antimicrobial effects of hydrophilic extracts of *Pelargonium* species (Geraniaceae). *Lett. Appl. Microbiol.*, 23, 205–207.

Lis-Balchin, M. and Deans, S.G. (1997) Bioactivity of selected plant essential oils against *Listeria monocytogenes*. *J. Appl. Microbiol.*, 82, 759–762.

Lis-Balchin, M. and Deans, S.G. (1998) Studies on the potential usage of mixtures of plant essential oils as synergistic antibacterial agents in foods. *Phytother. Res.*, 12, 472–475.

Lis-Balchin, M., Hart, S., Deans, S.G. and Eaglesham, E. (1995) Potential agrochemical and medicinal usage of essential oils of *Pelargonium* species. *J. Herbs, Spices and Medicinal Plants*, 3, 11–22.

Lis-Balchin, M., Hart, S., Deans, S.G. and Eaglesham, E. (1996a) Comparison of the pharmacological and antimicrobial action of commercial plant essential oils. *J. Herbs. Spices Med. Plants*, 4, 69–86.

Lis-Balchin, M., Ochocka, R.J., Deans, S.G. and Hart, S. (1996b) Bioactivity of the enantiomers of limonene. *Med. Sci. Res.*, 24, 309–310.

Lis-Balchin, M., Deans, S.G. and Hart, S. (1996c) Bioactivity of Geranium oils from different commercial sources. *J. Essent. Oil Res.*, 8, 281–290.

Lis-Balchin, M., Deans, S.G. and Eaglesham, E. (1998a) Relationship between bioactivity and chemical composition of commercial essential oils. *Flav. Frag. J.*, **13**, 98–104.

Lis-Balchin, M., Ribisch, K., Wenger, M.-T. and Buchbauer, G. (1998b) Comparative antibacterial effects of novel *Pelargonium* essential oils and solvent extracts. *Lett. Appl. Microbiol.*, **27**, 135–141.

Manolov, P., Petkov, V.D. and Ivancheva, S. (1977) Studies on the central depressive action of methanol extract from *Geranium macrorrhizum* L. *Dokladi na Bolgarskata Akademiya na Naukite*, **30**, 1657–1659.

Martindale, W.H. (1910) Essential oils in relation to their antiseptic powers as determined by their carbolic coefficients. *Perfumery and Essent. Oil Res.*, **1**, 266–296.

Moleyar, V. and Narasimham, P. (1986) Antifungal activity of some essential oil components. *Food Microbiol.*, **3**, 331–336.

Naigre, R., Kalck, P., Roques, C., Roux, I. and Michel, G. (1996) Comparison of antimicrobial properties of monoterpenes and their carbonylated products. *Planta Medica*, **62**, 275–277.

Nenoff, P., Haustein, U.F. and Brandt, W. (1996) Antifungal activity of the essential oil of *Melaleuca alternifolia* (tea tree oil) against pathogenic fungi *in vitro*. *Skin Pharmacol.*, **9**, 388–394.

Okuda, T., Yoshida, T. and Hatano, T. (1992) Polyphenols from Asian plants – Structural diversity and antitumour and antiviral activities. *ACS Symposium Series*, **507**, 160–183.

Pattnaik, S., Subramanyam, V.R., Kole, C.R. and Sahoo, S. (1995) Antibacterial activity of essential oils from *Cymbopogon*: Inter- and intra-specific differences. *Microbios*, **84**, 239–245.

Pattnaik, S., Subramanyam, V.R. and Kole, C.R. (1996) Antibacterial and antifungal activity of ten essential oils *in vitro*. *Microbios*, **86**, 237–246.

Pauli, A. and Knobloch, K. (1987) Inhibitory effects of essential oil components on growth of food-contaminating fungi. *Zeitschrift für Lebensmittel Untersuchung und Forschung*, **185**, 10–13.

Pelczar, M.J., Chan, E.C.S. and Krieg, N.R. (1988) Control of microorganisms by physical agents. In *Microbiology*, pp. 469–509. McGraw-Hill International, New York.

Pélissier, Y., Marion, C., Casadebaig, J., Milhau, M., Djenéba, K., Loukou, N.Y. and Bessière, J.M. (1994) A chemical, bacteriological, toxicological and clinical study of the essential oil of *Lippia multiflora* (Verbenaceae). *J. Essent. Oil Res.*, **6**, 623–630.

Petkov, V., Ivancheva, S., Tsonev, I., Klouchek, E. and Rainova, L. (1974) Chemistry and pharmacology of flavonoid fractions with hypotensive action extracted from geranium *Eksp. Med. Morfol.*, **13**, 29–36.

Rios, J.L., Recio, M.C. and Villar, A. (1987) Antimicrobial activity of selected plants employed in the Spanish Mediterranean area. *J. Ethnopharmacol.*, **21**, 139–152.

Rios, J.L., Recio, M.C. and Villar, A. (1988) Screening methods for natural products with antibacterial activity: a review of the literature. *J. Ethnopharmacol.*, **23**, 127–149.

Serkedjieva, J. (1995) Inhibition of influenza virus protein synthesis by a plant preparation from *Geranium sanguineum* L. *Acta Virol.*, **39**, 5–10.

Serkedjieva, J. (1996) A polyphenolic exract from *Geranium sanguineum* L. inhibits influenza virus protein expression. *Phytother. Res.*, **10**, 441–443.

Serkedjieva, J. (1997) Antiinfective activity of a plant preparation from *Geranium sanguineum* L. *Pharmazie*, **52**, 799–802.

Serkedjieva, J. and Manolova, N. (1988) On the antiviral activity of a polyphenolic complex, isolated from *Geranium sanguineum* L. *Antiviral R.*, **9**, 110–113.

Serkedjieva, J. and Hay, A.J. (1998) *In vitro* anti-influenza virus activity of a plant preparation from *Geranium sanguineum* L. *Antiviral Res.*, **37**, 121–130.

Serkedjieva, J. and Ivancheva, S. (1999) Antiherpes virus activity of extracts from the medicinal plant *Geranium sanguineum* L. *J. Ethnopharmacol.*, **64**, 59–68.

Shapiro, S., Meier, A. and Guggenheim, B. (1994) The antimicrobial activity of essential oils and essential oil components towards oral bacteria. *Oral Microbiol. Immunol.*, **9**, 202–204.

Shelef, L.A. (1984) Antimicrobial effects of spices. *J. Food Saf.*, **6**, 29–44.

Sherif, A., Hall, R.G. and El-Amamy, M. (1987) Drugs, insecticides and other agents from *Artemisia*. *Med. Hypotheses*, **23**, 187–193.

Ueda, S., Yamashita, H., Nakajima, M. and Kumabara, Y. (1982) Inhibition of microorganisms by spice extracts and flavouring compounds. *Nippon Shokuhin Kogyo Gakkaishi*, 29, 111–116.

Ushiki, J., Hayakawa, Y. and Tadano, T. (1996) Medicinal plants for suppressing soil-borne plant diseases. I. Screening for medicinal plants with antimicrobial activity in roots. *Soil Sci. Plant Nutr.*, 42, 423–426.

Ushiki, J., Tahara, S., Hayakawa Y. and Tadano, T. (1998) Medicinal plants for suppressing soil-borne plant diseases. II. Suppressive effect of *Geranium pratense* L. on common scab of potato and identification of the active compound. *Soil Sci. Plant Nutr.*, 44, 157–165.

Vokou, D., Vareltzidou, S. and Katinakis, P. (1993) Effects of aromatic plants on potato storage: sprout suppression and antimicrobial activity. *Agric. Ecosyst. Environ.*, 47, 223–225.

Youdim, K.A., Dorman, H.J.D. and Deans, S.G. (1999) The antioxidant effectiveness of thyme oil, α-tocopherol and ascorbyl palmitate on evening primrose oil oxidation. *J. Essent. Oil Res.*, 11, 643–648.

Zgorniak-Nowosielska, I., Zawilinska, B., Manolova, N. and Serkedjieva, J. (1989) A study on the antiviral action of a polyphenolic complex isolated from the medicinal plant *Geranium sanguineum* L. VIII Inhibitory effect on the reproduction of herpes simplex virus type 1. *Acta Microbiologia Bulgaria*, 24, 3–8.

14 Essential oils from different *Pelargonium* species and cultivars: their chemical composition (using GC, GC/MS) and appearance of trichomes (under EM)

Maria Lis-Balchin

INTRODUCTION

Commercial 'Geranium oil' is obtained from various *Pelargonium* cultivars (derived mainly from *P. graveolens*, *P. capitatum* and *P. radens*) growing mainly in Reunion, China and Egypt. The chemical composition of the oils is variable due to the difference in cultivars used, the climate (including sunlight, rainfall and temperature), the time of the harvest, fertilizers applied etc). Geranium oil is used mainly in perfumery, but also has potent antimicrobial potential (Lis-Balchin *et al.*, 1996). It also has relaxant pharmacological properties on smooth muscle e.g. guinea-pig ileum and the rat uterus (Lis-Balchin and Hart, 1997a) as well as striated muscle (Lis-Balchin and Hart, 1997b). The mode of action of commercial Geranium oil appears to act through cAMP (cyclic adenosine monophosphate) as the secondary messenger (Lis-Balchin and Hart, 1999).

Other scented *Pelargonium* species, hybrids and cultivars yield essential oils of many different chemical compositions, many of which also have antimicrobial potential (Lis-Balchin, 1991a; Lis-Balchin *et al.*, 1998) and some, with floral scents also act in a similar pharmacological way to that of commercial Geranium oil (Lis-Balchin and Hart,1997b, 1999). Others, with camphoric, mentholic or more medicinal scents did not show this mechanism of action (Lis-Balchin, 1991b; Lis-Balchin and Hart, 1999).

ODOUR CLASSIFICATION

Pelargonium essential oils, with their many distinctive odours, can be grouped in many different ways. One of the problems is that the odour can change dramatically during flowering e.g. *P. graveolens* becomes rosy during flowering, whilst minty at other times leading up to this; similarly the cultivars, 'Attar of Roses' and others. An example of a particular arbitrary grouping is shown together with the chemical composition of different *Pelargonium* species and cultivars, as assessed using gas chromatography (GC) with or without mass spectrometry (MS) in Tables 14.1–14.5. All samples were run on an OV 101 column or equivalent.

Table 14.1 The chemical composition of 'Rosy' pelargoniums

Component	*%*	*KI*
P. capitatum		
α-Pinene	0.1–47.5	929
Myrcene	0–5.5	980
Limonene	0–0.7	1020
Linalool	0–0.1	1082
cis-Rose oxide	0–1.4	1097
trans-Rose oxide	0–0.7	1107
Menthone	0–0.7	1142
Isomenthone	0–0.5	1159
Citronellol	0.2–10.7	1216
Geraniol	0.1–2.8	1233
Citronellyl formate	0.1–36.8	1255
Geranyl formate	0.1–9.8	1279
Citronellyl acetate	0–0.9	1333
Geranyl acetate	0–0.3	1359
β-Caryophyllene	0.9–9.1	1408
Gauia-6,9-diene	0.2–15.0	1431
Germacrene D	0.3–14.3	1467
Source: Demarne (1989).		
Attar of Roses		
Linalool	1.3–10	1086
cis-Rose oxide	0–0.7	1101
Isomenthone	2.7–7.8	1168
Citronellol	17.8–43.4	1211
Geraniol	7.4–42.8	1236
Citronellyl formate	2.6–11.2	1259
Geranyl formate	1.1	1283
β-Caryophyllene	0.2–3.1	1434
Guaia-6,9-diene	4.5–10.6	1451
+ many sesquiterpenes unidentified		
Source: Lis-Balchin (Master's Thesis Supervisor).		
P. cv. 'Rose'		
Linalool	4.6	1086
cis-Rose oxide	0.2	1101
Menthone	0.4	1129
Isomenthone	7.8	1168
Citronellol	19.0	1208
Geraniol	21.5	1233
Citronellyl formate	8.5	1259
Geranyl formate	9.5	1279
β-Caryophyllene	0.8	1408
Guaia-6,9-diene	7.2	1432
Geranyl propionate	1.6	1450
Germacrene D	2.1	1467

Citronellyl butyrate + XX	1.1	1507
Geranyl butyrate	1.2	1537
Phenylethyl tiglate	0.7	1547
Geranyl tiglate	1.4	1673
+ many sesquiterpenes unidentified		
Source: Demarne, 1989.		
P. radens		
α-Pinene	0.1–0.4	928
Myrcene	0.2–3.7	976
α-Phellandrene	0.3–1.1	1002
p-Cymene	0.2–0.7	1017
Limonene	0.6–1.7	1022
Linalool	0.3–0.9	1082
cis-Rose oxide	0.8–3.2	1097
trans-Rose oxide	0–0.4	1107
Menthone	0.3–4.3	1142
Isomenthone	32.8–81.8	1159
Citronellol	0.2–53.7	1216
Source:Lis-Balchin (Master's Thesis Supervisor).		
P. graveolens		
Linalool	1.2–12.8	1090
Menthone	0.2–4.5	1145
Isomenthone	7.0–81.5	1155
Citronellol	0–17.6	1208
Geraniol	19.1–41.9	1237
Citronellyl formate	11.2	1259
Guaia-6,9-diene	6.0	1450
Geranyl formate	0.7–4.2	1539
10-Epi-eudesmol	0–4.0	1626
+ many sesquiterpenes unidentified		
Unusual components identified by GC/MS		
Hexyl butyrate	0.2	1176
Hexenyl butyrate	0.2	1178
Source: Lis-Balchin (Master's Thesis Supervisor).		
'Rober's Lemon Rose'		
Linalool	1.5–12.8	1087
cis-Rose oxide	1.7	1102
trans-Rose oxide	0.7	1126
Isomenthone	5.8–8.1	1170
Citronellol	19.1–56.9	1212
Geraniol	4.3–9.8	1238
Citronellyl formate	3.1–12.6	1260
β-Caryophyllene	0–25.8	1434
Guaia-6,9-diene	0–2.7	1448
+ many sesquiterpenes unidentified		
Source: Lis-Balchin, 1988.		

Table 14.2 The chemical composition of 'Minty' pelargoniums

Component	*%*	*KI*
P. tomentosum		
α-Pinene	0.4	942
β-Pinene/myrcene	0.6	991
p-Cymene	2.2	1014
Limonene	2.2	1038
Menthone	45.6	1155
Isomenthone	46.6	1165
Terpinen-4-ol	0.9	1174
Lis-Balchin (Master's Thesis Supervisor).		
P. tomentosum (2)		
Tricyclene	1.0	921
α-Pinene	1.0	931
β-Pinene	tr	973
Myrcene	1.4	983
α-Phellandrene	2.8	1011
p-Cymene	0.2	1013
Limonene + X	3.9	1022
cis-Ocimene	0.5	1026
trans-Ocimene	0.3	1037
Linalool + X	0.1	1085
Menthone	25.1	1134
Isomenthone	60.8	1148
Menthol	0.1	1161
Citronellol	0.1	1210
Piperitone	0.8	1224
Geraniol	0.1	1235
Citronellyl formate	0.1	1257
β-Caryophyllene	0.3	1414
α-Humulene	0.1	1447
Germacrene D	0.2	1472
Source: Demarne *et al.*, 1986.		
Chocolate tomentosum		
α-Pinene	0–7.7	941
β-Pinene	0.1–2.9	989
Myrcene	0–0.7	989
α-phellandrene	2.5–17.5	1005
p-Cymene	1.4–4.7	1019
Linalool	0.9	1083
Menthone	0.1–39.1	1147
Isomenthone	0.4–22.1	1156
n-Hexyl isobutyrate	0.6–5.5	1153
Citronellol	0–5.0	1206
Geraniol	0–0.8	1233
Citronellyl formate	0–2.9	1256
Citronellyl acetate	0.3–0.7	1334
β-Caryophyllene	1.3–3.1	1431
Guaia-6,9-diene	0–5.6	1446
+ numerous sesquiterpenes unidentified		

Source: Lis-Balchin (Master's Thesis Supervisor).

'Lady Plymouth'		
α-Pinene	0.6–1.8	942
Myrcene	0.2–0.6	984
α-Phellandrene	0.4–0.9	1008
α-Terpinene	0.4–0.8	1022
p-Cymene	0.4–1.0	1022
Limonene	0.5–1.9	1032
Linalool	0.2–0.3	1087
Isomenthone	71.9–81.9	1158
α-Terpineol	0–0.4	1181
Octyl acetate	0.2	1185
Citronellol	0.2–0.5	1206
Geraniol	0.2–0.6	1241
+ sesquiterpenes unidentified		

Source: Lis-Balchin (Master's Thesis Supervisor).

Table 14.3a Chemical composition of 'Peppery-pungent' pelargoniums

Component	*%*	*KI*
P. vitifolium		
α-Phellandrene	0–0.9	997
p-Cymene	0–0.7	1011
Limonene	0–0.2	1019
cis-Ocimene	0.1–1.5	1024
trans-Ocimene	0–0.2	1035
Linalool	0.1–0.4	1083
Menthone	0.2–0.9	1129
Isomenthone	0.3–0.6	1139
Citronellol	2.6–6.3	1208
Geraniol	0–0.1	1233
Citronellic acid	76.9–83.9	1318
β-Caryophyllene	0.4–0.9	1408
Guaia-6,9-diene	0–0.1	1432
+ many sesquiterpenes unidentified		

Source: Demarne, 1989.

Table 14.3b Comparison of the main components of *P. papilionaceum* and 'Sweet Rosina'

Component	*P. papilionaceum*	*Sweet Rosina*
	%	*%*
Citronellic acid	85.9–89.3	27.8–43.1
Citronellol	0.2–1.6	14.5–43.1
Geranic acid	0.1–5.0	10.3–12.5
Isomenthone	0.1–0.4	15.2–24.8

Source: Lis-Balchin and Roth, 1999.

Table 14.4 The chemical composition of 'Citrusy' pelargoniums

Component	%	*KI*
P. citronellum		
Myrcene	0.5	987
p-Cymene	0.2	1018
γ-Terpinene	0.1	1054
Linalool	3.7	1084
Isomenthone	0.9	1161
Citronellol	30.5	1221
Neral	2.6	1224
Geraniol	0.1	1235
Geranial	42.6	1240
+ many sesquiterpenes unidentified		
Source: Lis-Balchin (Master's Thesis Supervisor).		
P. crispum 'variegatum'		
α-Pinene	0.4	939
Limonene	2.1	1028
γ-Terpinene	0.7	1053
Fenchone	2.2	1078
Citronellal	2.5	1134
Isomenthone	0–0.6	1170
Hexyl butyrate	0–2.8	1174
Citronellol	21.6–32.9	1223
Neral	2.5	1236
Geranial	27.9–48.7	1247
Sesquiterpene	55.3	1551
+ many sesquiterpenes unidentified		
Source: Lis-Balchin (Master's Thesis Supervisor).		
P. scabrum		
Myrcene	0.5–1.7	987
p-Cymene	0.9–1.3	1018
γ-Terpinene	0.3–1.3	1054
Linalool	0.2–1.5	1084
Isomenthone	0.4–1.1	1161
Terpinen-4-ol	0–13.9	1173
Citronellol	55.7–18.1	1221
Neral	1.4–5.0	1224
Geraniol	0.1–3.9	1237
Geranial	32.5–44.6	1256
Sesquiterpene	0–55.3	1551
+ many sesquiterpenes unidentified		
Lis-Balchin (Master's Thesis Supervisor).		
'Lemon Fancy'		
p-Cymene	1.3	1018
γ-Terpinene	0.5	1054
Linalool	2.8	1084
Isomenthone	1.1	1161
Citronellol	2.5	1221

Neral	28.8	1224
Geraniol	3.9	1237
Geranial	38.2	1252
Sesquiterpene	55.3	1551
+ many sesquiterpenes unidentified		

Source: Lis-Balchin (Master's Thesis Supervisor).

Table 14.5 The chemical composition of 'Camphoraceous-pungent' pelargoniums

Component	*%*	*KI*
P. glutinosum		
Myrcene	1.1–3.7	985
α-Phellandrene	2.0–16.4	1008
p-Cymene	16.0–23.0	1026
Limonene	2.8–13.4	1033
Hexyl butyrate	0.1–4.0	1170
Hexenyl butyrate	8.3–15.2	1176
Citronellol	0.8–4.3	1224
Sesquiterpene	23.5–25.0	1450
Sesquiterpene	20.7–26.4	1613
+ many sesquiterpenes unidentified		
Source: Lis-Balchin (Master's Thesis Supervisor).		
P. × *fragrans*		
α-Thujene	9.7	928
α-Pinene	19.7	940
β-Pinene	6.0	982
p-Cymene	1.5	1017
Limonene	7.5	1028
Fenchone	8.4	1079
α-Fenchyl alcohol	0.7	1106
Methyl eugenol	10.2	1373
β-Caryophyllene	2.6	1431
Source: Lis-Balchin and Roth, 2000.		
P. odoratissimum		
α-Pinene	1.2	944
β-Pinene	0.3	985
p-Cymene	3.5	1021
Limonene	1.3–4.3	1030
Fenchone	2.5–9.3	1081
Menthone	1.0	1144
Isomenthone	4.6–19.4	1155
Piperitone	1.4–8.7	1238
Methyl eugenol	31.7–79.8	1373
β-Caryophyllene	1.5–2.2	1432
+ many sesquiterpenes unidentified		

Source: Lis-Balchin and Roth, 2000.

Table 14.5 (Continued)

Component	%	*KI*
P. exstipulatum		
α-Thujene	25.0	921
α-Pinene	12.3	931
Benzaldehyde	4.3	949
Sabinene	0.5	968
Myrcene	2.7	978
Limonene	18.5	1029
γ-Terpinene	0.3	1053
Fenchone	4.5	1078
Linalool	0.2	1083
Camphor	4.3	1134
Menthone	0.2	1146
Isomenthone	0.6	1150
Terpinen-4-ol	3.7	1170
Hexyl butyrate	0.5	1174
α-Terpineol	0.9	1179
Sesquiterpene	0.2	1500
Sesquiterpene	10.3	1577
+ many sesquiterpenes unidentified		
Source: Lis-Balchin and Roth, 2000.		
P. quercifolium		
α-Thujene	0.2	933
α-Pinene	0.6	941
β-Pinene	19.2	982
α-Phellandrene	0.1	1005
p-Cymene	5.1	1025
Limonene	0.2	1030
trans-Ocimene	1.6	1046
Isomenthone	0.8	1159
α-Terpineol	30.5	1166
Nerol	1.1	1217
Neral	1.7	1221
Unknown	3.8	1365
+ many sesquiterpenes unidentified		
Source: Lis-Balchin (Master's Thesis Supervisor).		
'Clorinda'		
α-Pinene	1.0	940
β-Pinene	15.9	980
α-Phellandrene	9.6	1004
p-Cymene	3.1	1018
Limonene	2.7	1029
Fenchone	4.0	1084
Linalool	3.1	1189
+ many sesquiterpenes unidentified		
Source: Lis-Balchin (Master's Thesis Supervisor).		
'Copthorne'		
α-Thujene	4.5	931
α-Pinene	1.8	940

β-Pinene	5.5	980
p-Cymene	1.7	1017
Limonene	2.5	1027
γ-Terpinene	1.1	1053
Fenchone	4.0	1082
β-Caryophyllene	9.2	1482
Sesquiterpene	5.1	1445
+ many sesquiterpenes unidentified		

Source: Lis-Balchin (Master's Thesis Supervisor).

'Village Hill Oak'

α-Thujene	0.5	931
α-Pinene	0.6	940
β-Pinene/myrcene	0.8	997
α-Phellandrene	5.0	1004
p-Cymene	25.4	1022
Limonene	2.9	1030
γ-Terpinene	1.8	1054
Fenchone/linalool	2.3	1085
Terpinen-4-ol	11.0	1174
Nerol/citronellol	15.3	1226
+ many sesquiterpenes unidentified		

Source: Lis-Balchin (Master's Thesis Supervisor).

'Sweet Mimosa'

α-Thujene	8.0	939
α-Pinene	2.8	948
β-Pinene/myrcene	16.0	988
β-Phellandrene	5.5	1011
p-Cymene	1.8	1025
Limonene	2.6	1035
Linalool	0.6	1090
Isomenthone	7.3	1159
+ many sesquiterpenes unidentified		

Source: Lis-Balchin (Master's Thesis Supervisor).

P. grossularioides

Myrcene	1.8
p-Cymene	3.0
Limonene	1.0
Fenchone	8.2
Linalool	4.6
Menthone	1.8
Isomenthone	12.8
α-Terpineol	0.7
Citronellol	11.6
Neral	1.2
Geraniol	15.9
Geranial	3.7
Phenylethylisobutyrate	0.5
Methyl eugenol	11.2
β-Caryophyllene	0.0
Guaia-6,9-diene	0.0
Geranyl propionate	0.0
Geranyl butyrate	0.7

Table 14.5 (Continued)

Component	%	*KI*
Phenylethyl tiglate	0.8	
Caryophyllene oxide	1.8	
Geranyl tiglate	1.1	

Source: Lis-Balchin, (1991b).

Odour classification
1 Rosy
2 Minty
3 Peppery-pungent
4 Citrusy
5 Camphoraceous-pungent

Rosy: In the 'rosy' group (Table 14.1) there is *P. capitatum* L., found on the coastline of the Cape in South Africa, with many different varieties occurring from the point of view of appearance and odour (Demarne, 1989). 'Attar of Roses' is an old cultivar, which is probably the same as 'Otto of Roses', a form of *P. capitatum* mentioned and illustrated by Andrews (1805). This is the common form of the rose-scented *Pelargonium* in cultivation and is almost identical to the cultivar 'Rosé' grown on the Island of Reunion for its essential oil known as 'Geranium Bourbon'. There are many cultivars of this type from hybridizations between *P. capitatum* and *P. radens* H.E. Moore (Demarne and van der Walt, 1990) which produce essential oils of different odour and chemical composition (Demarne, 1989). *P. graveolens* L'Herit., is quoted most often as the species producing commercial 'Geranium oil' and could be considered the parent of some of the Egyptian and Moroccan Geranium oils due to its content of 10-epi-eudemol. A common hybrid is 'Rober's Lemon Rose', with rather jagged, irregular leaves and a distinctive rose-like odour with a hint of lemon. The parentage is said to involve *P. graveolens* (Abbott, 1994). *P. radens*, also known as as the cultivar 'Radula' could be placed in this group or the 'minty' group, (similarly to *P. graveolens*) depending on the time the sample is taken: if the plant is in flower, then the rosy odour is superior to the mintiness of the essential oils at other times.

Minty: The 'minty' group (Table 14.2) includes *P. tomentosum* Jacq. with over 90 per cent of menthone and isomenthone and also one of its hybrids (with *P. quercifolium* (L.f) L'Herit.) giving the more pungent 'Chocolate tomentosum'. A cultivar with a preponderance of isomenthone is 'Lady Plymouth', but it also sometimes has a rosy odour. It was said to be the Geranium capitatum variegatum (Abbott, 1994) illustrated and described by Andrews (1805–1806).

Peppery-pungent: The 'peppery-pungent' group (Table 14.3) includes *P. vitifolium* (L.) L'Herit., *P. papilionaceum* (L.) L'Herit. and the cultivar 'Sweet Rosina' whose parent was stated by the hybridiser to be *P. graveolens* (Lis-Balchin and Roth, 1999). All three contain a very distinctive, pungent and lasting component called citronellic acid, which has also beeen found in one essential oil sample from *P. 'filicifolium'* (Lis-Balchin *et al.*, 1999).

Citrusy: The 'citrusy' group (Table 14.4) include *P. scabrum* (Burm.f.) L'Herit., *P. citronellum* J.J.A. v.d.Walt and *P. crispum* Berg with a variegated form being often commercially available. The *Pelargonium* species differ in the size and leaf shape but have a uniquely similar citrus odour. The main components are geranial and citronellol. *P. scabrum* has very rough, large tri-lobed leaves and forms a bush sometimes reaching over a metre in height; *P. citronellum* has five-lobed leaves, smaller than previously and *P. crispum* has very tiny leaves on a woody, short bush which can be grown upright in various shapes by professionals. A common cultivar is 'Lemon fancy', which was raised by Helen Bowie in 1974 from 'Prince of Orange' and 'Mabel Grey' (Abbott, 1994), the latter being regarded as synonymous with *P. scabrum*.

Camphoraceous-pungent: The 'camphoraceous-pungent' pelargoniums (Table 14.5) include *P. exstipulatum* (Cav.) L'Herit., *P. odoratissimum* (L.) L'Herit. and *P.* × *fragrans* Willd. The latter is said to be a hybridization product of the first two species (Sweet, 1826) and the chemical composition seems to add weight to this proposal (Lis-Balchin and Roth, 2000, 2002). *P. quercifolium* (L.f.) L'Herit. and its various hybrids from 'Village Hill Oak' to 'Clorinda' and 'Copthorne' (which included also Regals in the hybridization) are put into this hotch-potch group of pungent odouriferous *Pelargonium*, many of which have similarities with Vick's ointment and which also includes *P. glutinosum* Jacq. The cultivar 'Sweet Mimosa' derived from *P. graveolens* is included. The very unusual P. *grossularioides* L. is also included, as it has a pungent, lemony smell; this is a tiny-leaved species which sprawls across the ground and is the only Pelargonium essential oil found to have a spasmogenic action on smooth muscle and the uterus (Lis-Balchin and Hart, 1994) and was used as an abortifacient by Zulus and Boers. The main components in this group include monoterpenes, with methyl eugenol and fenchone in *P. odoratissimum* and *P.* × *fragrans* as well as *P. grossularioides*, and, *p*-cymene featuring in *P. quercifolium* and the 'Oaks' (its hybrids).

TRICHOMES IN *PELARGONIUM* SPECIES AND CULTIVARS

As the essential oils are stored in trichomes on the leaves of the scented pelargoniums, there was a strong possibility that the shape, size and density of trichomes could be indicative of the resultant odour of each *Pelargonium* and a simple means of classification both of odour and also inherent biological properties.

A study of 133 species and subspecies of *Pelargonium* trichomes by Oosthuizen (1983) using a dissecting microscope showed numerous classes of indumentum, density and trichome types. The author described the indumentum as ciliated (thin hairs of equal size), glabrous (without trichomes), glandular (covered with glandular hairs), hirsute (covered with stiff, long, straight hairs), pubescent (short, thin, soft hairs) and other intermediate categories. The density varied from sparse to very dense. Eleven types of trichomes were described and illustrated by drawings. Both non-glandular and glandular trichomes were described. Oosthuizen (1983) stated that the glandular hairs had a uniserial stalk, of various lengths, and a unicellular head of various shapes. The heads were either globular, bulb-shaped or pear-shaped. Some of the non-glandular trichomes had a short, straight, stiff hair with or without a basal podium, while others were described as soft, with/without podium etc.

Oosthuizen (1983) stated that in the Hoarea Section, the petioles and laminae are mainly strigose (covered with stiff hairs with a sharp point, the hairs being compressed to the surface and orientated in a distal or proximal direction). In *Pelargonium*, *Glaucophyllum* and *Eumorpha*, it was hispid to setose and in *Cortusina* it was pilose, whilst in *Otidia* it was sericeous. The author stated that long glandular hairs occurred especially in the sections *Hoarea*, *Ciconium*, *Pelargonium* and *Polyactium* and were lacking in *Seymouria*, *Otidia* (except *P. carnosum*), *Myrrhidium*, *Peristera* (except *P. harveyanum*), *Dibrachya* and *Glaucophyllum*. Glandular hairs with pear-shaped heads occurred in sections *Ligularia*, *Jenkinsonia* and *Myrrrhidium*. Glandular hairs were said to be absent from *P. laevigatum*, *P. lanceolatum*, *P. grandiflorum*, *P. lateripes*. In some species in the section *Polyactium*, *Peristera* and *Glaucophyllum*, non-glandular hairs are completely absent with only glandular ones appearing. Oosthuizen (1983) concluded that there was some degree of classification possible using the trichomes. However, looking at the individual results of trichomes for each species in each section, some significant discrepancies occur and generalizations can only really be made regarding a few species in each section. No clear correlation could be made between the actual general odour of the plant and the actual trichomes present. In particular, the non-odourific leaves of e.g. *Hoarea* still contained glandular hairs, which would presumably have indicated the storage of scented exudates.

A study by van der Walt and Demarne (1988) on the origins of the 'Geranium oil' – producing cultivars included a comparison of the leaves of *P. graveolens* and *P. radens* using the scanning electron microscope. The leaves were dehydrated in a graded series of ethanol, critical point dried with liquid CO_2 and sputter-coated with gold. The authors showed slight differences in the size and shape of two types of glandular hairs in the two species, and also that one type only of non-glandular hairs occurring in *P. graveolens*, whilst two types were present in *P. radens*. Oosthuizen (1983) had described three types of non-glandular hairs in *P. radens*, with two types of glandular hairs but had not studied *P. graveolens*.

One of the problems of studying trichomes in *Pelargonium* is that changes seemingly occur in the morphology of glandular hairs during development and that in *P. scabrum* the different morphological types represent different developmental stages of a single glandular hair type (Oosthuizen and Coetzee, 1983). Their studies, using scanning electron microscopy and transmission electron microscopy. The results therefore throw into disarray, the results obtained by Oosthuizen (1983) regarding classification of the numerous trichomes in the different sections.

Oosthuizen and Coetzee (1983) also showed that the secretion of essential oil from the young trichome occurs repeatedly as a new cuticle develops each time beneath the ruptured one. Glandular hairs are initiated throughout the different stages of leaf development, whilst the non-glandular hairs are only initiated in the young leaf (Oosthuizen and Coetzee, 1983). Because the rate of glandular hair initiation is lower than that of epidermal cell differentiation and enlargement, the indumentum becomes less dense with leaf expansion and therefore age (Oosthuizen and Coetzee, 1984). This again makes comparison between petioles difficult, even within the same species, let alone between different sections as slight changes in development in leaves would give substantial variations in trichome density as well as appearance.

Personal studies of over 500 species and cultivars of *Pelargonium* using the scanning electron microscope, after processing the leaves in glutaraldehyde (to prevent any distortion) and sputter-coating with gold, indicated no precise segregation of pelargoniums based on their trichomes. Both scented and unscented *Pelargonium* leaves contained both glandular and non-glandular hairs (Figures 14.1–14.12). These varied largely during development of the leaf, more so than differences in different species.

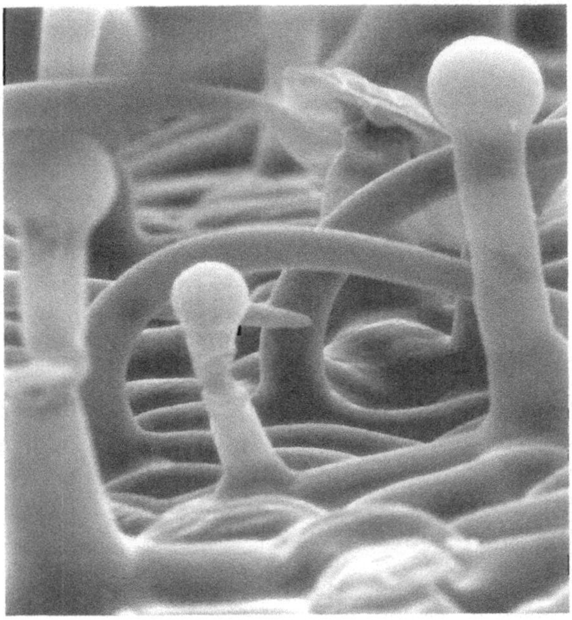

Figure 14.1 P. abrotanifolium – scented. Both glandular and non-glandular trichomes are illustrated.

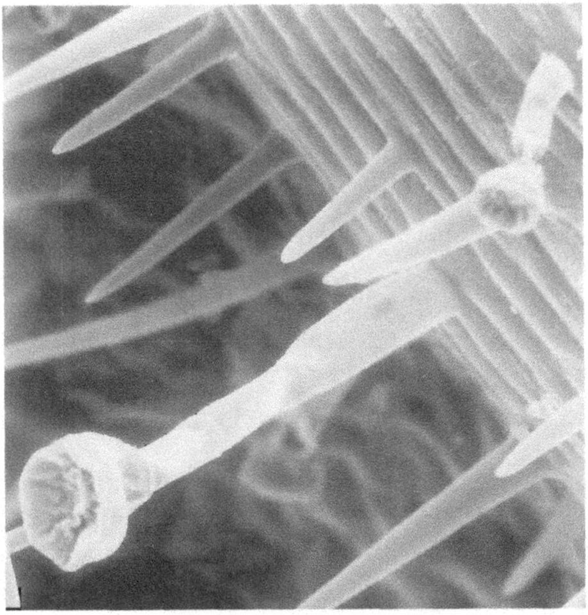

Figure 14.2 P. ribifolium – scented. Both glandular and non-glandular trichomes are illustrated.

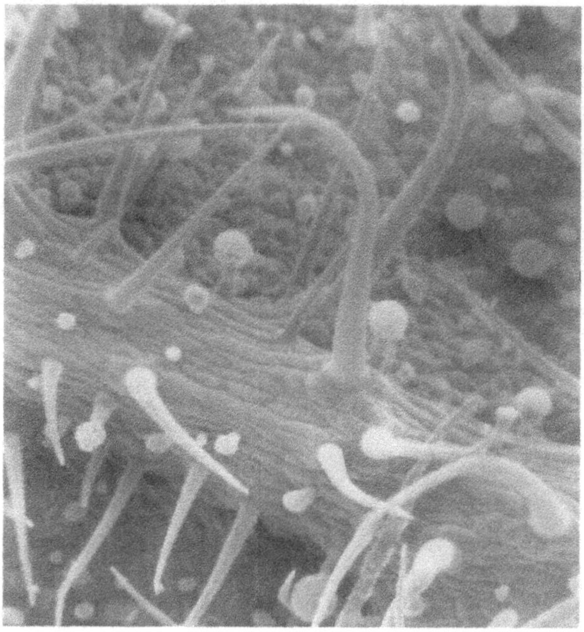

Figure 14.3 *P. tetragonum* – scented. Both glandular and non-glandular trichomes are illustrated.

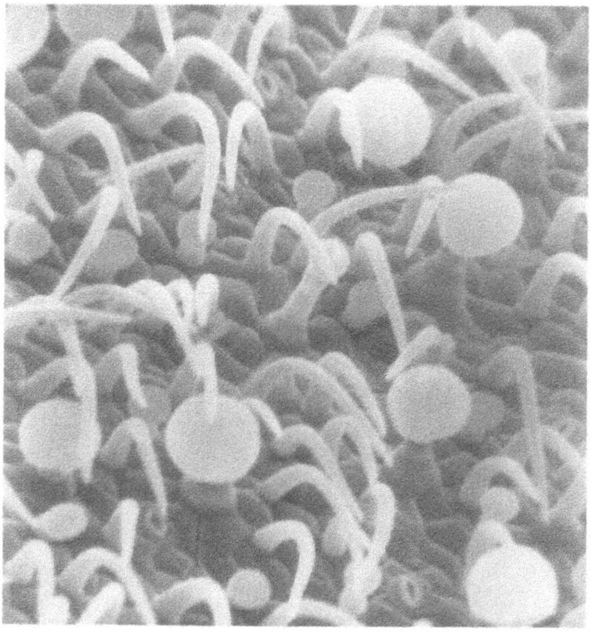

Figure 14.4 *P. graveolens* – scented. Both glandular and non-glandular trichomes are illustrated.

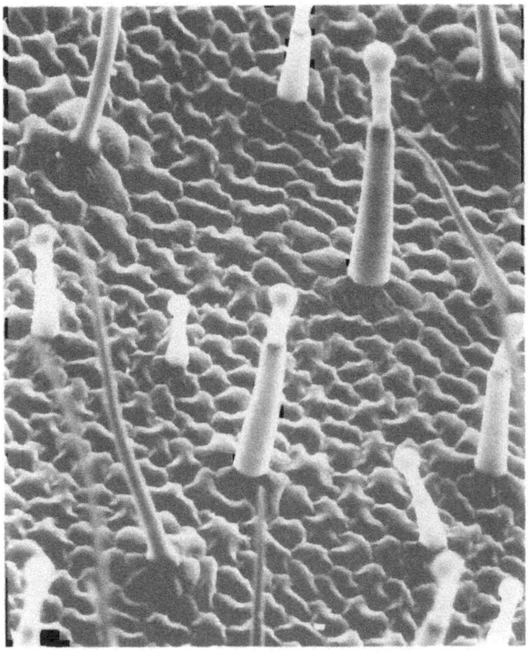

Figure 14.5 'Rober's Lemon Rose' – scented. Both glandular and non-glandular trichomes are illustrated.

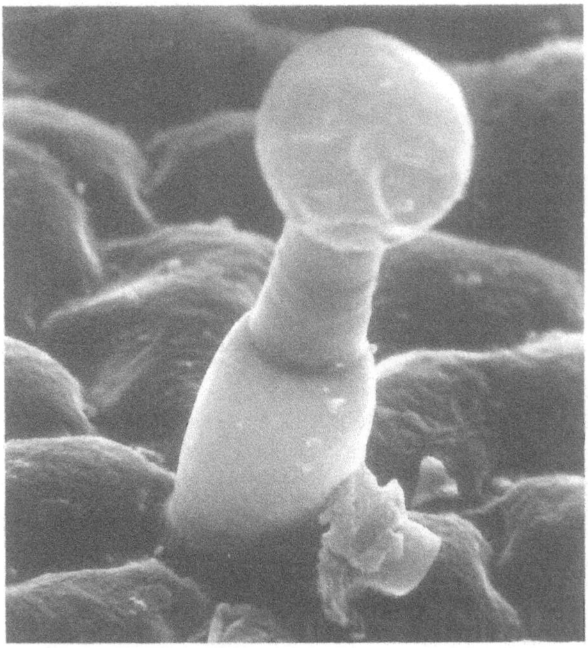

Figure 14.6 *P. barklyi* – unscented. Both glandular and non-glandular trichomes are illustrated where only the glandular hair is shown on the micrograph, although non-glandular hairs were also present.

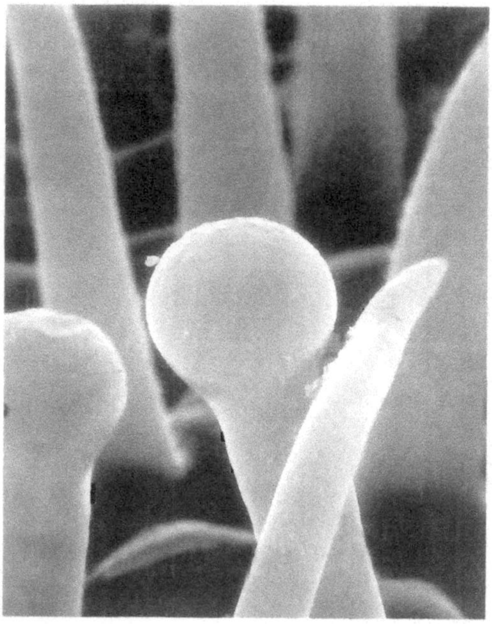

Figure 14.7 *P. triste* – unscented. Both glandular and non-glandular trichomes are illustrated.

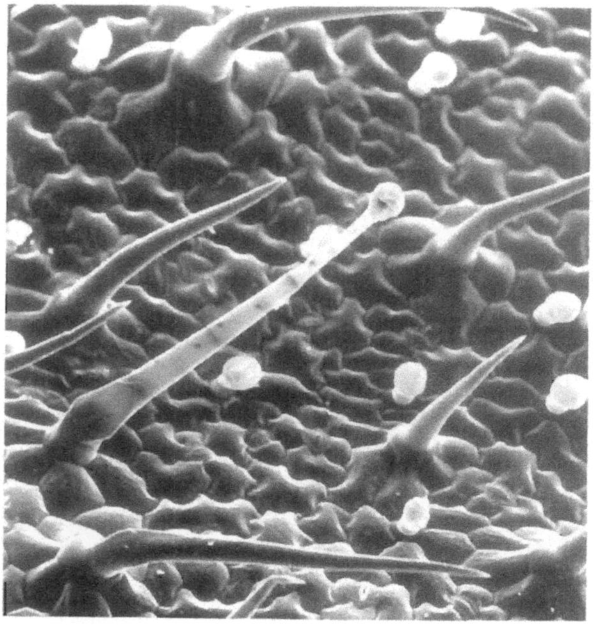

Figure 14.8 *P. tongaense* – unscented. Both glandular and non-glandular trichomes are illustrated.

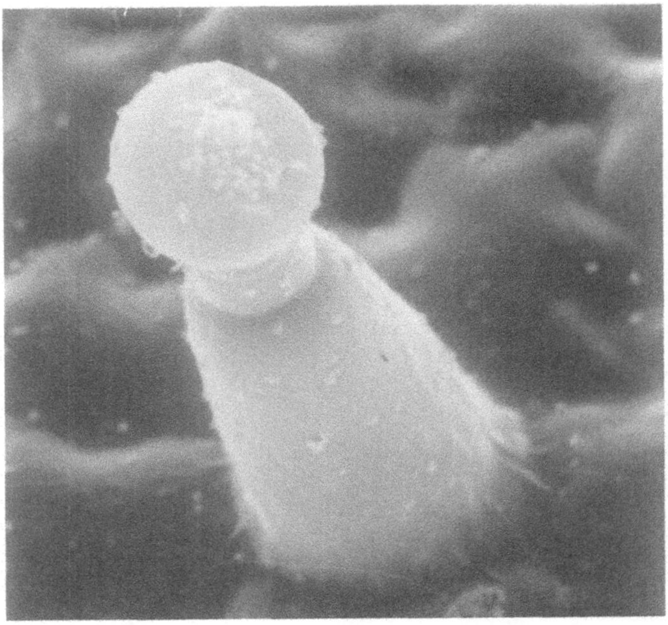

Figure 14.9 *P. rapaceum* – unscented. Both glandular and non-glandular trichomes are illustrated where only the glandular hair is shown on the micrograph, although non-glandular hairs were also present.

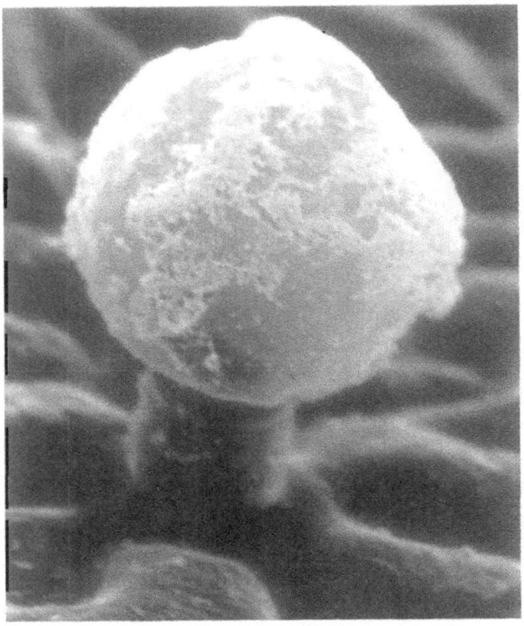

Figure 14.10 *P. grossularioides* – scented. Both glandular and non-glandular trichomes are illustrated where only the glandular hair is shown on the micrograph, although non-glandular hairs were also present.

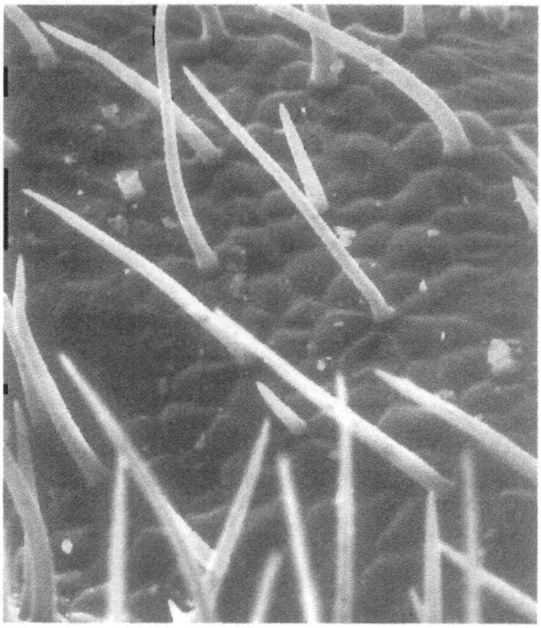

Figure 14.11 P. echinatum – unscented. Both glandular and non-glandular trichomes are illustrated.

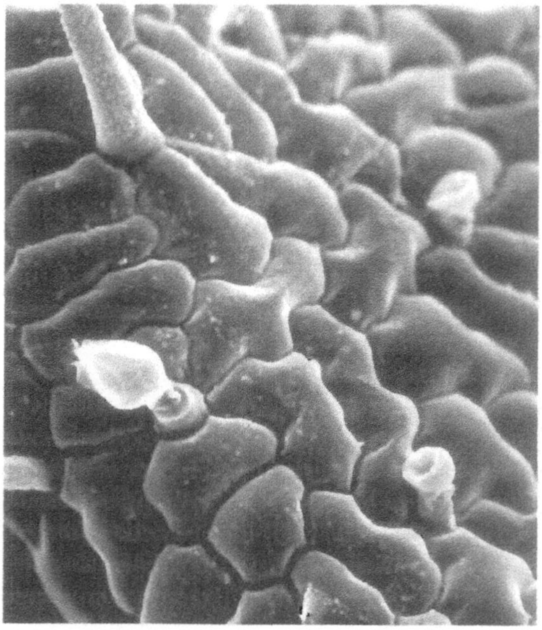

Figure 14.12 P. praemorsum – unscented. Both glandular and non-glandular trichomes are illustrated.

REFERENCES

Abbott, P. (1994) *A guide to Scented Geraniaceae*. Hill Publicity Services, West Sussex.

Andrews, H.C. (1805–1806) *Geraniums*, vol. 1 & 2, London.

Demarne, F. (1989) *L'Amelioration varietale du 'Geranium rosat'* (Pelargonium *sp.*). *Contribution systematique, caryologique et biochimique*. Thesis: Doctor of Science. Universite de Paris-Sud, Centre D'Orsay.

Demarne, F., Garnero, J. and Mondon, J.-M. (1986) L'Huile essentielle de *Pelargonium tomentosum* Jacquin (Geraniaceae). *Parf. Cos. Aromes.*, 70, Aug–Sept., 57–60.

Demarne, F. and van der Walt, J.J.A. (1990) Origin of the rose-scented *Pelargonium* cultivar grown on Reunion Island. *S. Afr. J. Bot.*, 55, 184–191.

Lis-Balchin, M. (1991a) Essential oil profiles and their possible use in Hybridization of some common scented Geraniums. *J. Essent. Oil Res.*, 3, 99–195.

Lis-Balchin, M. (1991b) The essential oil of *Pelargonium grossularioides* and *Erodium cicutarium* (Geraniaceae). *J. Essent. Oil Res.*, 5, 317–318.

Lis-Balchin, M. and Hart, S. (1994) A pharmacological appraisal of the folk medicinal usage of *Pelargonium grossularioides* and *Erodium cicutarium* (Geraniaceae). *Herbs, Spices Med. Plants*, 2, 41–48.

Lis-Balchin, M., Deans, S.G. and Hart, S. (1996) Bioactivity of commercial Geranium oil from different sources. *J. Essent. Oil Res.*, 8, 281–290.

Lis-Balchin, M. and Hart, S. (1997a) Pharmacological effect of esssential oils on the uterus compared to that on other different tisue types. In: Ch. Franz, A. Mathé and G. Buchbauer (eds), *Proc. 27th Int. Symp. Ess. Oils*, Vienna, Austria, 8–11 Sept., 1996.

Lis-Balchin, M. and Hart, S. (1997b) A preliminary study of the effect of essential oils on skeletal and smooth muscle *in vitro*. *J. Ethnopharmacol.*, 58, 183–187. Allured Pub. Corp., Carol Stream, III., pp. 24–28.

Lis-Balchin, M., Hirtenlehner, T. and Resch, M. (1998) Antimicrobial activity of novel Pelargonium essential oils added to a quiche filling as a model food system. *LAM*, 27, 207–210.

Lis-Balchin, M. and Hart, S. (1999) Studies on the mode of action of the essential oils of scented-leaf *Pelargonium* (Geraniaceae). *Phytother. Res.*, 12, 215–217.

Lis-Balchin, M. and Roth, G. (1999) Citronellic acid: a major component in two *Pelargonium* species (Geraniaceae). *J. Essent. Oil Res.*, 11, 83–85.

Lis-Balchin, M., Hart, S. and Roth, G. (1999) The pharmacological activity of the essential oils of scented *Pelargonium* (Geraniaceae). *Phytother. Res.*, 11(8), 83–840.

Lis-Balchin, M. and Roth, G. (2000) Composition of the essential oils of *P. odoratissimum*, *P. exstipulatum* and *P.* × *fragrans* (Geraniaceae). *Flav. Fragr. J.*, 15, 391.

Lis-Balchin, M. and Roth, G. (2002) The chemical composition of selected *Pelargonium* essential oils with reference to their pharmacological action. *Flav. Fragr. J.*, To be published.

Lis-Balchin, M. Master's Thesis Supervisor (Magister der Pharmazie) Pharmacy Department, University of Vienna, Austria [Gerhild, Roth (July, 1997), Ribisch, K. (September, 1998a), Wenger, M.-T. (September, 1998b), Brandstetter, A. (September, 2001), Groiss, S. (January, 2002)].

Oosthuizen, L.-D. (1983) The taxonomic value of Trichomes in *Pelargonium* L'Herit. (Geraniaceae). J. *S. Afr. Bot.*, 49, 221–242.

Oosthuizen, L.-D. and Coetzee, J. (1983) Morphogenesis of trichomes of *Pelargonium scabrum*. *S. Afr. J. Bot.*, 2, 305–310.

Oosthuizen, L.-D. and Coetzee, J. (1984) Trichome initiation during leaf growth in *Pelargonium scabrum*. *S. Afr. J. Bot.*, 3, 50–54.

Sweet, R. (1820–1830) *Geraniaceae*, vol. 1–5, London.

van der Walt, J.J.A. and Demarne, F. (1988) *Pelargonium graveolens* and *P. radens*: A comparison of their morphology and essential oils. *S. Afr. J. Bot.*, 54, 617–622.

15 Chemotaxonomy of *Pelargonium* based on alkaloids and essential oils

Peter Houghton and Maria Lis-Balchin

INTRODUCTION

There have been few chemotaxonomic studies of the genus *Pelargonium* and the results of studies on tannins and flavonoids (Asen and Grisbach, 1983; Bate-Smith, 1973, 1981; Marszewski, 1990) and unidentified phenolic compounds (Harney, 1966, 1976) were of limited value in the allocation of species into sections. Other components studied included tartaric acid (Stafford, 1961), which proved to be unsatisfactory as chemotaxonomic indices. Essential oils were studied in *Pelargonium* species and cultivars by Demarne and van der Walt (1989), Lis-Balchin (1991), Lis-Balchin *et al.* (1998a,b) and Lis-Balchin and Roth (1999) and proved of value in classification, but only according to odour. The detection of alkaloids in some *Pelargonium* species suggested that this criterion could be used as a chemotaxonomic marker (Lis-Balchin, 1993, 1996). Prior to the latter work, alkaloids had only been detected in some *Erodium* species (Medina *et al.*, 1977; Mossa *et al.*, 1983; Lis-Balchin and Guittonneau, 1995).

Alkaloid Chemotaxonomy

The *Geraniaceae* is not known as an alkaloid-producing family although alkaloids have been detected in phytochemical screening of *Erodium glaucophyllum* from Saudia Arabia (Mossa *et al.*, 1983) and *E. malacoides* from Argentina (Medina *et al.*, 1977). Traces of caffeine and choline had previously been found in the aerial parts of *Erodium cicutarium* (van Eijk, 1952a), a herb used by the Sotho in South Africa as a uterine tonic but caffeine was not detected in *Geranium molle* (van Eijk, 1952b).

Lis-Balchin and Guittoneau (1995) screened the leaves of 38 species of *Erodium* for the presence of alkaloids and found different alkaloidal thin-layer chromatography (TLC) profiles for the sections recognised within the genus. A single major compound was detected in the section *Malacoidea* although no alkaloids were observed in extracts from the subsection *Gruina*. The pattern in other sections was more variable and no clear correlations could be established.

Similar screening studies on *Pelargonium* species had revealed the presence of alkaloids, especially in 'zonal' cultivars. The identity of the alkaloids detected was not determined until studies by Lis-Balchin *et al.* (1996) on the cultivar 'Appleblossom'. The simple amines tyrosine 1 and tryptamine 2 were detected by TLC comparison with authentic samples but of more interest was the isolation of three other, more complex, indole alkaloids. The structures of two of these were determined by spectroscopic methods and they were shown to be the two isomers elaeocarpidine 3 and isoelaeocarpidine 4.

These were known compounds which had previously been isolated from *Elaeocarpus* (*Elaeocarpaceae*) (Johns and Lamberton, 1973) and *Tarenna vanprukii* (*Rubiaceae*) (Takayama *et al.*, 1992). Neither of these two families are considered to be closely related to the *Geraniaceae* in classical systematic botany so the presence of these alkaloids is likely to be the product of some process of convergent evolution.

Method of extraction and detection

The method used involved: extraction of the fresh leaves with boiling water, extracting the initial acidic chloroform extract and subsequently after basifying with saturated sodium carbonate solution to pH 10, extracting the alkaline chloroform fraction. The extracts were separated on silica gel plates using two solvent systems; and sprayed with Dragendorf's reagent or iodoplatinate reagent. Three to four major spots were obtained, which were very stable, remaining bright yellow for months on the plates; other spots were less stable. The general pattern proved consistent for all the species which contained the alkaloids.

Chemotaxonomy of the genus *Pelargonium*

The study of representative species of all Sections of the genus *Pelargonium* indicated that there was a main chemotaxonomic division in the genus. (Table 15.1), based on the presence or absence of the alkaloids (Lis-Balchin, 1995). The latter were found in just a few of the 44 species studied from all the sections: many of the species in the *Ciconium* Section but otherwise only in the root of *P. antidysentericum* ssp. *inerme*, Section *Jenkinsonia*, root of *P. radulifolium*, Section *Polyactium* and leaves of *P. fulgidum*, Section *Ligularia*.

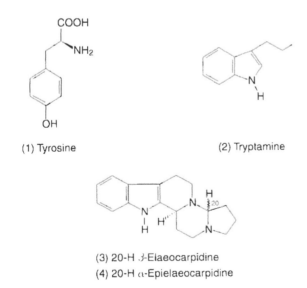

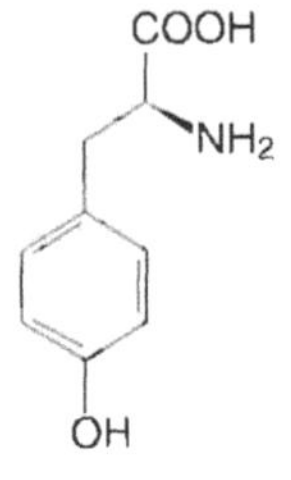

(1) Tyrosine

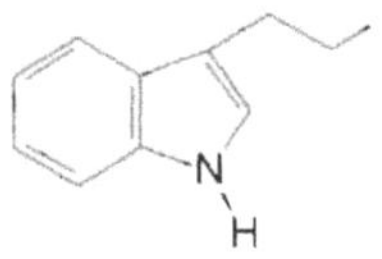

(2) Tryptamine

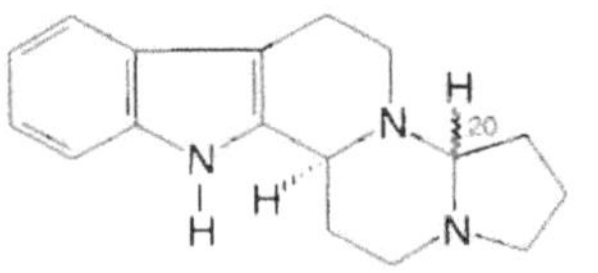

(3) 20-H β-Elaeocarpidine

(4) 20-H α-Epielaeocarpidine

Figure 15.1 The structure of elaeocarpidine and related compounds.

Table 15.1 A survey of the presence of alkaloids in representative species in different sections of *Pelargonium*

Section	*Species*
Campylia (Sweet) D.C.	*P. elegans* (Andr.) Willd
Ciconium (Sweet) Harv.	*P. acetosum* (L.) L'Hérit. + *P. alchemilloides* (L.) L'Hérit. sensu lato *P. inquinans* (L.) L'Hérit. + *P. tongaense* Vorster *P. transvaalense* Knuth *P. zonale* (L.) L'Hérit. +
Cortusina D.C.	*P. odoratissimum* (L.) L'Hérit. *P. reniforme* Curt. sensu lato
Dibrachya (L.) L'Hérit.	*P. peltatum* (L.) L'Hérit.
Glaucophyllum Harv.	*P. grandiflorum* (Andr.) Willd *P. lanceolatum* (Cav.) Kern
Hoarea D.C.	*P. appendiculatum* (L.f.) Willd *P. punctatum* (Andr.) Willd
Isopetalum (Sweet) D.C.	*P. cotyledonis* (L.) L'Hérit.
Jenkinsonia (Sweet) Harv.	*P. antidysentericum* ssp. inerme + (root) (Eckl. and Zeyh.) Costel *P. antidysentericum* ssp. antidysentericum (root) (Eckl. and Zeyh.) Costel *P. tetragonum* (L.f.) L'Hérit.
Ligularia (Sweet) Harv.	*P. fulgidum* (L.) L'Hérit. + *P. pulchellum* Sims- *P. mollicomum* Fourcade
Myrrhidium D.C.	*P. multicaule* Jacq.
Otidia (Sweet) G. Don	*P. carnosum* (L.) L'Hérit. sensu lato *P. dasyphyllum* E. Mey. ex Knuth
Pelargonium (D.C.) Harvey	*P. betulinum* (L.) L'Hérit. *P. capitatum* (L.) L'Hérit. *P. citronellum* J.J.A.v.d. Walt *P. cucullatum* subsp. tabulare (L.) L'Hérit. *P. denticulatum* Jacq. *P. graveolens* L'Hérit. *P. hispidum* (L.f.) Willd *P. tomentosum* Jacq. *P. vitifolium* (L.) L'Hérit. *P. 'viscossimum'* *P. 'filicifolium'*
Peristera D.C.	*P. australe* Willd *P. grossularioides* (L.) L'Hérit. sensu lato
Polyactium (Eckl. and Zeyh.) D.C.	*P. luridum* (Andr.) Sweet sensu lato- *P. radulifolium* (root) (Eckl. and Zeyh.) Steud. + *P. triste* (L.) L'Hérit.

Chemotaxonomic revision of the *Ciconium* Section

Studies on the Section *Ciconium* species (Lis-Balchin, 1996) showed a 50 per cent division in the presence of alkaloids (Table 15.2). The results strongly suggested that species which were recently re-allocated from other sections into *Ciconium* do not

Table 15.2 The presence of alkaloids in Section *Ciconium*

Alkaloid-positive species	*Alkaloid-negative species*
P. zonale	*P. alchemilloides*
P. acraeum	*P. elongatum*
P. scandens	*P. transvaalense*
P. frutetorum	*P. multibracteum*
P. salmoneum	*P. quinquelobatum*
P. ranunculophyllum	*P. caylae*
	P. mutans
	P. tongaense

comply with the chemotaxonomy of the original *Ciconium* (Sweet) Harv., members. These include *P. alchemilloides*, *P. elongatum* and *P. transvaalense* (all lately in Eumorpha); *P. multibracteum*, *P. quinquelobatum*, *P. caylae*, *P. mutans* and *P. tongaense* also did not comply. The remaining species with true Ciconium characteristics (and containing alkaloids) are: *P. zonale*, *P. inquinans*, *P. acraeum*, *P. scandens*, *P. frutetorum*, *P. salmoneum* and *P. ranunculophyllum*. The fact that *P. zonale*, *P. inquinans* and *P. frutetorum* were alkaloid-positive is consistent with their use in the development of *P.* × *hortorum* cultivars which are all alkaloid-positive.

The chemotaxonomic separation of *P. alchemilloides* from *P. ranunculophyllum* supports the morphological findings and cytological differences observed (Gibby, 1989). The alkaloid-negative species have a wider distribution than the alkaloid-positive species, which are found in the Cape area, e.g. *P. caylae* is found as far as Madagascar. The chemotaxonomic division found is supported by recent botanical studies (Vorster, 1993). The presence of alkaloids suggests a more evolved characteristic, possibly in response to the presence of non-beneficial insects or other predators, as alkaloids are largely unpalatable.

Chemotaxonomy of modern cultivars

The modern cultivars, known as geraniums (Table 15.3) showed the presence of the alkaloids in *P.* × *hortorum* Bailey cultivars (zonals), but absent from *P.* × *domesticum* Bailey (regals) and *P.* × *peltatum* L'Heritier (ivy-leaf). The zonal × ivy-leaf hybrids also contained the alkaloids in their leaves regardless of whether they had an ivy-leaf appearance (e.g. *Millfield Rose*, *Jack of Hearts*, *Elsi*) or zonal appearance (e.g. Deacons), indicating a dominant trait (Lis-Balchin, 1997).

All zonals showed a similar alkaloid pattern regardless of the leaf colouration etc. including the various golden-leaf cultivars, bronze-leaf cultivars, dark-leaf stellar cultivars, the Deacons and Highfield cultivars, the mesh-leaf cultivar 'Wantirna', the Boar, 'Magic Lantern', 'Red Kewense' and the white-stemmed 'Freak of Nature' and shiny-leaf 'Skelly's Pride'.

The results support the theory that *P.* × *domesticum* cultivars and *P.* × *hortorum* cultivars arose from different species. The lack of alkaloids in the *P.* × *domesticum* cultivars supports the theory that *P. cucullatum* and *P. capitatum* were the probable parents, together with a large number of other possible parents from a number of sections e.g. *P. fulgidum* (Ligularia).

Table 15.3 Modern *Pelargonium* cultivars assessed for presence of alkaloids

P. × *hortorum* cultivars (all +ve)
(Zonals)
Highfield's Symphony
Highfield's Delight
Highfield's Perfection
Penny (Irene)
Christopher Lee
Burgenland Girl
Mr Wren
Magic Lantern (*P. frutetorum* cultivar)
White Boar (*P. frutetorum* cultivar)
Crystal Place gem (ornamental)
Distinction (ornamental)
Bridesmaid (golden-leaf)
Dovedale (golden-leaf)
Appleblossom (Rosebuds)
Skelly's Pride (shiny-leaf)
Wantirna (mesh-leaf)
Bird Dancer (steller)
Redondo (dwarf, dark-leaf)
Fire Dragon (cactus-flower)
Els (stellar)
Red Star (stellar)
P. × *domesticum* cultivars (all −ve)
(Regals)
Judith Thorpe
Fringed Aztec
Mosaic Belle Notte
Noche
P. peltatum cultivars (all −ve)
(Ivy-leaved)
Crocketta
Madame Crousse
Feuerriesse
Sugar Baby
Hybrid ivy-leaf (all +ve)
Ivy-leaf appearance
Millfield rose
Jack of Hearts
Elsi
Auden Ken an Emil Eschbach
Zonal appearance
Deacon Moonlight
Deacon Birthday
Deacon Ragatta
Deacon Regalia

The parents of *P.* × *hortorum* cultivars were obviously derived from the alkaloid-positive members of the *Ciconium* Section e.g. *P. acraeum*, *P. inquinans* and *P. zonale*. These all originate from South Africa, mainly from the Cape, and had presumably evolved the ability to synthesise alkaloids in response to either the original geographical location or more likely to a biological effect. A local infestation by predators for example could have provoked the synthesis of phytoalexins: in this case alkaloids. With time, alkaloids could have become a regular secondary metabolite and therefore the information was incorporated into the DNA. It is known that the earliest species of Ciconium exported from the Cape was *P. zonale* and this was presumably one of the major hybridisation parents of *P.* × *hortorum* cultivars.

The ability to synthesise alkaloids is apparently a strong characteristic as illustrated by the inheritance of alkaloids from the *P.* × *hortorum* parent in the hybrid ivy-leaf pelargoniums – both those with a zonal appearance and those with an ivy-leaf appearance.

Possible insecticidal function of alkaloids

The presence of these alkaloids, identified as elaeocarpidine alkaloids (Lis-Balchin *et al.*, 1996) in zonal *Pelargonium* suggested an insecticidal action, as the latter are more resistant to whitefly than the regals and the alkaloids were distributed at the zones, associated with tannins, and whitefly appeared to desist from staying on these areas (Woldemariam *et al.*, 1997).

Citronellic acid

This unusual component has been found so far in just two *Pelargonium* species, *P. papilionaceum* L. L'Herit. and *P. vitifolium* L. L'Herit. and one cultivar, P. 'Sweet Rosina' (Lis-Balchin and Roth, 1999). Both the species and the cultivar have a very acrid, persistent odour, mildly reminiscent of the he-goat or 'Rambossie', which is the vernacular name given to *P. papilionaceum* in South Africa. The cultivar is said to be a cross between *P. capitatum* and *P. graveolens*: (Keys, 2000) both of which are implicated in the parentage of the rose-like Geranium oil of commerce (Dermarne and van der Walt, 1989).

P. papilionaceum is related both to *P. vitifolium* and *P. capitatum* (van der Walt and Vorster, 1988). The former was used as a tobacco substitute in South Africa and smoked, possibly for medicinal purposes, as well (Watt and Breyer-Brandwijk, 1962). The percentage of citronellic found in *P. vitifolium* and *P. papilionaceum* essential oils was about 79–86 per cent. This component was also obtained in similar quantities by solvent extractions of the species using hexane or petroleum spirits, therefore it was shown not to be an artefact of steam distillation (Lis-Balchin and Roth, 1999). The percentage in 'Sweet Rosina' was 28–43 per cent depending on extraction technique.

Citronellic acid was also identified in considerable quantities in one sample of steam-distilled '*P. filicifolium*', but this remained unconfirmed in other extractions (Lis-Balchin personal research).

Citronellol and geranic acid

Citronellol was found at about 1 per cent in *P. vitifolium* and *P. papilionaceum*, but was between 15 and 43 per cent in the cultivar. Geranic acid was found at 0–5 per cent in *P. papilionaceum*, it was higher in the cultivar (10–13 per cent), and absent in *P. vitifolium*.

The results offer strong support for the implication of *P. papilionaceum* and *P. vitifolium* as a parent of the cultivar, although the actual breeder was adamant that only *P. capitatum* (the rose-scented parent of commercial Geranium oil) and *P. graveolens* (another possible parent of a different cultivar for Geranium oil production) was involved in the cross. It is noteworthy that many *Pelargonium* crosses bear no resemblance to either parent.

The citronellic containing *Pelargonium* species and cultivars could possibly be used to produce a very strong fixative for perfumery, as other possibilities of use, including antimicrobial and pharmacological investigations have not shown any other attributes to date.

Further differences in the essential oil composition of scented *Pelargonium* species and cultivars are found in Chapter 14.

REFERENCES

Asen, S. and Grisbach, R. (1983) High pressure liquid chromatographic analysis of flavonoids in geranium florets as an adjunct for cultivar identification, *J. Am. Soc. Hort. Sci.*, 108, 845–850.

Bate-Smith, E.C. (1973) Chemotaxonomy of Geranium. *Bot. J. Linn. Soc.*, 67, 347–359.

Bate-Smith, E.C. (1981) Astringent tannins of the leaves of *Geranium* species. *Phytochem.*, 20, 211–216.

Demarne, F. and van der Walt, J.J.A. (1989) Origin of the rose-scented *Pelargonium* cultivar grown on Reunion Island, *S. Afr. J. Bot.*, 55, 184–191.

Gibby, M. (1989) *Pelargonium ranunculophyllum* (Geraniaceae) in Southern Africa. *S. Afr. J. Bot.*, 55, 539–542.

Harney, P.M. (1966) A chromatographic study of species presumed ancestral to *Pelargonium* × *hortorum* Bailey. *Canad. J. Genet. Cytol.*, 8, 780–787.

Harney, P.M. (1976) The origin, Cytogenetics and Reproductive Morphology of the zonal Geranium: A Review. *Hort. Sci.*, 11, 189–194.

Johns, S.R. and Lamberton, J.A. (1973) *Elaeocarpus* alkaloids. *Alkaloids*, 14, 325–346.

Keys, H. (2000) National Pelargonium Collection, Fibrex Nurseries, Pebworth, Warwickshire. Personal communication.

Lis-Balchin, M. (1991) Essential oil profiles and their possible use in hybridisation of some common scented Geraniums. *J. Essent. Oil Res.*, 3, 99–105.

Lis-Balchin, M. (1993) A chemotaxonomic study of the *Pelargonium* (Geraniaceae) species and their modern cultivars. *J. Hort. Sci.*, 72, 791–795.

Lis-Balchin, M. (1996) A chemotaxonomic reappraisal of the Section Ciconium *Pelargonium* (Geraniaceae). *S. Afr. J. Bot.*, 62, 277–279.

Lis-Balchin, M. and Guittonneau, G.-G. (1995) Preliminary investigations on the presence of alkaloids in the genus *Erodium* L'Herit. (Geraniaceae). *Acta Bot. Gallica*, 141, 31–35.

Lis-Balchin, M. and Roth, G. (1999) Citronellic acid: a major compound in two *Pelargonium* species (Geraniaceae) and a cultivar. *J. Essent. Oil Res.*, 11, 83–85.

Lis-Balchin, M., Houghton, P. and Woldermariam, T. (1996) Elaeocarpidine alkaloids from *Pelargonium* (Geraniaceae). *Nat. Prod. Lett.*, 8, 105–112.

Lis-Balchin, M., Buchbauer, G., Hirtenlehner, T. and Resch, M. (1998a) Antimicrobial activity of novel Pelargonium essential oils and solvent extracts, *Lett. Appl. Microbiol.*, 27, 135–141.

Lis-Balchin, M., Buchbauer, G., Hirtenlehner, T. and Resch, M. (1998b) Antimicrobial activity of novel Pelargonium essential oils added to a quiche filling as a model food system. *Lett. Microbiol.*, 27, 207–210.

Marszewski, D.E. (1990) Chemotaxonomy of *Pelargonium*, *Proceedings of the International Geraniaceae Symposium*, Stellenbosch University, South Africa 24–26 Sept., (Vorster, P. ed.), pp. 211–214.

Medina, J.E., Rondina, R.V.D. and Coussio, J.D. (1977) Phytochemical screening of Argentine plants with potential pharmacological activity (Part VII). *Planta Med.*, 31, 136–140.

Mossa, J.S., Al-Yahya, M.A., Al-Meshal, I.A. and Tariq, M. (1983) Phytochemical and biological screening of Saudi medicinal plants Part 5 *Fitoterapia*, 54, 147–152.

Stafford, H.A. (1961). Distribution of tartaric acid in the Geraniaceae, *Am. J. Bot.*, 48, 699–701.

Takayama, H., Katsura, M., Seki, N., Kitajima, M., Aimi, N., Sakai, S., Santiarworn, D. and Liawruangrath, B. (1992). Elaeocarpidine, A Naturally Occurring Racemate From *Tarenna vanprukii. Planta Med.*, 58, 289–290.

van Eijk, J.L. (1952a) *Pharm.. Weekblad*, 87, 425 (cited in Hegnauer, R. (1966) *Chemotaxonomie der Pflanzen*, Vol. 4. Birkhauser, Basle, p. 200).

van der Walt, J.J.A. and Vorster, P.J. (1988) *Pelargoniums of Southern Africa*, National Botanic Gardens, Kirstenbosch, S.Africa.

van Eijk, J.L. (1952b) *Pharm. Weekblad*, 87, 70 (cited in Hegnauer, R. (1966) *Chemotaxonomie der Pflanzen*, Vol. 4. Birkhauser, Basle, p. 200).

Vorster, P.J. (1994) Taxonomy of *Pelargonium*, Austrian Pelargonium Soc. Symposium, Burgenland, Aug. 12–14.

Watt, J.M. and Breyer-Brandwijk, M.G. (1962) *The Medicinal Plants of Southern Africa.* Livingston Ltd, Edinburgh.

Woldermariam, T., Houghton, P.J., Lis-Balchin, M. and Simmonds, M.S.J. (1997) Alkaloid and tannin distribution in the leaves of *Pelargonium* zonale with reference to insect behaviour. *Pharmaceut. J.*, 259, 481.

16 Phylogenetical relationship within the genus *Pelargonium* based on the RAPD-PCR method of DNA analysis correlated with the essential oil composition

J. Renata Ochocka, Adam Bogdan, Arkadiusz Piotrowski and Maria Lis-Balchin

INTRODUCTION

DNA markers have recently been extensively used for genetic studies and plant identification. With the introduction of new techniques such as polymerase chain reaction (PCR), restriction fragment length polymorphism (RFLP) or sequencing of DNA molecule, taxonomy based on DNA is one of the most developed approaches to identification of biodiversity (Schierwater *et al*., 1997; Demeke and Adams, 1994, Caetano-Annoles, 1996; Hadrys and Schierwater, 1992). Such studies were also carried out on the *Pelargonium* genus, and also other genera of the *Geraniaceae* family (*Erodium*, *Geranium*, *Monsonia*, *Sarcocaulon*) as well. Price and Palmer (1993) examined chloroplast gen *rbc*L in order to assess kinship within the *Geraniaceae* family and likewise among other families from the *Geraniales*. The same approach was employed by Pax *et al*. (1997) in the analysis of relatedness' among endemic Hawaiian geraniums. Apart from chloroplast DNA (fragment between *trn*L and *trn*F), internal transcribed spacer ITS sequences from rDNA, was the basis for work by Bakker *et al*. (1998) to analyse the phylogenetics of the *Peristera* section of *Pelargonium*.

One of the approaches used during the last few years is PCR-based amplification of arbitrary DNA sequences (Schierwater *et al*., 1997). A technique derived from this is randomly amplified polymorphic DNA RAPD which was used in this study (Demeke and Adams, 1994; Hadrys and Schierwater, 1992; Caetano-Annoles, 1996). The main advantage of RAPD is that no prior knowledge of the genome subjected to analysis is required. The reaction can also be done with nanogram amounts of total genomic DNA. In standard PCR protocols, one pair (two different) primers (one stranded, oligonucleotide fragment of DNA) is used. After shifting the temperature of the assay to that above the denaturation temperature of template DNA and subsequently reducing to an annealing temperature, primers are attached to the template but only at sites whose sequences are complementary to that in the primers. If primers are situated in an inverted direction, after elevating the temperature to about 72 °C, new fragments of

DNA, the same as between sites of annealing primers, are amplified in an exponential fashion. In contrast to standard PCR protocol, researchers do not have to know what the sequences flanking the amplified fragment are in order to construct appropriate primers. The RAPD amplification protocol differs from standard protocols in that only a single primer is employed. Using short (i.e. 10 nucleotides in length) primers, there is a high probability that genomic, template DNA contains several sites close to one another, complementary to the sequence of the primer. The amplification products are resolved according to length by means of gel electrophoresis, stained, and then the visible bands are scored. Bands on the electrophoretic gel are marked as 0 (absence) in some positions, and 1 (present) in others, i.e. converted the data to the binary form. To explain genetic relationships, about ten primers, yielding a total of 100 bands are needed. The data are subsequently subjected to numerical analysis with one of the selected algorithms and plotted as a tree. RAPD analysis provides the knowledge about events studied at various taxonomical levels. Despite limitation regarding reproducibility and sometimes scoring of bands, RAPD is often used because of its simplicity.

Plant material used in the study

The material used was dried leaf tissue of representative species of *Pelargonium: P. 'filicifolium'*, *P. tomentosum*, *P. scabrum*, *P.* 'Lady Plymouth', *P. glutinosum*, *P. 'Chocolate Peppermint'*, *P. 'Atomic Snowflake'*, *P. graveolens*, *P. cucullatum*, *P. quercifolium* and *P. trifidum* which are shown in the table below with their appropriate numbers.

Name	*No.*
P. 'filicifolium' var. of *P. denticulatum* Jacq.	1
P. tomentosum Jacq.	5
P. scabrum L'Herit.	6
P. 'Lady Plymouth'	7
P. glutinosum (Jacq.) L'Herit.	10
P. 'Chocolate Peppermint'	11
P. 'Atomic Snowflake'	12
P. graveolens L'Herit.	13
P. cucullatum subsp. *tabulare* L'Herit.	15
P. quercifolium L'Herit.	16
P. trifidum Jacq.	18

All the species bar *P. trifidum* belong to the Section *Pelargonium*, the latter is in the Section *Ligularia*.

Essential Oil analysis, Genomic DNA Isolation and PCR protocol

Essential Oil analysis by GC and MS was carried out on all the essential oil samples and the total Genomic DNA Isolation was isolated according to the CTAB method (Palmarczyk *et al.*, 1995) from eleven samples as described in the table above. The PCR protocol using the following sequences of primers: RA07-GAAACGGGTG, RA09-GGGTAACGCC, RA17-GACCGCTTGT, RA18-AGGTGGACCGT, RA20-GTTGCGATCC, RN01-CTCACGTTGG resulted in 220 of the total bands.

Gel electrophoresis and data analysing

After the PCR reaction, products were resolved in agarose gel and their positions formed the basis in constructing a matrix consisting of one (when the band was present) or zero (when there was no band, but if it existed at the same position along any of tracks of the other samples). The original data were converted to dissimilarity matrices according to algorithms contained in RAPDistance Package Version 1.04. using various algorithms (given below):

1 2*n11/((2*n11) + n01 + n10) – Dice (Czekanowski, 1913; Dice, 1945; Nei and Li, 1979; Sorensen, 1948).
2 n11/(n – n00) – Jaccard (1901, 1908).
3 n11/(n01 + n10) – Kulczynski 1 (Kulczynski, 1927).
4 0.5*((n11/(n11 + n01)) + (n11/(n11 + n10))) – Kulczynski 2 (Kulczynski, 1927).
5 (n11*n00) – (n01*n10)/sqrt((n11 + n01)*(n01 + n00)*(n11 + n10)*(n10 + n00)) the Phi coefficient or Pearson's Phi coefficient (Sokal and Sneath, 1963).
6 n11/n – Russell and Rao (1940).
7 n11/(n11 + 2*(n10 + n01)) – Sokal and Sneath 1 or Anderberg (Sokal and Sneath, 1963).
8 0.25*((n11/(n11 + n10)) + (n11/(n11 + n01)) + (n00/(n00 + n10)) + (n00/(n00 + n01))) Sokal and Sneath 2 (Sokal and Sneath, 1963).
9 n11/sqrt((n11 + n10)*(n11 + n01)) – Ochiai (1957).
10 n11*n00/sqrt((n11 + n10)*(n11 + n01)*(n00 + n10)*(n00 + n01)) – Sokal and Sneath 3 (1963).
11 ((n11*n00)–(n10*n01))/((n11*n00) + (n10*n01) – Yule and Kendall (1950).
12 0.5*(sqrt((F*F) + (8*F))–F))**(1/n), where F = 2*n11/(nx + ny) – Upholt (1977).
13 'Evolutionary distance estimate' (K) (Li and Graur, 1991).
14 (n11 + n00)/n – Simple Matching (or Apostol) (Apostol *et al.*, 1993).
15 n*(1 – (n11/n)) – Excoffie (Excoffier *et al.*, 1992).
16 (n11 + n00)/(n11 + 2*(n10 + n01) + n00) – Rogers and Tanimoto (1960).
17 (n11 + n00)/(n11 + 0.5)*(n10 + n01) + n00 – Sokal and Sneath (1963).
18 (n11 – (n10 + n01) + n00)/n – Hamman (Spath, 1980).

where: n = the number of band positions
nx = the number of bands present in the track with bands from one sample
ny = the number of bands present in the track with bands from compared sample
n11 = the number of positions where x = 1 AND y = 1
n00 = the number of positions where x = 0 AND y = 0
n01 = the number of positions where x = 0 AND y = 1
n10 = the number of positions where x = 1 AND y = 0.

Trees reflecting relationships of analysed samples were drawn according to the neighbour joining method. Validity of trees was tested by means of permutation tail probability (PTP) test. Unweighted pair-group method using arithmetic averages (UPGMA) was used with the average distance between all pairs of objects in the two different clusters according to Sneath and Sokal (1973) and also weighted pair-group method using arithmetic averages (WPGMA) with the size of the respective clusters used as a weight (Sneath and Sokal, 1973).

The results were compared to the results of factor analysis by means of principal components analysis (PCA) (Dunteman, 1989; StatSoft, Inc., 1995), presented on a 3D plot (unrotated) using the first three principal components that were chosen using the Kaiser criterion (Kaiser, 1960) as well as the scree test (Cattell, 1966).

COMPARISON OF RESULTS OF GENETIC ANALYSIS AND PHYTOCHEMICAL ANALYSES

With algorithms included in the RAPDistance program two kinds of trees were obtained. Algorithms from 1 to 13 (without No. 6) gave similar trees; the length of branches of particular trees were variable, but the positions of the branches were the same. An example of such a tree is shown in Figure 16.1a, drawn according to algorithm 4. Algorithms from 14 to 18 resulted in another group of trees; Figure 16.1b represents the tree based on algorithm 14.

P. '*filicifolium*', *P. glutinosum*, *P.* 'Chocolate Peppermint' and *P. quercifolium* are clustered in all trees so their reciprocal positions seem to be convincing. The same conclusion is drawn as to *P.* 'Atomic Snowflake' and *P. graveolens* with *P. scabrum* and *P. cucullatum* as well. High values of the PTP test was observed both with the first group of algorithms (1–13) or the second (14–18) so none of the plots were preferable. The tree was therefore also drawn using the original set of data (0;1) by the UPGMA method (Fig. 16.2a), calculating the dissimilarity matrix, by the WPGMA method (Fig. 16.2b) and the PCA method (Fig. 16.3). The results obtained with the UPGMA and PCA method were more consistent than with those obtained using the neighbour-joining method based on algorithms from 1 to 13. Taking into account these facts a model of genetic relationships from Figure 16.1a is therefore proposed.

When analysing the contents of 51 compounds in *Pelargonium* essential oils with the application of the PCA method (Fig. 16.4), it is possible to point out a clustering of samples into three subgroups. In one cluster there are: *P.* '*filicifolium*' (P1), *P. glutinosum* (P10) and *P. quercifolium* (P16); in the second and the third there are: *P. scabrum* (P6), *P. cucullatum* (P15) and then *P. tomentosum* (P5), *P.* 'Lady Plymouth' (P7), *P.* 'Chocolate Peppermint' (P11) and *P. graveolens* (P13) respectively (Table 16.1).

The genetic results presented in the PCA diagram (Fig. 16.3) also point out the clustering of the samples investigated. Similarity of the genetic and phytochemical profiles can be seen for samples P6 (*P. scabrum*) and P15 (*P. cucullatum*).

The results obtained initially for *Pelargonium*, indicate that the RAPD–PCR method may be very useful for comparison of genetic changes with chemical composition of plants.

COMPARISON OF RESULTS WITH THE PHYLOGENY OF PELARGONIUM USED

There was a consistent clustering of *P.* 'Atomic Snowflake' and *P. graveolens* on all the cladograms. This indicates the close genetic relationship, as the former is a hybrid of *P. graveolens*. The chemical analysis of *P.* 'Atomic Snowflake' was not available, but the odour partly resembles the mintiness and rosyness of *P. graveolens* with an extra persistent acrid odour which resembles that of *P. papilionaceum* containing citronellic acid (Lis-Balchin *et al.*, 1999).

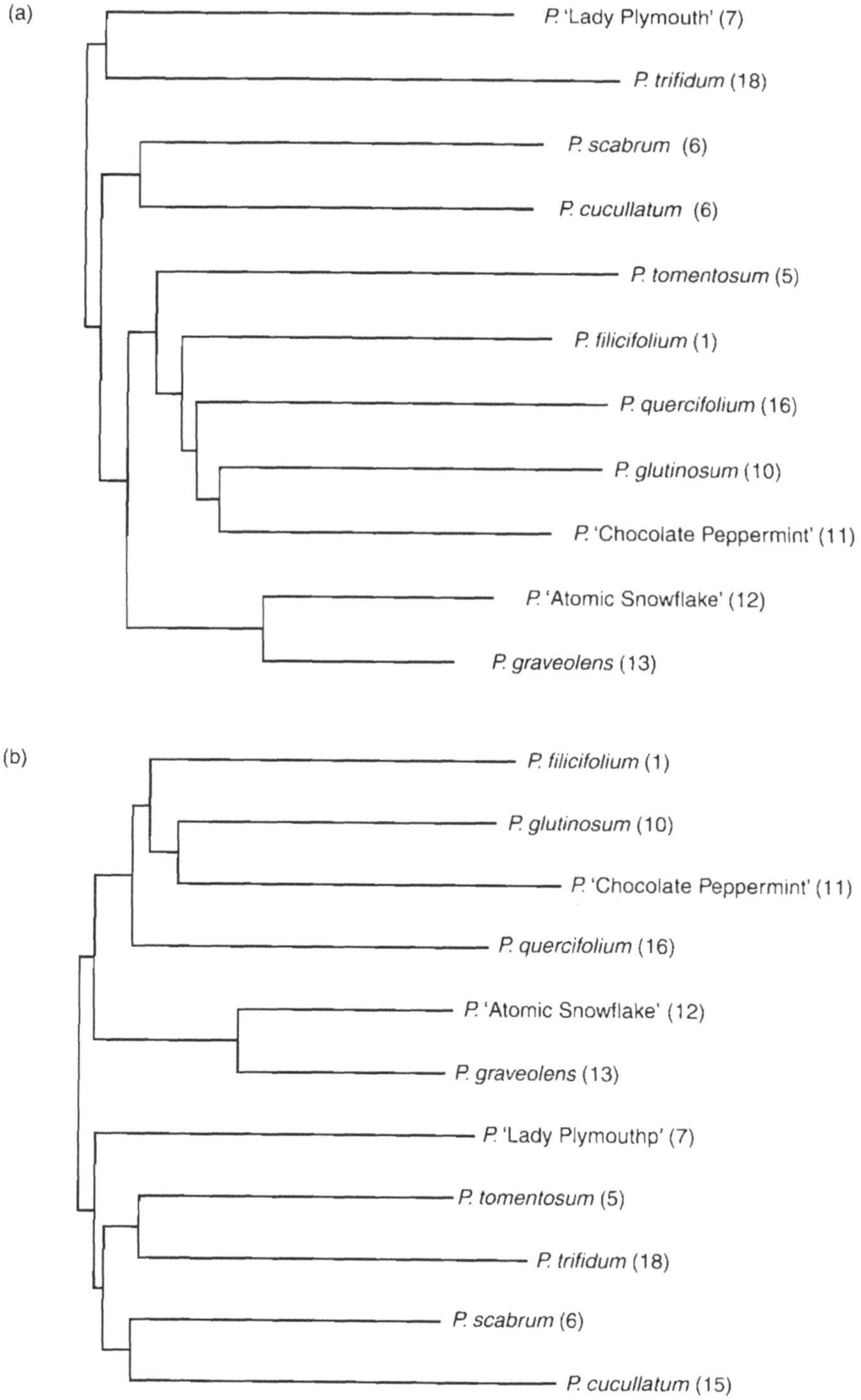

Figure 16.1 a,b The tree diagram obtained by the neighbour joining method for RAPD dataset: (a) using algorithm 4 included in the RAPDistance program (b) using algorithm 14 included in the RAPDistance program.

Table 16.1 Concentration of Essential oil components in various *Pelargonium* species

Compounds	*Essential oils concentration (%)*								
	1	5	6	7	10	11	13	15	16
cis-3-Hexanol	0	0	1.7	0	0	0	0	0.63	0.37
α-Thujene	0	0	0	0	0	0	0	0	0.17
α-Pinene	0	0.4	0	6.00	0	7.7	0	0	0
β-Pinene	0	0.6	0	0.6	0	0.1	0	0	0
Benzaldehyde	0.70	0	0	0	0	0	0	0	0.62
Sabinene	0.12	0	0	0	0	0	0	0	3.16
Myrcene	0.07	0	0.26	0	1.1	0	0	0	0
α-Phellandrene	0.58	0	0	0.7	2.0	0	0	0	0.10
Hexyl acetate	0	0	0	0	0	0	0	0	17.32
1,4-Cineole	1.89	0	0	0	0	0	0	0	19.19
α-Terpinene	0.39	0	1.16	0	0	0	0	0	0.81
p-Cymene	0.23	2.2	0	2.2	23.7	0	0	0	5.01
Limonene	0	2.2	0	2.5	2.8	0	0	0.30	0.24
Ocimene	0	0	0	0	0	0	0	0	1.56
γ-Terpinene	0	0	0.27	0	0	0	0	0	0
(−)-Fenchone	0.15	0	0	0	0	0	0	0	0
Linalool	0	0	1.53	0	0	0.9	12.8	0	0
Camphor	0.14	0	0	0	0	0	0	0	0
Menthone	0	45.6	0	0	0	0	0.2	0	0.81
Isopulegol	0.36	0	0	0	0	0	0	0	0
Isoborneol	0	0	0.28	0	0	0	0	0	0
p-Cymene-8-ol	31.29	0	0	0	0	0	0	0	0
Isomethone	0	46.6	0.44	73.4	0	0.4	7.0	0	30.50
Terpinen-4-ol	0	0.9	13.85	0	0	0	0	0	0
α-Terpineol	0	0	0.68	0	0	0	0	0	0
Hexyl butyrate	0	0	0	0	1.0	0	0.2	0	0
Hexenyl butyrate	0	0	0	0	8.3	0	0	0	0
Octyl acetate	0	0	0	0	0	0	0	18.55	0
trans-Carveol	0.30	0	0	0	0	0	0	0	0
Citronellol	0.14	0	2.60	17.6	0.9	5.0	17.6	0	0.21
Nerol	0	0	17.96	0	0	0	0	0	0
Neral	0	0	4.95	0	0	0	0	0	0
Geraniol	0	0	32.48	0	0	0.8	41.9	0	0
Citronellyl formate	0	0	0	0	0	2.9	11.2	0	0
n.i.	1.18	0	0	0	0	0	0	0	0
Geranial	0	0	0.70	0	0	0	0	0	0
Cuminyl alcohol	0	0	2.29	0	0	0	0	0	0
n.i.	47.47	0	0	0	0	0	0	0	0
Citronellyl acetate	0	0	0	0	0	0	0	0	0.87
2-Propenyl benzene	0.53	0	0	0	0	0	0	0	0
β-Caryophyllene	0	0	0.38	0	0	0	0	0.99	0
Guaia-6,9-diene	0	0	0	0.2	0	5.6	6.0	0	0
Geranyl butyrate	0	0	0	0	0	24.0	0.7	0	0
Citronellyl ester	0	0	2.20	0	0	0	0	0	0
10-Epi-eudesmol	0	0	0	0	0	0	4.0	0	0
Geranyl tiglate	0	0	0	0	0	0	2.7	0	0

Notes

(1) – *P. filicifolium*, (5) – *P. tomentosum*, (6) – *P. scarbum*, (7) – *P.* 'Lady Plymouth', (10) – *P. glutinosum*, (11) – *P.* 'Chocolate Peppermint', (13) – *P. graveolens*, (15) – *P. cucullatum*, (16) – *P. quercifolium*.

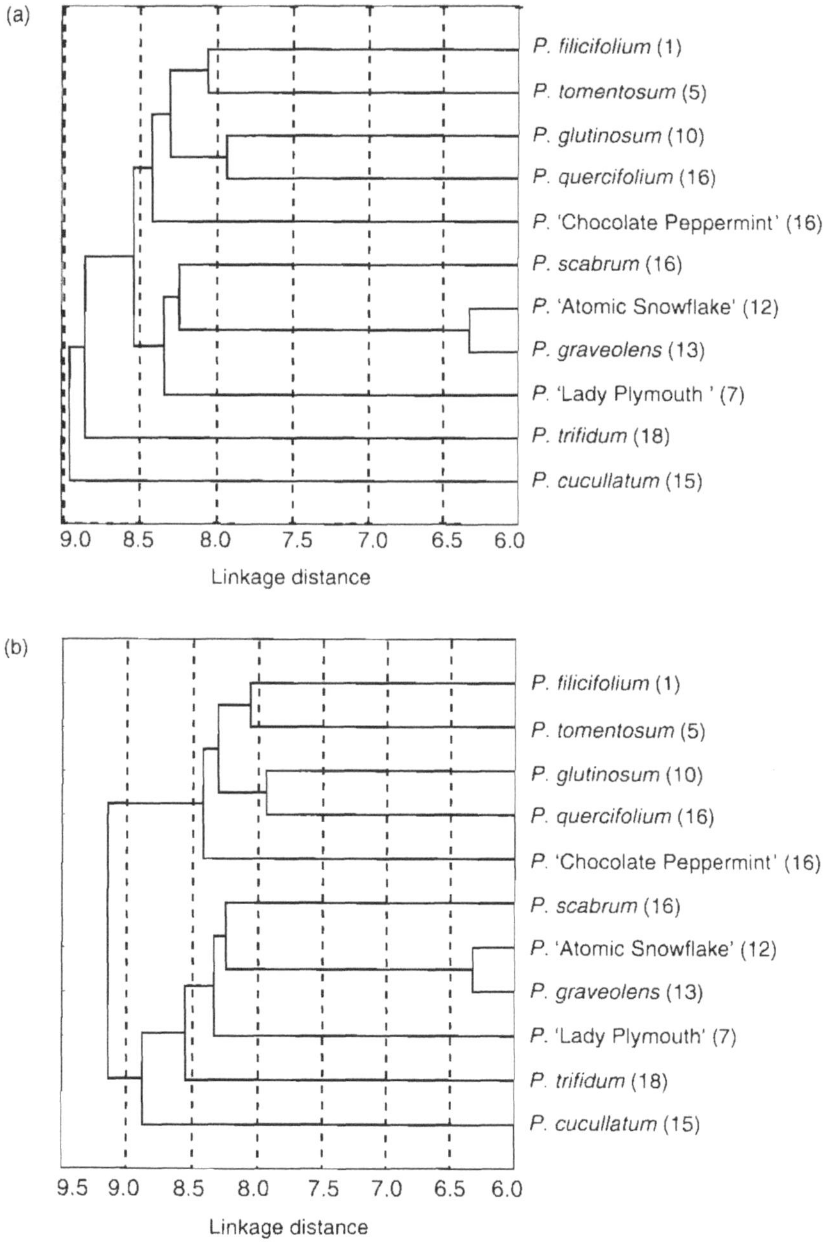

Figure 16.2 a,b The tree diagram obtained for the RAPD dataset (a) by the UPGMA method (b) by the WPGMA method.

The linking together of *P. filicifolium*, *P. glutinosum*, *P. quercifolium*, *P.* 'Chocolate Peppermint' and *P. tomentosum* can be explained almost entirely on the basis of their odour, although there are some verifications also from the genetic point of view as *P. tomentosum* is the parent of the hybrid, *P.* 'Chocolate Peppermint' and the other parent is derived initially from *P. quercifolium*.

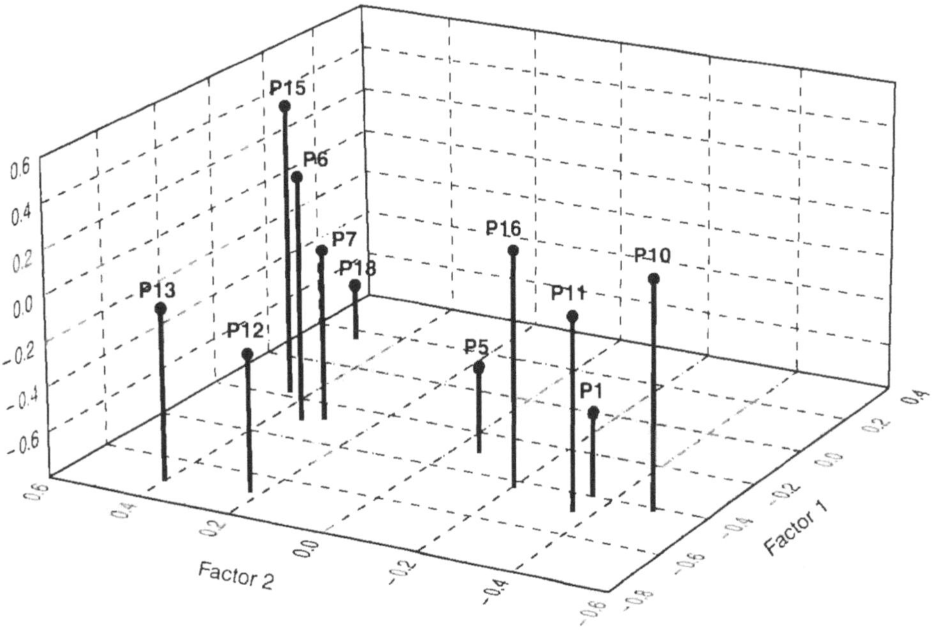

Figure 16.3 Genetic results: The 3D plot obtained by the PCA method for RAPD dataset.

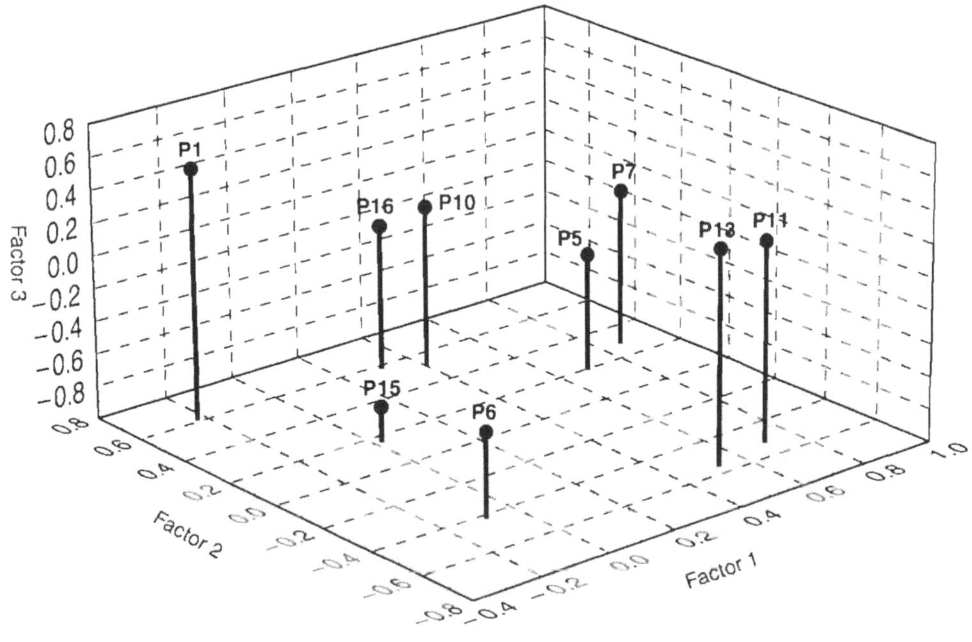

Figure 16.4 Phytochemical results: The 3D plot obtained by PCA method for phytochemical dataset.

GC data for *P. trifidum* indicates that it contains a substantial percentage of menthone, some isomenthone, and monoterpenes including camphene, myrcene, γ-terpinene, limonene (MLB personal analysis). This explains the clustering with *P. 'Lady Plymouth'* (Figures 16.1a, 16.2a, b); with *P. tomentosum* (Figure 16.1b). Unfortunately the analytical results were not available for the essential oil correlations in Figure 16.4. The results also support the hypothesis that the DNA profiles obtained by the RAPD-PCR method correlates with the chemical composition rather than with the genetic angles, as *P. trifidum* is in a completely different section to that of the other two *Pelargonium* with which it is clustered.

The clustering of *P. cucullatum* with *P. scabrum* in Figure 16.1a and 16.1b can only be explained on the basis of similarity in their chemical composition as the two species are very different in most botanical parameters.

REFERENCES

Apostol, B.L. *et al.* (1993) Estimation of the number of full sibling families at an oviposition site using RAPD-PCR markers: applications to the mosquito Aedes aegypti. *Theor. Appl. Genet.*, 86, 991–1000.

Bakker, F.T., Helbrugge, D., Culham, A. and Gibby, M. (1998) Phylogenetic relationships within *Pelargonium* sect. Peristera (Geraniaceae) inferred from nrDNA and cpDNA sequence comparision. *Pl. Syst. Evol.*, 211, 273–287.

Caetano-Annoles, G. (1996) Scaning of nucleic acids by *in vitro* amplification: New developments and applications. *Nat. Biotech.*, 14, 1668–1674.

Cattell, R.B. (1966) The scree test for the number of factors. *Multivariate Behav. Res.*, 1, 245–276.

Czekanowski, J. (1913) Zarys metod statystycznych w zastosowaniu do antropologii. Travaux de la Societe des Sciences de Varsovie III. *Classes des sciences mathematiques et naturelles, No. 5.*

Demeke, T. and Adams, R.P. (1994) The use of PCR-RAPD analysis in plant taxonomy and evolution, In: H.G. Griffin and A.M. Griffin (eds), *PCR Technology: Current Innovations*, CRC Press, Boca Raton, FL.

Dice, L.R. (1945) Measures of the amount of ecologic association between species. *Ecology*, 26, 297–302.

Dunteman, G.H. (1989) *Principal Components Analysis*. Sage publications.

Excoffier, L., Smouse, P.E. and Quattro, J.M. (1992) Analysis of molecular variance inferred from metric distances among DNA haplotypes: application to human mitochondrial DNA restriction data. *Gen. Soc. Amer.*, 131, 479–491.

Hadrys, H. and Schierwater, B. (1992) Applications of random amplified polymorphic DNA (RAPD) in molecular ecology. *Mol Ecol.*, 1(1), 55–63.

Jaccard, P. (1901) Etude comparative de la distribution florale dans une portion des Alpes et des Jura. Bull. Soc., *Vaudoise Sci. Nat.*, 37, 547–579.

Jaccard, P. (1908) Nouvelles recherches sur la distribution florale. *Bull. Soc. Vaud. Sci. Nat.*, 44, 223–270.

Kaiser, H.F. (1960) The application of electronic computers to factor analysis. Educational and Psychological Measurement, 20, 141–151.

Kulczynski, S. (1927) Die Pflanzenassoziationen der Pieninen. *Bull. Intern. Acad. Pol. Sci. Lett. Cl. Sci. Math. Nat.. B(Sci. Nat.)*, (Suppl. 2), 57–203.

Li, W.-H. and Graur, D. (1991) *Fundamentals of Molecular Evolution.* Sinauer, pp. 61–63.

Lis-Balchin, M., Hart, S. and Roth, G. (1999) The pharmacological activity of the essential oils of scented *Pelargonium* (Geraniaceae). *Phytotherapy Res.*, 11(8), 83–84.

Nei, M. and Li, W.H. (1979) Mathematical model for studying genetic variation in terms of restriction endonucleases. *Proc. Natl. Acad. Sci. USA*, 76, 5269–5273.

Ochiai, A. (1957) Zoogeographic studies on the soleoid fishes found in Japan and its neighbouring regions. *Bull. Jap. Soc. Sci. Fish*, 22, 526–530.

Palmarczyk, G., Rytka, J. and Skoneczny, M. (1995) *Inzynieria genetyczna i biologia molekularna. Metody. Podrecznik labolatoryjny.* Instytut Biochemii i Biofizyki Polskiej Akademii Nauk, Warszawa.

Pax, D.L., Price, R.A. and Michaels, H.L. (1997) Phylogenetic position of the Hawaiian Geraniums based on rbcL sequences. *Am. J. Bot.*, 84(1), 72–78.

Price, R.A. and Palmer, J.D. (1993) Phylogenetic relationships of the Geraniaceae and Geraniales from rbcL sequence comparison. *Ann. Missouri Bot. Gard.*, 80, 661–671.

Rogers, D.J. and Tanimoto, T.T. (1960) A computer program for classifying plants. *Science*, 132, 1115–1118.

Russell, P.F. and Rao, T.R. (1940) On habitat and association of species of anopheline larvae in south-eastern Madras. *J. Malar. Inst. India*, 3, 153–178.

Schierwater, B., Ender, A., Schroth, W., Holzmann, H., Diez, A., Streit, B. and Hadrys, H. (1997) Arbitrarily amplified DNA in ecology and evolution. In: Gustavo Caetano-Annoles and Peter Gresshoff, *DNA Markers: Protocols, Applications and Overviews*, Wiley & Sons. Inc.

Sneath, P.H.A. and Sokal, R.R. (1973) *Numerical Taxonomy*. San Francisco, W.H. Freeman & Co.

Sokal, R.R. and Sneath, P.H.A. (1963) *Principles of Numerical Taxonomy*, Freeman p. 134.

Sorensen, T. (1948) A method of establishing groups of equal amplitude in plant sociology based on the similarity of species content and its application to analyses of the vegetation on Danish commons. *K. Dan. Vidensk. Selsk. Biol. Skr.*, (Copenhagen) 5, 1–34.

Spath, H. (1980) *Cluster Analysis Algorithms*. Trans. Ursula Bull. Ellis Horwood (Halstead/Wiley), Chichester, England.

StatSoft, Inc. (1995) STATISTICA for Windows (Computer program manual). Tulsa, OK: StatSoft, Inc., 2300 East 14th Street, Tulsa, OK, 74104–4442, (918) 749–1119, e-mail: info@statsoft.com, WEB: http://www.statsoft.com

Upholt, W.B. (1977) Estimation of DNA sequence divergence from comparison of restriction endonuclease digests. *Nucl. Acid Res.*, 4, 1257–1265.

Yule, G.U. and Kendall, M.G. (1950) *An Introduction to the Theory of Statistics*. 14th Edition. Hafner.

17 Geranium essential oil: standardisation, ISO; adulteration and its detection using GC, enantiomeric columns, toxicity and bioactivity

Maria Lis-Balchin

INTRODUCTION

Definition of Geranium oil and specification

The International Organisation for Standardization or ISO, defines Geranium oil as 'The oil obtained by steam distillation of the fresh or slightly withered herbaceous parts of *Pelargonium graveolens* L'Heritier ex Aiton, *Pelargonium roseum* Willdenow and other undefined hybrids which have given rise to differing ecotypes in the various geographical areas' (International standard 4731: 1972). The colour is various shades of amber-yellow to greenish-yellow. The odour is given as characteristic of the origin, rose-like with a varying minty note.

The specification does not include the Bulgarian Geranium oil distilled from *Geranium macrorrhizum*, known as Zdravetz oil, containing mainly sesquiterpenes of which half is apparently germacrone (Ognyanov, 1985).

ISO 4731 has set the concentration for citronellol content at a minimum 42 per cent/maximum 55 per cent for Bourbon Geranium oil; 35/58 for Moroccan; 40/58 for Egyptian and 40/58 per cent for Chinese oils . Other physico-chemical values are given in Table 17.1.

ORIGIN AND NOMENCLATURE OF GERANIUM OIL

There are numerous misnomers given for the origin of 'Geranium oil' in both scientific books (Fenaroli, 1997) and most aromatherapy books. The worst are: *Geranium maculatum*, *G. robertianum* and other *Geranium* species, as the medicinal properties of 'Geranium oil' were mistakenly taken from Herbals (Culpeper, 1653; Grieve, 1937).

Some of the other misnomers most frequently used are: *Pelargonium odoratissimum*, *P. graveolens*, *P. asperum* and *P. roseum* Willd. However, the main source of the commercial oil is from a cultivar known as *P.* cv. *'rose'* giving the 'Geranium oil, Bourbon' from Reunion and also lately from China. The 'Rose' cultivar has been found to be, most probably, a hybrid between *P. capitatum* × *P. radens* (Demarne and van der Walt, 1989). The cultivars used for the production of Geranium oil in many parts of the world

Table 17.1 Physico-chemical characteristics of Geranium oil 'Bourbon' and from other sources

	Bourbon	*Morocco*	*Egypt*
Relative density at 20 °C	0.884–0.892	0.883–0.900	0.887–0.892
Refractive Index at 20 °C	1.462–1.468	1.464–1.472	1.466–1.470
Optical rotation at 20 °C	−8 to −14	−8 to −13	−8 to −12
Acid value maximum	10	10	6
Ester value	52–78	35–80	42–58
Ester value after Acetylation	205–230	192–230	210–235
Carbonyl value Expressed as iso-menthone	58	58	Not given
Apparent citronellol (rhodinol) content	42–55	35–58	40–58

remain confused; in some papers originating in India, the cultivar is stated to be that obtained as a cutting from the cultivar 'Rose' from Reunion, however, many papers state that their Geranium oil source is from *P. graveolens*.

PRESENT SOURCE

Numerous parts of the world have produced Geranium oil in the past (Weiss, 1997) but even in the country of origin of *Pelargonium* i.e. South Africa, this has proven to be non-economical. Small amounts, for own country consumption is produced in India, Morocco and Algeria. In Kenya it is known as Mawah oil, and is stated to be from *P. fischeri* (Weiss, 1997). Nowadays, there is no production of Geranium oil in Grasse, which simply acts as a market base for imported oil from all over the world and where some adjustments are made to the oil, demanded for its various uses. African Geranium oil usually refers to the Egyptian oil but also to that of Morocco and Algeria where the same or similar *Pelargonium* cultivar is used.

China produced a type of Bourbon oil, but Weiss (1997) refers to it as an Morocco-style oil, which makes it more like the Egyptian type. The area of Yunnan, Binchuan, Shiping and Yuxi Provinces are gradually being decreased in favour of other essential oil crops.

The descriptions of the unique but different odour qualities found in the essential oil 'bible' by Arctander (1960) probably no longer apply to the modern produce. The different chemical compositions of Geranium oils from different geographical areas (Lawrence, 1992, 1994; Weiss, 1997) are also not found to conform to norm in practice, as adulteration of commercial oils is profuse (Lis-Balchin *et al.*, 1996).

GERANIUM CONCRETE AND ABSOLUTE

Geranium concrete and absolute are made in small amounts for certain perfumes and are produced mainly in Egypt. The concrete, extracted with petroleum spirits or hexane is dark green or brownish-green with a foliage-like odour and great tenacity. The geranium absolute, made from the concrete by dissolving in absolute alcohol and then chilling to precipitate the insoluble components, followed by evaporation of the solvent, is also greenish with a somewhat leaf-earthy and powerful odour (Arctander, 1960). Terpeneless Geranium oil can be produced from the Geranium oil and the absolutes, by vacuum distillation; this makes the oil more soluble in diluted alcohol and is useful in foods as well as cosmetics.

The name 'geranium rose' oil was formerly used to indicate a special French Geranium oil produced in Grasse, which was distilled over rose petals. Nowadays the Geranium rose refers to the *Geranium* c.v. 'Rose' cultivated in Reunion as the Bourbon Geranium oil.

One of the main products of Geranium oil and absolutes in the past was rhodinol, which is composed mainly of the citronellol fraction. This was used extensively in the 'poor-man's' rose perfumes and cosmetics, including soaps, creams, etc. (Arctander, 1960). Nowadays, rhodinol is produced synthetically, as the price of geranium has soared.

ADULTERATION

Geranium oil contains mainly citronellol and geraniol and their esters, therefore can be easily concocted from cheaper essential oils and adjusted to the recommended ISO standards. The antimicrobial activity of such essential oils is much greater than that of some authentic oils but has a similar pharmacological effect on smooth muscle (spasmolytic) and the actual odour can be even more appreciated by perfumers than the real essential oil. The essential oil composition of this Geranium oil differs completely from that of a true *G. robertianum* oil (Pedro *et al.*, 1992) or that of *G. maccrorhizum* (Ognyanov, 1985).

The most expensive Geranium oil was always the Bourbon, but over the years this has inexplicably increased in tonnage (which is surprising for a small volcanic island with relatively limited *Pelargonium*-growing areas) as well as value; this could be partly due to the increase in Geranium oil production in China, which being very similar to that of Bourbon would often get accepted as such (Verlet, 1992).

Recent Geranium oil production in China is restricted to the region of Binchuan 450 km from Kunming (Cu, 1996) and there are two harvests, a summer one which yields an oil which is relatively similar to Bourbon and the winter harvest which gives a low-grade oil with only 4 per cent geraniol, compared to the summer 7 per cent and the Bourbon with 14 per cent. The citronellol content is however much greater than that of Bourbon geranium and is virtually doubled. The characteristic sesquiterpene is guaia-6,9-diene as in Bourbon oil.

Adulteration of Geranium oil is perhaps encouraged by the ISO requirements themselves and the comparatively low price of synthetics. The yield of Geranium oil is less than 0.3 per cent, and is usually 0.2 per cent. There are excellent recipes for a synthetic Geranium oil, one of which is shown in Table 17.2, which was given to the author by a perfumer 15 years ago.

Adulteration of all essential oils occurs to a considerable extent with diluents like propylene glycol, triacetin, triethyl citrate or benzyl alcohol, ethyl alcohol and in the case of aromatherapy oils with fixed oils like almond oil, which are added in excessive amounts. Adulteration also implies giving the wrong source on the labelling e.g. Bourbon, if it came from another country or was synthetic, or even when Body Shop had a Geranium oil leaflet stating that it originated from *G. maculatum* (which is not only the wrong species but has no odour).

Table 17.2 Recipe for a synthetic 'Geranium oil'

Component/chemical	*Parts*
Dimethylsulphoxide (DMSO) 0.1%	10
Citral	200
Citronellol	2800
Citronellyl formate	1000
Geraniol	1500
Geranyl formate	800
Geranyl propionate	150
Geranyl tiglate	150
guaia-6,9-diene	500
iso-menthone	500
menthone	150
L-rose oxide	150
Linalool	1000
Diluent	1090
	100%

DETECTION OF ADULTERATION

Carrier or fixed oil or solvents

Geranium oil Bourbon is frequently adulterated and the real or preferred oil can only be detected by experienced noses and stringent chemical analysis, unless solvents are used in any quantity, where simple gas chromatography (GC) can detect adulteration, if one knows where to look for it. Adulteration with carrier oils (fixed oils like almond oil) is easily detected by putting a drop of the sample on blotting paper or a piece of cloth and looking for signs of a halo of grease remaining after a few hours (as pure essential oils would evaporate completely, leaving no residual mark.

Chiral or enantiomeric columns

However, ordinary GC, with or without mass spectrometry (MS) or other identification facilities, like Infra-red (IR) etc. are not sophisticated enough to find most adulterations when fractions of other oils or synthetic components are used. The determination of such adulteration of essential oils was perfected by the use of special enantiomeric or chiral columns, mainly composed of an α-cyclodextrin phase (Ravid *et al.*, 1992; Lis-Balchin *et al.*, 1999). One of the major components, citronellol occurs in the (−)-form in geranium and rose oils and has a finer rose odour than the (+) enantiomer, and a sweet, peach-like flavour. The (+)-citronellol enantiomer has been found in citronella oils from Ceylon and Java, *Cymbopogon winterianus*, *Boronia citridora*, *Eucalyptus citriodora*, Spanish verbena and other essential oils.

The two enantiomers are starting materials for numerous chiral pheromones and flavours (Ravid *et al.*, 1992) and are prepared commercially by partial or total synthesis, sometimes involving particular yeast strains. This abundance of the citronellol lends itself to adulteration on a grand scale. Initial analyses of commercial Egyptian Geranium oil yielded almost a racemic mixture of citronellol enantiomers, whilst a true Bourbon oil gave a highly concentrated S(−)-citronellol (Ravid *et al.*, 1992); Further analyses using chiral columns showed that some cultivars of Egyptian type geranium leaves (distilled in the laboratory) had an almost racemic mixture, while others had more of the (−) enantiomer. Bourbon Geranium oil distilled in the laboratory had in contrast a predominance of the (−) enantiomer of 73–78 per cent, whilst commercial samples had a more racemic content, as did *Eucalyptus citriodora* and Lemon mint oil. Citronella had only 26 per cent of the (−) enantiomer. The rose-like quality of the (−)-citronellol determines the odourific value of the Geranium oil and synthetic rose compounds and therefore addition of synthetic citronellol or extracts from plant oils with a low content of (−)-citronellol would be detrimental to the odour and quality of the product.

Recent studies on Australian Geranium oils grown from specific *Pelargonium* clones showed that using 10 key chiral components, and calculating a so-called 'chiral excess', it was possible to distinguish geographically different and seasonally different essential oils and also adulteration (Doimo *et al.*, 1999).

It is worth noting that chiral columns can also be used by synthetic chemists and those involved in adultering esential oils, as the same type of column can be used to separate out the enantiomers, which could then be added in the correct proportion for a given essential oil, i.e. for Geranium oil Bourbon, the proportion of the (−)-citronellol could be adjusted to 75 per cent for maximum odour quality and 'authenticity'. It is always difficult to keep up with forgers, let alone be a step ahead of them!

Like Geranium oil itself, Rhodinol ex Geranium is often adulterated with synthetic rhodinol, fractions of citronella or palmarosa oils and synthetic components.

G. *macrorrhizum* (Zdravetz oil), produced almost solely in Bulgaria, has been used for adulterating Geranium oil (Guenther, 1950; Pedro *et al.*, 1992). The essential oil composition of this Geranium oil differs completely from that of a true *G. robertianum* oil or that of commercial Geranium oil from *Pelargonium* cultivars.

G. robertianum contains mainly α-terpinene, linalool, α-terpineol and an assortment of monoterpenes in contrast to commercial Geranium oil with citronellol and geraniol as its main components (Pedro *et al.*, 1992).

The percentage of linalool in *G. robertianum* is considerably higher than that in commercial Geranium oil, which is about 3–10 per cent (Lis-Balchin, 1995; Lis-Balchin *et al.*, 1996), which was based on the actual analytical data of over 40 commercial Geranium oils from different geographical sources (as on labels) bought from many different commercial outlets). Adulteration with *G. robertianum* oil would therefore be easily detected using conventional GC as well as simply by its smell.

Geographical source and chemical composition of commercial Geranium oil

The apparent geographical source had on the whole no correlation with the chemical composition of commercial Geranium oil (Lis-Balchin *et al.*, 1996) except for the presence or absence of the relevant sesquiterpene: i.e. 10-epi-g-eudesmol in Egyptian oils (3–7 per cent) and guaia-6,9-diene (1–7 per cent) in the Bourbon and China oils; a Moroccan oil contained both these sesquiterpenes. The proportion of the main components i.e. citronellol, geraniol, linalool, iso-menthone, citronellyl formate and geranyl formate was not consistent for any geographical source. The bioactivity, as determined by the action of the oils against 25 different bacterial species, 20 different *Listeria monocytogenes* cultivars, three different fungi and also their anti-oxidant action was not correlated with either the geographical source of the Geranium oil specimens or their chemical composition. The activity of the main components, citronellol and geraniol, was assessed against all the bioactivity parameters either singly or in combination, in the percentages listed by the ISO for different Geranium oils. The bioactivity was very potent for both the components, and the mixtures. However, a sample of Australian oil extracted using field-distillation and obtained directly from its source, was comparatively inactive, suggesting possible adulteration of commercial oils with synthetic components (Lis-Balchin *et al.*, 1996).

The effects of different samples of Geranium oil was also investigated pharmacologically using guinea-pig ileum *in vitro* (Lis-Balchin *et al.*, 1996). There was again a variation in the bioactivity as shown by the relaxation produced in the smooth muscle. There was insufficient variation to warrant this to be a sensitive method for geranium, but other work using enantiomers (Lis-Balchin *et al.*, 1996, 1999) have indicated that there is scope for seeing differences in activity due to individual enantiomers which react differently in different tissues.

In conclusion, due to the high sensitivity, biological evaluation using several different parameters, can be a useful tool in evaluating essential oils and checking for their adulteration, as it is more sensitive than ordinary GC.

PRICE OF GERANIUM OIL

The price of Geranium Bourbon has recently rocketed due to poor crops (through cyclones and other calamities) and therefore scarcity of oil production. Although, it is almost impossible to find out the production of pure Geranium oil, the imports into USA from 1992, 1993, 1994 was increasing, showing about 53,000 kg, 64,000 kg, 82,000 kg respectively and in the last year having a market value of $57 per kg, having increased from $37 per kg from 1992.

The price/kg in 1999–2000 has reached between $160 for Chinese '*P. graveolens* oil' to $360 for 'Geranium Rose Maroc' (CH-Imports Ltd., Greensboro, USA catalogue). Other sources of the oil are therefore eagerly sought, as China has decreased production (from 80 ton in 1990 to 9.4 ton in 1991, (Quinhua, 1993)), and this is still falling; India may perhaps be able, in future, to bridge the production gap.

TOXICITY OF *PELARGONIUM* SPECIES

Toxicology of the essential oil

Status

Geranium oil Bourbon, Algerian, Moroccan were granted generally recognised as safe (GRAS) status by Flavoring Extract Manufacturer's Association (FEMA) (1965) and approved by the Food and Drug Administration (FDA) for food use. The Council of Europe (1970) included Geranium oil in the list of spices, seasonings, etc., deemed admissible for use with a possible limitation of the active principle in the final product.

Biological toxicity studies

Acute toxicity

Oral LD50 in rats, >5 g/kg; dermal in rabbits, 2.5 g/kg (Moreno, 1973). Irritation: applied undiluted to abraded or intact rabbit skin for 24 h under occlusion was found to be moderately irritant (Moreno, 1973), but applied to backs of hair-less mice, it was not irritating (Urbach and Forbes, 1972). Human patch test (closed) to 10 per cent Geranium oil in petrolatum produced no irritation after 48 h (Kligman, 1966).

Sensitisation

a maximisation test on 25 volunteers, using 10 per cent in petrolatum produced no sensitisation (Kligman, 1966).

Phototoxicity

Was not found for Geranium oil (Urbach and Forbes, 1972).

Toxicity of* Pelargonium *species

Toxicity of *Pelargonium* is usually found under the heading of 'geranium' toxicity. There are very few, scattered, references to any toxicity, and all references are due to contact dermatitis and sensitisation. Most of the references are to the Geranium oil and the main components geraniol (Lovell, 1993).

Toxicity of Geranium oil Components

Patch tests to geraniol proved negative but dermatitis to perfumes containing Geranium oil has been shown in a few cases (Klarmann, 1958). Ointments containing geraniol e.g. 'Blastoestimulina' were reported to cause sensitisation when used in the treatment of chronic leg ulcers (Romaguera *et al.*, 1986, Guerra *et al.*, 1987), although the patients were also sensitive to other ointments which contained no essential oils. Blastoestimula contained:

	Conc.
Glyco-D-116-F	as is
Corn oil	as is
Neomycin sulphate	20 per cent pet.
Geraniol	2 per cent pet.
Propylene glycol	10 per cent aq.
Centella asiatica extract	1 per cent o.o.
Lavender	2 per cent pet.

The patients were obviously sensitised prior to this by other chemicals. Sensitisation to geraniol using a maximisation test proved negative (Opdyke, 1975), but the allergen may be geraniol as cross-reactions often occurred with citronella (Keil, 1947), however, the main sensitiser in citronella is citronellal, with citronellol less reactive; geraniol was even weaker, as was citral. A patient who had used citronella oil to smear over his windows against mosquitoes developed a skin reaction and four other patients were found with sensitivity to citronella. In two cases, strong reactions were obtained with 1 per cent solutions of citronellal and weaker ones with citronellol, geraniol, geranyl acetate. In 23/23 cases no response was found using lemon oil, therefore suggesting specificity of the response. However, sensitisation to geraniol using a maximisation test proved negative (Opdyke, 1975). In a lemon oil sensitisation case, α-pinene gave a greater response than β-pinene: this is due to the close similarity between limonene and β-pinene (due to an exposed methylene radical).

Latest reports from Japanese studies, using patients with ordinary cosmetic dermatitis and pigmented cosmetic dermatitis, who showed a positive allergic responses to a wide range of fragrances (Nakayama, 1998), gave a list of Class A fragrances which were termed common cosmetic sensitisers and primary sensitisers. This Class included Geranium oil, geraniol, sandalwood oil, artificial sandalwood, musk ambrette, jasmine absolute, hydtroxycitronellal, Ylang ylang oil, cinnamic alcohol, cinnamaldehyde, eugenol, balsam of Peru and lavender oil.

Geraniol was found to give a positive patch test in over 1.2 per cent cases when used at 1 per cent in white petrolatum with 5 per cent sorbitan sesquioleate (Frosch, 1998).

Toxicity of Pelargonium plants

A vesicular hand dermatitis in a young man who removed dead leaves from an unknown *Pelargonium* was reported (Anderson, 1923). This was not identified, and the paper mentions it could have been one of hundreds of varieties. The patient had been suffering from tuberculosis, therefore his immunity was very low. The dermatitis was successfully treated with 'black mercurial lotion'!

Both zonals and forms of the scented rose geranium have caused sensitisation in a few cases (Rook, 1961; Hjorth, 1969).

REFERENCES

Anderson, J.W. (1923) Geranium dermatitis. *Archives Dermatol. Syphilology*, 7, 510–511.

Arctander, S. (1960) *Perfume and Flavor Materials of Natural Origin*, Elizabeth, New Jersey, USA.

Cu, J.-Q. (1996) Geranium oil from Yunnan, China. *Perf. Flav.*, 21, 23–24.

Culpeper, N. (1653) *The English Physitian Enlarged*, George Sawbridge, London.

Demarne, F.E. and van der Walt, J.J.A. (1989) Origin of rose-scented *Pelargonium* cultivar grown on Reunion Island. *S. Afr. J. Bot.*, 55, 184–191.

Doimo, L., Fletcher, R.J. and D'Arcy, B.R. (1999) Chiral excess: measuring the chirality of geographically and seasonally different Geranium oils. *J. Essent. Oil Res.*, 11, 291–299.

FEMA (Flavoring Extract Manufacturer's Association) (1965) Survey of flavoring ingredient usage levels. No. 2508. *Food Technol. Champaign*, 19, part 2, 155.

Frosch, P.J. (1998) Are major components of fragrances a problem? In: P.J. Frosch, J.D. Johansen, and I.R. White, (eds), *Fragrances: Beneficial and Adverse Affects*, Springer Verlag, Berlin pp. 92–99.

Fenaroli, G. (1997) *Handbook of Flavour Ingredients*. 3rd ed. CRC Press, London. Vol.1.

Grieve, M. (1937) *A Modern Herbal*. Reprinted 1992. Tiger books International London.

Guerra, P., Aguilar, A., Urbina, F., Cristobal, M.C. and Garcia-Perez, A. (1987) Contact dermatitis to geraniol in a leg ulcer. *Contact Dermat.*, 16, 298–299.

Guenther, E. (1950) *The Essential Oils*, Vol. 4, van Nostrand Co., New York.

Hjorth, N. (1969) Plant dermatitis. *Contact Dermat. Newsletter*, 6, p. 126.

Keil, H. (1947) Contact dermatitis due to oil of Citronella. *J. Investig. Dermatol.*, 8, 327–334.

Klarmann, E.G. (1958) Perfume dermatitis. *Ann. Allergy*, 16, 425–434.

Kligman, A.M. (1966) Report to RIFM, 31 October.
Lawrence, B.M. (1992) Progress in Essential Oils. *Perf. Flav.*, 17(2), 46–49; (6), 59–60.
Lawrence, B.M. (1994) Progress in Essential Oils. *Perf. Flav.*, 19(1), 40–42.
Lis-Balchin, M. (1995) *Aroma Science: The chemistry and activity of Essential Oils*, Amberwood Pub. Ltd., Surrey.
Lis-Balchin, M., Deans, S.G. and Hart, S. (1996) Bioactivity of commercial Geranium oil from different sources. *J. Essent. Oil Res.*, 8, 281–290.
Lis-Balchin, M., Ochocka, R.J., Deans, S.G. and Hart, S. (1999) Differences in bioactivity between the enantiomers of α-pinene. *J. Essent. Oil Res.*, 11, 393–397.
Lovell, C.R. (1993) *Plants and the skin*, Blackwell Scientific. Publ., Oxford.
Moreno, O.M. (1973) Report to RIFM, 25 July.
Nakayama, H. (1998) Fragrance Hypersensitivity and its control. In: P.J. Frosch, J.D. Johansen, and I.R. White (Eds) *Fragrances: Beneficial and Adverse Affects*, White, Springer Verlag, Berlin pp. 83–91.
Ognyanov, I. (1985) Bulgarian Zdravetz oil. *Perf. Flav.*, 10(6), 38–44.
Opdyke, D.L.T. (1975) Monographs on fragrance raw materials. *Food Cosmet. Toxicol.*, 13, 451.
Pedro, L.G., Pais, M.S.S. and Scheffer, J.J.C. (1992) Composition of the essential oil of *Geranium robertianum* L. *Flav. Fragr. J.*, 7, 223–226.
Quinhua, Z. (1993) China's perfumery industry picks up. *Perf. Flav.*, 18, 47–48.
Ravid, U., Putievsky, E., Katzir, I., Ikan, R. and Weinstein, V. (1992) Determination of the enantiomeric composition of citronellol in essential oils by chiral GC analysis on a modified γ-cyclodextrin phase. *Flav. Fragr. J.*, 7, 235–238.
Romaguera, C., Grimalt, F. and Vilaplana, J. (1986) Geraniol dermatitis. *Contact Dermat.*, 14, 185–186.
Rook, A. (1961) Plant dermatitis-botanical aspects. *Trans. St. John's Dermatol. Soc.*, 46, 41–47.
Urbach, F. and Forbes, P.D. (1972) Report to RIFM, 22 September.
Verlet, N. (1992) Geranium Bourbon: quel avenir? *Parf Cosmet. Aromes*, 108, 49–51.
Weiss, E.A. (1997) *Essential Oil Crops*, CAB International, Oxon.

18 'Rose-scented geranium' a *Pelargonium* grown for the perfume industry

Frédéric-Emmanuel Demarne

INTRODUCTION

The interest shown in growing scented-leaf geranium dates from the mid nineteenth century. At that time, the real 'Rose of the Levant', *Rosa damascena* Mill. (Rosaceae), became rare and its essential oil reached excessive levels of price in Paris (Héricart De Thury, 1845). This situation forced the perfumers to look for new essential oils with a rose scent and they remembered the distillation trials of rose geranium carried out by Rochez in Lyon (France) in 1819. In 1844, Demarson established the first plantations in the sunshine of Provence, near Grasse (France).

But, from a physiological point of view, the geranium cultivation requires generous sunshine and well-drained soils, rich in organic matter. The crop dreads frost and temperatures below 2 °C as they are harmful to the growth and even to the survival of the plant.

And, from an agro-economic point of view, the traditional cultural practices are very labour consuming: preparation of cuttings for plantation, muck-spreading, hoeing for weed control, manual harvesting and firewood duty for distillation are costly operations. Furthermore, in the countries with pronounced winter season, the risk of frost leads to an annual crop management and needs to shelter the cuttings till the next plantation.

For these reasons and because of the lack of mechanised solutions, the production of rose Geranium oil did not take a long time to abandon the South of France and to move to more merciful skies, where an abundant and cheap labour existed. From 1850, the French colonies of northern Africa and Indian Ocean seem predestined for growing geranium. The crop is introduced to Algeria in 1847 and developed on an industrial scale by Léon Chiris and Monk towards 1865 (Naves, 1934; Igolen, 1941). In 1880 the plant is introduced to Reunion Island where it will be distilled for the first time in 1882. After the turn of the twentieth century, the crop was attempted in Corsica, Italy, Morocco, Tunisia, Egypt, Southern Russia, Congo, Kenya, Tanganyika, Madagascar, India, Spain, Portugal, Brazil, Comoro Islands...The most recent development was in the People's Republic of China in the fifties.

TAXONOMY AND GENETICS

Since the beginning, and as pointed out by Guenther (1951), 'the taxonomy of the plants which are cultivated in various parts of the world for the production of commercial Geranium oil has been a matter of much controversy and has given rise to considerable confusion'. Basically, the name geranium itself is incorrect, since the species involved in the production of rose Geranium oil are all *Pelargonium*, not *Geranium*. All these plants belong to the Geraniaceae family, but *Geranium* and *Pelargonium* form two different genera, beside three other which are *Erodium*, *Monsonia* and *Sarcocaulon*.

In the same way, the cultivars which, by a great misnomer, are referred to as *P. capitatum* or *P. graveolens* or *P. radens* are not natural species. According to the revision of the genus *Pelargonium*, carried out since 1977 under the supervision of J.J.A. van der Walt at the University of Stellenbosch (Republic of South Africa) (van der Walt, 1985), we now know that the true *P. radens* H.E. Moore (Moore, 1955) and *P. graveolens* L'Hérit. (L'Héritier Dom De Brutelle, 1789) are mint-scented species (van der Walt and Demarne, 1988) with 88 chromosomes (Albers and van der Walt, 1984), while *P. capitatum* (L.) L'Hérit. is a 66 chromosomes species very poor in essential oil and with a very faint rose scent (Demarne *et al.*, 1993; Viljoen *et al.*, 1995). All these natural species have originated from Southern Africa.

According to past and recent researches in genetics (Ducellier, 1933; Kuchuloriya, 1964, 1968; Demarne, 1989; Demarne and van der Walt, 1989), we also know that all the cultivars grown worldwide for rose Geranium oil production are interspecific hybrids between *P. capitatum*, on one hand, and *P. graveolens* or/and *P. radens*, on the other hand. Most of these cultivars have 77 chromosomes (Tamai *et al.*, 1958, 1963; Payet, 1981; Demarne, 1989) and their main features (chromosome number, habit, leaf and flower shape, essential oil yield and composition...) are in-between those of their natural parents. The oil composition in particular depends on the *P. capitatum* parent which transmits the ability to synthesise geraniol and citronellol rather than isomenthone. Furthermore, the presence of 6,9-guaiadiene and/or 10-epi-γ-eudesmol, which usually distinguishes between the African type and Bourbon type oils (Vernin and Metzger, 1983), is also inherited from the *P. capitatum* parent.

Thus, the cultivars of rose-scented geranium should be named *Pelargonium* hybrids of *P. capitatum* × *P. radens* or *Pelargonium* hybrids of *P. capitatum* × *P. graveolens*. The choice of the proper variety, when establishing a new production anywhere in the world, is fundamental: it will directly influence the yield and the quality of the oil.

CULTIVATION

Growing rose geranium for essential oil production is not easy. For a profitable crop, this fragile bush must be cultivated as a perennial on a 3–5 year basis, and every harvest can question its survival. This is especially true when the plantations are established in tropical countries, where fungal and bacterial diseases are numerous and particularly virulent.

Probably because of that fragility, rose geranium is usually grown on small plots of about 1–2 ha, sometimes less. Most of the cultural practices are done by hand, making the cultivation resemble horticulture and keeping the farmer busy throughout the year.

Propagation

All the cultivars of rose geranium are heterozygotic and highly polyploid ($x = 11$; '2n' $= 7x = 77$ chromosomes). Due to their hybrid character and their unusual chromosome number, the plants are usually male-sterile and do not set seed (Ducellier, 1933; Tokumasu, 1970). Their agronomic performances and the quality of their essential oils depend on complex genetic balances. Therefore, asexual propagation is the rule and the cultivars are usually propagated by terminal stem cuttings.

Rooting of the cuttings is influenced by the nutritional state (carbohydrates/nitrogen ratio), the sanitary state and the age of the mother-plants (Bricheteau *et al.*, 1980; Dartigues and Lemaire, 1980; Vidalie, 1980), by wilting and application of root-promoting hormones and fungicides (Michellon, 1975, 1976; Demarne, 1981, 1992c; Rajeswara Rao *et al.*, 1993), and by water supply, temperature and growing medium.

From a common experience, the best rooting is obtained from herbaceous terminal stem cuttings of 12–20 cm long (7–10 nodes) with 4–5 terminal leaves, grown in nursery under partial shade and mist irrigation. Day/night temperature should be 21–27/12–17 °C and the growing medium should provide a sufficient porosity as well as a good water-holding capacity to allow good aeration and good water supply. Treating the base of the cuttings with a mixture of indolebutyric acid (IBA/0.1–0.2 per cent) and captane (10 per cent) in talcum powder is also recommended. Under those conditions, the cuttings quickly develop a profuse root system and are ready for transplanting after 40–60 days.

However, beside these ideal conditions of propagation, a traditional method is still in use in most of the producing countries. It consists of planting unrooted woody cuttings, 30–40 cm long, directly into the field, just after the rainy season, but unfortunately, the mortality rate after planting is usually high.

Planting and cropping system

The profitability of a geranium crop is directly correlated to the number of plants per hectare. Depending on soil and climatic conditions, plant spacing varies between 80 × 30 cm and 100 × 60 cm; the optimum being of about 35,000 plants/ha. Rose geranium is generally cultivated as a pure crop installed on a cleared land. Cuttings are planted in holes or furrows dug manually, at the bottom of which the farmer usually puts a handful of manure (farm manure or composted distillation residues) and sometimes, some fertilizer, lime, and a slow-released insecticide to prevent the young roots from white grubs and other soil insect attacks.

During the first 6 months, geranium grow quite slowly, leaving large portions of the inter-row uncovered. Intercroping geranium with short duration legumes or maize allows a profitable use of this space and an efficient weed control (Lougnon, 1924; Prakasa Rao *et al.*, 1984a, 1986; Narayana *et al.*, 1986; Garin, 1987).

In the traditional cropping systems, rose geranium is thus a perennial weeded crop, and in the tropical countries, rainfall combined with hoeing for weed control causes severe erosion. Fields on a slope must be arranged in bench terraces with swathes, and new concepts of sustainable agriculture, based on planting geranium into a controlled cover crop, are now developed. These techniques favour the microbiological activity of the soil and lead to a better preservation of soil fertility and a better water management (Michellon *et al.*, 1994a,b, 1996; Perret *et al.*, 1996). The foliage production and the life span of the crop are thus improved, but the drawback is that this agriculture is more technical, especially in keeping the cover crop under control, without damaging the rose geranium.

Fertilization

In traditional cultivation, the fertilization of rose geranium usually combines chemical fertilization, restitution of organic matter and sometimes liming (Garin, 1987).

Because of their relatively high costs, the chemical fertilizers are sparingly used (Chabalier, 1992). The composted residues of distillation and/or an organic manure bring back to the field some fertilizer elements, but the time of application and the placement of these residues mainly benefit the intercrops. Generally speaking, the fertilization level is low and the traditional system is not stable.

Chemical fertilization

The first fertilization trials on rose geranium date from 1894. According to Fritz (1976), a high yield crop can produce 7 tons of dry matter/ha/year, which correspond to average exportations of 100 kg N, 32 kg P_2O_5, 165 kg K_2O, 250 kg CaO, 28 kg MgO, 15 kg Na and 10 kg S. The dry matter is about 18–20 per cent of the total biomass. These figures obtained on Reunion Island have been confirmed in India (Prakasa Rao *et al.*, 1988).

It is now well known that rose geranium responds to nitrogen applications but on the condition that balanced fertilization is used and other elements are not limited, especially phosphorus (Chabalier, 1992). On Reunion Island, Fritz has observed linear increases of herb yield with nitrogen applications up to 150 kg/ha/year (Fritz, 1976). Similar results were obtained in India with doses up to 225 kg/ha/year (Prakasa Rao *et al.*, 1986, 1988).

Rose geranium also responds to phosphate applications. Fritz (1976) observed that the essential oil yield is linked to the P content of the plant and, as far back as 1899, Dantes (1899) reported that massive applications of superphosphate (2 t/ha) doubled the herb yield. Weiss (1967) reported similar results in Kenya and advised the renewal of the application of 200 kg/ha of P_2O_5 after the fifth harvest of a geranium crop. Other hints are given by Chabalier (1992) to grow rose geranium in andic[1] soils and to prevent phosphorus deficiencies when this element is made unavailable by the absorbing complex of the soil.

Rose geranium can withstand low levels of potassium before displaying deficiency symptoms. Those symptoms and a yield drop appear when the K content of the plant drops down 0.3 per cent of the dry matter (Chabalier, 1992).

Lastly, spraying miconutrients (B, Cu, Zn, Mo...) is sometimes recommended (Ladaria, 1968; Arumugam and Kumar, 1979; Dhakshinamoorthy *et al.*, 1980; Prakasa Rao *et al.*, 1984b). Under certain conditions, those trace elements can favour the oil production.

In practice, the type of chemical fertilizer will depend on the soil conditions and on the cropping system; the doses and the applications' timetable will depend on the possible level of intensification. For an intensive crop, the theoretical fertilization, based on the restoration of the exported nutrients and on the normal losses is 150–200 kg N/ 60–80 kg P_2O_5/150–200 kg K_2O.

N, P and K can be combined by using a complete fertilizer at the time of planting and then once a year at the end of the rainy season. In intensive cultivation N (urea or calcium ammonium nitrate) can be applied in 2–4 split doses. In any case, when the fertilizer is put in the hole at the time of planting, it is important to avoid the contact between the cutting and the fertilizer to prevent burning.

1 Dark soil of volcanic origin, with particular physical and chemical properties.

Lime and organic manure

When rose geranium is traditionally grown without a cover crop, its erosive character often requires incorporated applications of lime and organic matter for restoring the soil fertility. In Reunion Island, an additional practice consists in a several years' growth of bush fallow of *Acacia decurrens* before planting.

In all cases, the use of organic manure in combination with fertilizers at the time of planting is advocated. The beneficial impact of organic matter to the microbial activity and soil structure is well known. Michellon (1987) observed that application of 5–10 tons of distillation residues on a geranium intercropped with legumes leads to a better root development and a two-fold increase of essential oil yield compared with geranium grown as a sole crop. Further to that observation, Chabalier (1992) gave the composition of different sources of organic matter suitable for rose geranium cultivation.

Irrigation

Rose geranium is rather drought resistant and dreads excess of water. Irrigation is usually left to nature and when a waterlogging risk exists, the cuttings are planted on ridges for insuring a good drainage, as in the Nile valley in Egypt.

But, attempts made to grow geranium in harsh weather conditions in Egypt (Koriesh and Atta, 1986), in Israel (Sanderovich *et al.*, 1983) or in India (Singh *et al.*, 1996; Sabina Aiyanna *et al.*, 1998; Singh, 1999), demonstrated that rose geranium can take advantage of different techniques of irrigation and fertigation to increase both its herb and essential oil yields per hectare. However, a negative correlation is often found between plant water content and the essential oil content, as a result of irrigation treatment (Sanderovich *et al.*, 1983).

In intensive cropping systems, both plasticulture and cover crops allow the reduction of the water needs of the crop.

Weed control

Weed control is important in geranium cultivation. Under tropical and sub-tropical climates, weeds grow very quickly and tend to invade the fields, competing with the crop for space, light, water and nutrients. When left uncontrolled, the weeds smother the young cuttings which become etiolated; they limit growth and branching leading to poor yields (Trémel, 1992; Rajeswara Rao and Bhattacharya, 1997). Moreover, some odouriferous weeds can be harvested and distilled with the crop and will affect the quality of the oil.

In most producing countries, manual weeding is practised. This operation is very labour-consuming and requires from 60 to 125 days of labourer/ha/year. Even when herbicides are used in the inter-row. weeding the row is always done manually. In practice, weed control is a combination of manual hoeing, herbicide application, cultural tricks (mulching, plasticulture...) and suitable rotations or co-cultivation with soil-cover plants.

In all cases, man must keep in mind that hoeing tends to deconstruct and to dry the soil, increasing the erosion. On the other hand, pre-emergence, systemic or contact herbicides can be used but they must be checked against their ability to be traced later in the essential oil. Different active ingredients have been tested on numerous weeds in geranium cultivation (atrazine, diuron, glyphosate, paraquat, metribuzine, nitrophen, alachlor...); their selectivity and the way to use them is available in the literature (Gravaud *et al.*, 1976; Srinivasan *et al.*, 1979; Didelot *et al.*, 1985; Garin, 1987; Trémel, 1992).

Pest and diseases

Rose geranium in cultivation is subject to attacks of insects, nematodes, fungi and bacteria. Pests and diseases are so numerous that it seems inconceivable to give the chemicals a miss and to produce organic Geranium oil in profitable conditions.

A little information is available on pests in Egypt and China, but in Reunion Island, a general survey (Quilici *et al.*, 1992) of the harmful insects revealed that geranium is attacked by at least 14 different species belonging to the *Hemiptera* (six species), *Coleoptera* (three species), and *Lepidoptera* (five species). Among the most important pests are the white grubs of *Hoplochelus marginalis* Fairmaire (*Coleoptera*, Fam. *Scarabaeidae*), the cockchafers *Cratopus humeralis* Boh., and *C. angustatus* Boh. (*Coleoptera*, Fam. *Curculionidae*), the whitefly *Trialeurodes vaporariorum* Westwood (*Hemiptera*, Fam. *Aleyrodidae*), the scale insect *Pseudaulacaspis pentagona* Targioni-Tozzetti (*Hemiptera*, Fam. *Diaspididae*) and the defoliating caterpillar of *Lobesia vanillana* (*Lepidoptera*, Fam. *Tortricidae*). An efficient integrated control of these insects exists as a combination of light trapping, agronomic controls (minimum tillage, cover-crop...), chemical controls (insecticides spraying, chemical trapping...) and biological controls (with insects and fungi which parasitize the larva) (Quilici *et al.*, 1992).

Nematodes have also been reported as important pests, especially in India. According to Rajeswara Rao (personal communication) several species of *Criconemoides*, *Helicotylenchus*, *Meloidogyne*, *Pratylenchus*, *Scutellonema* and *Xiphenema* damage the crop and inflict yield losses up to 75.8 per cent (Doraswamy *et al.*, 1979; Kumar and Nanjan, 1985; Anita and Vadivelu, 1997). Chemical control seems possible as well as biological control with a nematophagous fungus and/or companion cropping with nematicidal plants (periwinkle, marigold...) (Anita and Vadivelu, 1997).

Fungal and bacterial diseases are also numerous and very virulent, and considerable literature is available on these subjects. Several lists were established by different plant pathologists who have studied and referenced the pathogens in the producing countries (Bouriquet, 1946; Sinaretty and Trémel, 1992) (Rajeswara Rao) personal communication). Wilt, dieback, leaf blight, leaf spot, root and stem rot, anthracnose, are common and can be ascribed to several species of fungus (*Colletotrichum*, *Botrytis*, *Septoria*, *Cercospora*, *Armillaria*, *Rosellinia*, *Phomopsis*, *Pythium*, *Fusarium*...) and bacteria (*Pseudomonas solanacaerum*). Certain of those pathogens cause severe damages leading sometimes to the total destruction of the crop and the impossibility of growing geranium again on the same plot (e.g. wilt caused by *Pseudomonas solanacearum*). Generally speaking, the suggested control measures of these pathogens are only partially satisfactory, even if anthracnose and botrytis can be controlled by spraying specific fungicides. Despite the need for resistant varieties, a genetic improvement program does not seem reasonable or economically viable regarding the economic importance of the crop in the individual oil-producing countries.

HARVESTING

The Geranium oil is contained in glandular trichomes (Demarne and van der Walt, 1989) which are mainly located on both surfaces of the young leaves, on the young stems, on the buds and on different parts of the inflorescences. Oil is thus obtained from the top young parts of the plant and, because of the perennial character of the crop, harvesting geranium must fulfil two opposing objectives: the maximum production of herb and the preservation of the subsequent shoot development ability (Demarne, 1992d).

Therefore, the performance of a particular harvesting technique depends on the actual herb and essential oil yields, but also on the re-growth of the plants and the rate of mortality. The quality of the shoot development will govern the frequency and the yield of the subsequent harvests on individual plants, while the rate of mortality will drastically influence the number of plant per hectare, the time spent to fill in the gaps and hence the survival and the medium-term profitability of the crop.

From that point of view, manual harvesting with secateurs, knives or sickles is the best way for harvesting geranium as it prioritizes an individual approach of the plant architecture and tends to optimize both the herb yield and the re-growth ability. On the contrary, mechanical harvesting carries out an intermediate trimming of the whole field and only optimizes the labour productivity. Whatever principles they are based on (Korezawa *et al.*, 1967; Boyko, 1977; Paillat, 1987; Ducreux, 1993), the mechanical harvesters developed up till now partly solve the question of plant re-growth in cutting the utmost top of plants only (Demarne, 1992a) and in the different trials conducted mainly in Reunion, the loss of earnings was offset in increasing the harvesting frequency. Details on the way to prune rose geranium in different conditions are available in the literature (Doraswamy and Sundaram, 1982; Demarne, 1992d).

For all these reasons, geranium is still largely harvested by hand. The crop is ready for the first harvest 6–8 months after planting. Subsequent harvests are made at 3–5 months intervals, depending on plant development, weather conditions, crop management, labour availability... In all cases, harvests must be performed on clear and sunny days. The plant material is usually left one day on the inter-row to wilt. It is then transported to the distillery. At that time, heaping up or chopping the plant material is inadvisable to avoid fermentation, and distillation must be done quickly.

DISTILLATION

Distillation is the ultimate step of geranium cultivation. Because the essential oil is located in secretory sacs at the end of glandular hairs, its extraction is easy. Therefore, all the techniques of distillation and organic solvent or supercritical fluid extraction (SFE) give results; the differences being only in the yield and the quality of the products. In practice, there is only a market for water-distilled essential oil. Only Egypt produces and markets on request small quantities of geranium concrete and/or absolute.

Different kinds of distillery can be found in the producing countries, which correspond to different technical, economical and social situations. The numerous direct-fired stills of about 1000 l capacity operating in the countryside in Reunion Island, in China, in Egypt or in India have nothing in common with the batteries of 3000 l stills equipped with a separate boiler and encountered in certain factories, in Egypt or in Reunion Island.

However, the major part of the world production of Geranium oil is distilled in small units, belonging to small holders or to village communities. Processing 300–350 kg of plant material in a 1000 l direct-fired equipment lasts 90–150 min and gives a 0.1–0.2 per cent yield of oil. A complete description of the traditional equipment, including the limiting factors, the possible improvements and the detailed field distillation procedure is available in the literature (Demarne, 1992b). However it is interesting to point out that under such conditions, the distillation of 1 ha can take 40–100 days/year, without counting the time necessary for firewood or charcoal duty.

ESSENTIAL OIL PRODUCTION

World production

After more than one century of attempts, the so-called 'rose Geranium oil' is now produced mainly in the People's Republic of China, Egypt, Reunion Island and India. A few kilos are also exported from Madagascar and smaller quantities are sometimes produced in Algeria, Morocco and Russia.

China is now by far the biggest producer of Geranium oil. The cultivation is concentrated in the Yunnan Province, in the district of Binchuan. From an agricultural point of view, geranium is grown as a perennial (3 years) and makes the best of the horticultural techniques of the Chinese agriculture; the fields can be harvested four times a year, and the average oil yield is pretty good (± 45 kg/ha). An accurate assessment of the Chinese production and domestic consumption is precluded by the lack of reliable statistical data. Nevertheless and for several years, the annual production seems to be about 80–110 t, of which 20–30 t are used on the Chinese domestic fragrance market and 60–80 t are exported.

Egypt is the second biggest producer. Major areas of production are primarily in upper Egypt, mainly Bani Sweif and Fayoum. In this country the crop is annual, with two harvests a year. The oil is of the African type, as it contains substantial quantities of 10-epi-γ-eudesmol. According to different sources, the production in the past 10 years has averaged 50–55 t per year, with important year to year fluctuations.

On Reunion Island, despite big efforts made by the French government and the local authorities to support the production through research and development programs, and the remarkable results obtained by the agronomists, the production of rose Geranium oil has drastically declined over the years. The unequalled quality of the real Bourbon oil, universally recognised in high class perfumery, has not been sufficient to maintain the price at levels that can balance the increasing standard of living of the population and nowadays, the annual production is about 6 t, all of which is exported.

India has a small production (±2 t) in south hills and plains (Nilgiri hills, Pulney hills) which is used for its internal market.

International trade and demand

The main importers of Geranium oil are France and USA. However, whereas the bulk of US imports are consumed domestically, there is a big re-export trade from France. The French exports consist, on one hand, in pure Geranium oil and, on the other hand, in a proportion of oil that has been re-processed and blended to users' specifications. In most cases these oils are of very high quality.

'The international trade appears to be relatively stable and it is very likely that the USA and France will continue to dominate the import trade for the foreseeable future' (Robbins, 1984). The average annual imports of Geranium oil into the USA are about 60–65 t, and into France about 90–95 t. Japan imports about 15 t and the remainder of the European Community an other additional 30 t. Other appreciable markets include Brazil, India and a few others, but the trade statistics often fail to show Geranium oil separately, and it is difficult to gauge the extent of these trade flows in the majority of cases.

For all these markets combined, the total need is about 220 t, when trade between the USA and European Community, and within the European Community, is excluded.

All in all, the supply no longer fits the demand, and despite a production adjustment between the main sources, there is still an extra demand for 20–25 t of a very high quality Geranium oil to replace the Bourbon type.

Geranium oil is purchased and assessed mainly on the basis of its odour character and, in a lesser extent on its suitability for the preparation of rhodinol. For many years, it was traded almost exclusively on the basis of samples and this procedure still applies, even if, for certain origins (Bourbon and certain Yunnan oils), it is possible to buy on the basis of product description only.

In the importing countries, Geranium oil is traditionally marketed through a comprehensive network of intermediaries of various types including import merchants, dealers, commissioned brokers, agents and the processing and compounding houses. Geranium oil prices vary widely depending on the quality, the origin and also unforeseen supply fluctuations. Bourbon oil from Reunion Island traditionally commanded a price premium over oils from other sources, but the shortage of this quality in the recent years has caused switches to oils from alternative sources, causing the price of the latter oils to rise sharply.

OIL COMPOSITION AND QUALITY

Geranium oil is one of the most important raw materials used in perfume and soap industries. This product is rather well known and has been studied for the end of the nineteenth century. Its physical properties and its chemical composition have given rise to a great number of publication, as the analytical techniques progressed.

The works of Naves (Naves, 1934, 1944, 1951, 1957; Naves *et al.*, 1935, 1961, 1963; Melera and Naves, 1961; Naves and Tullen, 1961; Naves and Ochsner, 1962; Benesova *et al.*, 1964; Romanuk *et al.*, 1964; Krepinsky 1966), Teisseire (Teisseire and Corbier, 1964; Pesnelle *et al.*, 1969, 1971), Giannotti (Giannotti, 1966; Giannotti and Schwang, 1968a, b), Wolff *et al.* (1963), Klein (Klein and Rojahn, 1977; Rojahn and Klein, 1977), have been amongst the important contributions to the knowledge of the chemical composition of the Geranium oil. Since 1960, the development of gas-chromatography (GC) and associated identification techniques (GC-MS, GC-FTIR…) have sped up this acquisition of knowledge with the works of Garnéro *et al.* (Garnéro and Buil, 1969; Garnéro, 1974; Garnéro *et al.*, 1980), Peyron (Peyron, 1962a, b), Teisseire (Corbier and Teisseire, 1966; Pesnelle *et al.*, 1969), Kapétanidis (Heuss *et al.*, 1969), Timmer (Timmer *et al.*, 1971; Ter-Heide *et al.*, 1975), Kami (Kami *et al.*, 1969), Kaiser (Kaiser, 1984), Demarne (Guérère and Demarne, 1985; Demarne, 1989), Vérin (Vérin *et al.*, 1998), and others. More recently, techniques of multidimensional chiral gas chromatography coupled with mass spectrometry (enantio-MDGC/MS) (Ravid *et al.*, 1992; Kreis and Mosandl, 1993; Casabianca *et al.*, 1996; Wüst and Mosandl, 1998, 1999; Wüst *et al.*, 1996, 1998a,b, 1999; Allemant, 1998; Doimo *et al.*, 1999) has led to a better knowledge of the enantiomeric distribution of chiral compounds in the oils of different origins and, in a certain extent, could be used as criteria for authenticity assessment.

Based on this mass of academic information, applied research programs were carried out on the variation in the quality of the oil and to determine the influence of different internal, external and management factors influencing this (Gailleton, 1959; Korezawa and Tanida, 1969; Naragund and Divakar, 1983; Obaladze and Kuchuloriya, 1986; Angadi and Vasantha Kumar, 1989; Demarne and van der Walt, 1989; Zobenko and

Arinshtein, 1989; Kulkarni *et al.*, 1997), the climate and the soil (Charapov, 1968; Kalix, 1968; Michellon, 1980; Sun Handong *et al.*, 1985; Rajeswara Rao *et al.*, 1990, 1996; Bhattacharya *et al.*, 1993; Yan Dongwei *et al.*, 1994; Southwell *et al.*, 1995; Chakib, 1998), the plant part (Pieribattesti, 1982; Mallavarapu *et al.*, 1997), the crop age (Kaul *et al.*, 1998), the shade (Kaul *et al.*, 1997b), the fertilization (Korezawa, 1961), the presence of weeds and diseases (Rajeswara Rao and Bhattacharya, 1997; Rajeswara Rao *et al.*, 1999a), the irrigation (Singh, 1999), the time of harvesting and the technique used (Yoshida *et al.*, 1968; Mani and Sampath, 1981; Ravid and Putievsky, 1983; Tiberghien, 1989; Baret, 1991; Prakasa Rao *et al.*, 1995; Rajeswara Rao *et al.*, 1996), the post-harvest treatments (Mani *et al.*, 1981; Rajeswara Rao, *et al.*, 1992, 1999b), the distillation (Guérère *et al.*, 1985), the conditions and duration of storage (Mani *et al.*, 1981; Mahmoud *et al.*, 1983; Kaul, *et al.*, 1997a).

All these studies would tend to confirm that as far as the production scale is concerned, the three main important factors influencing the oil quality are the variety, the soil and climate conditions and also the distillation procedure. The quality variations of the oil mainly concern the relative concentrations of eight important compounds which are (*E*)- and (*Z*)-rose oxides, linalool, geraniol, citronellol, isomenthone, guaia-6,9-diene and 10-epi-γ-eudesmol.

Chemical composition of the oil

Several reviews were published by Lawrence (Lawrence, 1976, 1978, 1984, 1985, 1986, 1988, 1992, 1994, 1996, 1999) during the last years, which amalgamated the information published in the scientific literature on the composition of different Geranium oils from different geographical origins, but also on the identification of new components of the oil and on the enantiomeric ratios of some chiral compounds (α-terpineol, rose oxides, linalool, citronellol, menthone, isomenthone). Rose Geranium oil is a very complex product that contains hundreds of compounds, some being hydrocarbons (aliphatic, aromatic, monoterpenic, sesquiterpenic with different skeletons), and the others being oxygenated with alcohol, phenol, oxide, aldehyde, ketone, acid, ester and ether functional groups.

Only 30 compounds are regularly and individually present at more than 0.3 per cent of the oil; together they represent about 90 per cent of the essential oil and are sufficient to form the basis of the essential oil quality. The other components appear as traces, the correct identification of which is sometimes difficult, especially when sesquiterpenic hydrocarbons are concerned.

Quality

So, the quality of rose Geranium oil used in the fragrance industries depend on several criteria, from analytical determinations to compliance with international standards. The standards classify the Geranium oils into three types, according to their origins: the Chinese type is applicable to the essential from the People's Republic of China, the African type is for the oils from Algeria, Morocco and Egypt, and the Bourbon type for the essential oil from Reunion Island. The other oils are produced in too small quantities to be the subject of particular standards; when they appear on the international market, they are bought on the basis of their similarities with one of the three above-mentioned types.

According to the standards (Anonymous, 1996), the Geranium oils are easily characterised by their physical properties (specific gravity, optical rotation, refractive index, flash point, solubility in ethanol and colour), chemical properties (acid value, ester value, carbonyl value) and above all by their organoleptic properties (aspect, colour, odour) and their GCs profiles. The rose Geranium oils are liquid, limpid and less dense than water ($0.882 \leq {}^{20}d_{20} \leq 0.892$). Their colour vary from amber-yellow (Egypt) to yellowish-green, brownish-green or green (Bourbon). They all have a strong rose scent with a more or less pronounced mint note depending on their origin.

Capillary GC using polar or non-polar phases reveals that the major components of the oils are the monoterpenic alcohols (citronellol, geraniol, nerol, linalool, α-terpineol) and their esters (citronellyl formate, citronellyl butyrate, citronellyl propionate, citronellyl tiglate, geranyl formate, geranyl butyrate, geranyl propionate, geranyl tiglate, neryl formate), the monoterpenic ketones (menthone, (−)-isomenthone), the monoterpenic aldehydes (neral, geranial), the monoterpenic oxides ((*E*)- and (*Z*)-rose oxides), monoterpenic hydrocarbons (α-pinene, myrcene, limonene...), an aromatic ester (phenylethyl tiglate), several sesquiterpenic hydrocarbons (guaia-6,9-diene, germacrene D, β-bourbonene, β-caryophyllene), two sesquiterpenic ketones (furopelargones A and B) and a sesquiterpenic alcohol (10-epi-γ-eudesmol).

Among these components, geraniol, citronellol, linalool, isomenthone, geranyl formate, citronellyl formate, guaia-6,9-diene and 10-epi-γ-eudesmol play a particular role, as they allow to distinguish between oils of different origins and different varieties of rose geranium. The Bourbon type oil from Reunion Island contains large quantities (ca. 5–7 per cent) of guaia-6,9-diene but does not contain 10-epi-γ-eudesmol. Chinese geranium does not contain 10-epi-γ-eudesmol as well. On the contrary, the African types contain 10-epi-γ-eudesmol (ca. 4–5 per cent) but do not contain guaia-6,9-diene (Pesnelle *et al.*, 1969, 1971). However, when the varieties and the growing conditions are not fixed for long, both guaia-6,9-diene and 10-epi-γ-eudesmol can be found together, as in some Indian oils (Kaul *et al.*, 1997b), but on the international market this lead to suspicions of blending between oils of different origins.

Linalool is a good indicator of the extraction process as it seems to be formed from geraniol during steam water distillation. When the herb is extracted with solvent, the concrete thus obtained contains only small quantities of linalool (<0.5 per cent). But when the herb is distilled for more than 2 h in a traditional field equipment, the oil contains 5–10 per cent of linalool; the sum (linalool + geraniol) being constant in both the solvent extract and the water-distilled oil (Guérère *et al.*, 1985). In the industrial process, because of a limited contact between the herb and hot water, the linalool percentage is in-between those limits (ca. 4–5 per cent).

Geraniol, linalool and citronellol together also typify the different qualities. Citronellol is usually higher in Chinese oil (38–40 per cent) and Egyptian oil (31–33 per cent) than in Bourbon oil (19–21 per cent). On the contrary, geraniol and consequently linalool are higher in Bourbon oil (16–19 per cent/7–10 per cent) than in Egyptian oil (13–15/5–6 per cent) and Chinese oil (7–10/2–3 per cent); the ratio (linalool + geraniol)/citronellol varies from 0.3 (Chinese oil) to 0.6 (Egyptian oil) and 1.4 (Bourbon oil). The amount of geranyl formate and citronellyl formate are proportional to the corresponding alcohol in the oils; isomenthone is slightly higher in Bourbon oil (8–9 per cent) than in the others oils (ca. 6 per cent).

As far as the Indian oils are concerned, these figures vary depending on the variety under cultivation. The oil of certain cultivars even contain more geraniol than citronellol.

REFERENCES

Albers, F. and van der Walt, J.J.A. (1984) 'Untersuchungen zur Karyologie und Mikrosporogenese von *Pelargonium* sect. *Pelargonium* (Geraniaceae).' *Plant Systematics and Evol.*, 147(3), 177–188.

Allemant, S. (1998) Etude comparative des huiles essentielles de géranium selon leur origine géographique. Report, Ets. MANE et Fils, Grasse (France).

Angadi, S.P. and Vasantha Kumar, T. (1989) 'A new record of seed setting in scented geranium (*Pelargonium graveolens* L'Hérit.) from India.' *Current Sci.*, 58(24), 1401–1402.

Anita, B. and Vadivelu, S. (1997) 'Management of root knot nematode, Meloidogyne hapla on scented geranium, *Pelargonium graveolens* L.' *Indian J. Nemat.*, 27, 123- 125.

Anonymous (1996) Norme NFT 95-212/Dec.1987: Huile essentielle de Géranium (*Pelargonium* × *asperum* Ehrhart ex Willdenow). in Recueil des Normes Françaises. AFNOR. Paris (France), Association Française de Normalisation. 2, 101–108.

Arumugam, R. and Kumar, N. (1979) '*Geranium* cultivation in Kodaikanal hills.' *Indian Perfumer*, 23(2), 128–130.

Baret, P.J.J. (1991) Comparaison de l'extraction d'huile essentielle de géranium à partir d'une coupe manuelle et d'une coupe mécanique. Report, Université de la Réunion, Saint-Denis (Réunion).

Benesova, V., Chou, P.N., Herout,V., Naves,Y.R. and Lamparasky, D. (1964) 'Sur les terpènes: CLX – structure d'un gaiène inedit présent dans la fraction sesquiterpénique de l'huile essentielle de Géranium Bourbon.' *Czechoslov. Chem. Commun.*, (29), 1042–1047.

Bhattacharya, A.K., Kaul, P.N., Rajeswara Rao, B.R., Ramesh, S.I. and Mallavarapu, G. (1993) 'Composition of the oil of rose-scented geranium (*Pelargonium* sp.) grown under the semiarid tropical climate of south India.' *J. Essent. Oil Res.*, 5(2), 229–231.

Bouriquet, G. (1946) Maladies du Géranium rosat. In Les Maladies des Plantes Cultivées à Madagascar, Paris (France), 386–398.

Boyko, I.Y. (1977) The mechanized line for geranium processing. In *7th. Int. Cong. Essential Oils*, Kyoto (Japan). pp. 95–96.

Bricheteau, J., Alegre, J. and Lecouls, D. (1980) Le problème de la production de boutures de *Pelargonium* à partir de pieds-mères cultivés sur substrat inerte. In Le *Pelargonium (acquisitions nouvelles)*. E.N.I.T.H. Angers (France). 1, 17–39.

Casabianca, H., Graff, J.B. and Guillamet, S. (1996) 'Analyses chirales et isotopiques des principaux constituants de roses et de géraniums.' Rivista Italiana EPPOS (Numero Speciale), 244–261.

Chabalier, P.F. (1992) La fertilisation. In Le géranium rosat à la Réunion. C.A. Hauts. Saint-Denis (Reunion Island), *Graphica*, 51–62.

Chakib, S. (1998) Les essences de géranium: influences pédoclimatique et saisonnière sur les essences marocaines, classification par les méthodes d'analyse de données. Doctorate, Chemistry, Université de Droit, d'Economie et des Sciences d'Aix-Marseille III, Marseille (France), p. 166.

Charapov, N.I. (1968) Influence du climat sur la productivité des plantes et la qualité de l'huile essentielle. *4th. Int. Cong. Essential Oils*, Moscou (USSR).

Corbier, B. and P. Teisseire (1966) Contribution à la connaissance de l'huile essentielle de Géranium Bourbon. *Recherches*, 89–90.

Dantes, E. (1899) 'Le géranium.' Revue Agricole de la Réunion, pp. 458–460.

Dartigues, A. and Lemaire, F. (1980) Besoins en élements minéraux du pieds-mère de *Pelargonium* × *hortorum* conséquences de la nature de l'azote sur la qualité des boutures. In Le *Pelargonium* (acquisitions nouvelles). Angers (France), E.N.I.T.H. 1, 41–50.

Demarne, F.-E. (1981) Essai sur le bouturage du Géranium rosat. Report, CIRAD, Saint-Denis (Reunion).

Demarne, F.-E. (1989) L'amélioration variétale du Géranium rosat (Pelargonium sp.) – contribution systématique, caryologique, et biochimique. Doctorate, Plant Genetics, Paris Sud, Orsay (France), p. 163.

Demarne, F.-E. (1992a) 'Influence du mode de récolte sur la distillation du géranium rosat.' *L'Agronomie Tropicale*, 46(2), 161–163.

Demarne, F.-E. (1992b) La distillation. In Le géranium rosat à la Réunion. C.A. Hauts. Saint-Denis (Reunion Island), *Graphica*, 97–102.

Demarne, F.-E. (1992c) Le bouturage. In Le géranium rosat à la Réunion. C.A. Hauts. Saint-Denis (Reunion Island), *Graphica*, 39–43.

Demarne, F.-E. (1992d) Les opérations de récolte. In Le géranium rosat à la Réunion. C.A. Hauts. Saint-Denis (Reunion Island), *Graphica*, 91–96.

Demarne, F.-E. and van der Walt, J.J.A. (1989) 'Origin of rose-scented *Pelargonium* cultivar grown on Réunion island.' *S. A. J. Bot.*, 55(2), 184–191.

Demarne, F.-E., Viljoen, A. M. and van der Walt, J.J.A. (1993) 'A study of the variation in the essential oil and morphology of *Pelargonium capitatum* (L.) L'Hérit. (Geraniaceae). Part 1: The composition of the oil.' *J. Essen. Oil Res.*, 5(5), 493–499.

Dhakshinamoorthy, M., Arumugam, R. and Mani A.K. (1980) 'Effect of copper and molybdenum on the herbage yield and oil of geranium (*Pelargonium graveolens* L'Hérit.).' *Indian Perfumer*, 24(3), 214–215.

Didelot, D., Garin, P. and Michellon, R. (1985) Essai d'herbicides dans l'association géranium-haricot. Report, CIRAD-IRAT, Saint-Denis (Réunion).

Doimo, L., Fletcher, R.J. and D'Arcy, B.R. (1999) 'Chiral excess: measuring the chirality of geographically and seasonally different Geranium oils.' *J. Essen. Oil Res.*, 11(3), 291–299.

Doraswamy, K. and Sundaram, M. (1982) 'Geranium cultivation in south India.' Cultivation & Utilization of Aromatic Plants, CSIR India Pub., 573–577.

Doraswamy, K., Sundaram, M. and Manian, K. (1979) 'Root knot nematode control in scented geranium (*Pelargonium graveolens* L'Hérit.).' *The Madras Agric. J.*, 66(3), 205–206.

Ducellier, L. (1933) 'Observations sur la descendance du Géranium Rosat.' *Bull. Soc. Hist. Nat.*, Afrique du Nord 24, 142–148.

Ducreux, A. (1993) La récolteuse de géranium: un outil indispensable pour la modernisation de la culture. Report, CIRAD-SAR REUNION, Saint-Denis (Reunion Island).

Fritz, J. (1976) 'Effet de la fertilisation azotée sur la production du Géranium rosat.' *L'Agronomie Tropicale*, 31(4), 369–374.

Gailleton, J.M. (1959) Le géranium Rosé, variété à propager. Report, Direction Départementale de l'Agriculture de la Réunion, Saint-Denis (Réunion).

Garin, P. (1987) 'Systèmes de culture et itinéraires techniques dans les exploitations à base de géranium dans les hauts de l'ouest de La Réunion.' *L'Agronomie Tropicale*, 42(4), 289–300.

Garnéro, J. (1974) Etude de deux huiles essentielles complexes: les essences de *Géranium* et les essences de Vétyver. Report, Ets. Robertet, Grasse (France).

Garnéro, J. and Buil, P. (1969) Contrôle par l'analyse instrumentale des huiles essentielles: 21 – sur les essences de *Géranium* Bourbon et Maroc. Report, Ets. Robertet, Grasse (France).

Garnéro, J., Joulain, D. and Buil, P. (1980) Etude de la composition chimique de l'huile essentielle de géranium citronelle de La Réunion. Report, Ets. Robertet, Grasse (France).

Giannotti, C. (1966) 'Isolement de deux nouveaux diols monoterpéniques de l'essence de Géranium Bourbon.' *C.R. Acad. Sci. Paris*, Série C, 262, 422–425.

Giannotti, C. and Schwang, H. (1968a) 'Sur la structure d'un nouveau nor-sesquiterpène isolé de l'essence de Géranium Bourbon: la 11-nor-bourbonanone-1.' Bulletin de la Société Chimique de France(6), 2452–2456.

Giannotti, C. and Schwang, H. (1968b) 'Sur l'isolement et la structure de deux dehydrofuropélargones, nouveaux sesquiterpènes de l'essence de Géranium Bourbon.' *Tetrahedron*, 24, 2055–2061.

Gravaud, A., Roura, A. and Bédier, A. (1976) Desherbage du Géranium. Report, Service Protection des Végétaux de La Réunion, Saint-Denis (Réunion).

Guenther, E. (1977) Essential oils of the family Geraniaceae: oil of Geranium. *The Essential oils*. Robert E. Krieger Pub. Co., pp. 671–737.

Guérère, M. and Demarne, F.-E. (1985) 'Caractérisation de l'huile essentielle de Géranium Bourbon par chromatographie en phase gazeuse sur colonne capillaire.' *Ann. Fals. Exp. Chim.*, 78(837), 183–188.

Guérère, M., Demarne, F.-E., Mondon, J.M. and Pajaniaye, A. (1985) 'Etude d'huiles essentielles de Géranium Bourbon obtenues de trois façons différentes.' *Ann. Fals. Exp. Chim.*, 78(836), 131–136.

Héricart De Thury (1845) 'Culture du Géranium rosat (*Pelargonium capitatum*), et emploi de ses produits.' *Revue Horticole*, 206–208.

Heuss, A., Kapétanidis, I. and Mirimanoff, A. (1969) 'Etude par chromatographie en phase gazeuse de l'huile essentielle d'un Géranium Rosat (*Pelargonium* × *asperum* Ehrhart ex Willdenow).' *Plantes médicinales et phytothérapie*, 3(1), 28–43.

Igolen, G. (1941) 'L'essence de géranium d'Algerie.' *Annales de Chimie Analytique*, Tome, 23(10), 260–262.

Kaiser, R. (1984) '(5R*,9S*)- and (5R*,9R*)-2,2,9-trimethyl-1,6-dioxaspirol [4.4] non-3-ene and their dihydro derivatives as new constituents of Geranium oil.' *Helvetica Chimica Acta.*, 67, 1198–1203.

Kalix, P. (1968) Culture en milieu synthetique d'une plante à essence (Géranium Rosat): étude préliminaire de l'influence du facteur climatique et du milieu nutritif sur la biosynthèse des terpénoïdes. In Pharmacy, University of Geneva, Geneva (Switzerland): pp. 80p.

Kami, T., Otaishi, S., Hayashi, S. and Matsuura, T. (1969) 'A study on low-boiling compounds of essential oil of geranium species.' *Agr. Biol. Chem.*, 33(4), 502–505.

Kaul, P.N., Rajeswara Rao, B.R., Bhattacharya, A.K., Mallavarapu, G.R. and Ramesh S.I. (1997a) 'Changes in chemical composition of rose-scented geranium (*Pelargonium* sp.) oil during storage.' *J. Essent. Oil Res.*, 9(1), 115–117.

Kaul, P.N., Rajeswara Rao, B.R., Bhattacharya, A.K. and Singh, K. (1998) 'Relationship between crop age, essential oil concentration and its composition in rose-scented geranium (*Pelargonium* sp.).' *J. Essent. Oil Bearing Plants*, 1, 88–92.

Kaul, P.N., Rajeswara Rao, B.R., Bhattacharya, A.K., Singh, K. and Singh, C.P. (1997b) 'Effect of partial shade on essential oils of three *Geranium* (*Pelargonium* sp.) cultivars.' *Indian Perfumer*, 41, 1–4.

Klein, E. and Rojahn, W. (1977) 'La synthèse de l'acide oxo-6-dihydro-6,7-citronellique, un constituant inédit de l'essence de Géranium Réunion.' Dragoco Report (3), 55–58.

Korezawa, N. (1961) 'Studies on the yield analysis of essential oil in *Pelargonium* species: (IV) relation between the vicissitude of the ingredients and quality of green herb, the oil yield and the percentage yield of oil.' *Jpn. J. Trop. Agr.*, (Nettai Nogyo), 4(4), 169–172.

Korezawa, N. and Tanida, M. (1969) 'On the characteristics and some problems in cultivation of *Pelargonium* new cultivar B1-N °10.' *Jpn. J. Trop. Agr.*, (Nettai Nogyo), 12(3–4), 153–157.

Korezawa, N., Tanida, M., Takeuchi, M. and Kubo, H. (1967) 'Studies on the practicability of the rotary grass cutter and the planting-form of the rose geranium field.' *Jpn. J. Trop. Agr.* (Nettai Nogyo), 11(1–2), 13–19.

Koriesh, E.M. and Atta, S.K.H. (1986) 'Relationship between irrigation intervals and soil conditioner application on growth and oil yield of geranium plant (*Pelargonium graveolens*, L'Hérit.).' *Annals of Agric. Sci.*, Moshtohor, 24(1), 447–454.

Kreis, P. and Mosandl, A. (1993) 'Chiral compounds of essential oils. Part XIII. Simultaneous chirality evaluation of Geranium oil constituents.' *Flav. Frag. J.*, 8(3), 161–168.

Krepinsky, J., Samek, Z., Sorm, F., Lamparsky, D., Ochsner, P. and Naves, Y.R. (1966) 'The structure and absolute configuration of alpha- and beta-bourbonene, sesquiterpenic hydrocarbons from the essential oil of Geranium Bourbon.' *Tetrahedron*, Supp. 8 (Part 1), 53–70.

Kuchuloriya, T.L. (1964) Sélection du géranium. *3rd. Int. Cong. Essential Oil*, Plodiv (Bulgaria).

Kuchuloriya, T.L. (1968) 'Résultats de la sélection du Géranium Rosat.' Izd. pyscevaja promyslennost, 39–47.

Kulkarni, R.N., Baskaran, K., Ramesh, S. and Kumar, S. (1997) 'Intra-clonal variation for essential oil content and composition in plants derived from leaf cuttings of rose-scented geranium (*Pelargonium* sp.).' *Industrial Crops and Products*, 6(2), 107–112.

Kumar, S. and Nanjan, K. (1985) 'Increasing Geranium oil yield.' *Indian Perfumer*, 29, 121–124.

Ladaria, L.N. (1968) 'Utilisation des oligo-élements en culture de Géranium rosat.' Efirnomaslichye rastenija ikh Kul'tura i pererabotka, 91–98.

Lawrence, B.M. (1976) 'Progress in essential oils.' *Perf. Flav.*, 1, 45–46.

Lawrence, B.M. (1978) 'Progress in essential oils.' *Perf. Flav.*, 3, 54–56.

Lawrence, B.M. (1984) 'Progress in essential oils.' *Perf. Flav.*, 9(5), 87.

Lawrence, B.M. (1985) 'Progress in essential oils.' *Perf. Flav.*, 10(5), 93.

Lawrence, B.M. (1986) 'Progress in essential oils.' *Perf. Flav.*, 11(3), 49.

Lawrence, B.M. (1988) 'Progress in essential oils.' *Perf. Flav.*, 13(5), 61.

Lawrence, B.M. (1992) 'Progress in essential oils.' *Perf. Flav.*, 17(6), 51.

Lawrence, B.M. (1994) 'Progress in essential oils.' *Perf. Flav.*, 19(1), 31.

Lawrence, B.M. (1996) 'Progress in essential oils.' *Perf. Flav.*, 21(6), 58–62.

Lawrence, B.M. (1999) 'Progress in essential oils.' *Perf. Flav.*, 24(1), 58–62.

L'Héritier Dom De Brutelle, C.L. (1789) In: Aiton (ed.), Hortus Kewensis. London, George Nicol., 2.

Lougnon, A. (1924) 'Le Géranium et les cultures vivrières.' Revue Agricole de la Réunion(3), 104–108.

Mahmoud, M.M., El-Gmal, E.L.A.H., Hassancin, A.F.M. and Mohamed, S.M. (1983) Géranium: effets sur l'huile essentielle des conditions de conservation. In 4th Symposium I.S.H.S.

Mallavarapu, G.R., Rajeswara Rao, B.R., Kaul, P.N. and Ramesh, S. (1997) 'Contribution of the essential oils of leaf, petiole and stem of scented geranium to the odour of Geranium oil.' *J. Med. Arom. Plant Sci.*, 19, 1020–1023.

Mani, A.K., Mohandass, S., Kumar, N. and Sampath, V. (1981) 'Effect of storage time of the herbage prior distillation on oil recovery and its quality in geranium (*Pelargonium graveolens* L'Hérit.).' *Indian Perfumer*, 25(3–4), 35–36.

Mani, A.K. and Sampath, V. (1981) 'Seasonal influence on the oil content and quality of oil in geranium (*Pelargonium graveolens* L'Hérit.).' *Indian Perfumer*, 25(3–4), 41–43.

Melera, A. and Naves, Y.R. (1961) 'Résonance magnétique nucléaire: spectres de résonance magnétique nucléaire et stéreochimie des oxydes $C_{10}H_{18}O$ de l'huile essentielle de Géranium.' *Comptes Rendus des Séances de l'Académie des Sciences (Paris)* Tome, 252, 1937–1938.

Michellon, R. (1975) Essais de bouturage du Géranium rosat. IRAT-REUNION, Saint-Denis (Réunion). pp. 1–11.

Michellon, R. (1976) Essaish de bouturage du Géranium rosat. IRAT-REUNION, Saint-Denis (Réunion). pp. 1–4.

Michellon, R. (1980) Influence de l'altitude sur le comportement et la composition de l'huile essentielle du Géranium rosat. IRAT-REUNION, Saint-Denis (Réunion).

Michellon, R. (1987) Amélioration des systèmes de culture à base de géranium. IRAT, Saint-Denis (Réunion).

Michellon, R., Dejante, P., Vincent, G. and Nativel, R. (1994a) Gestion d'une couverture de kikuyu (Pennisetum clandestinum) associée au géranium rosat. CIRAD-CA Réunion, Saint-Denis (Réunion). pp. 41 p.

Michellon, R., Dejante, P., Vincent, G. and Nativel, R. (1994b) Gestion d'une couveture de lotier (Lotus uliginosus) associée au géranium rosat. CIRAD-Réunion, Saint-Denis (Réunion). pp. 1–42.

Michellon, R., Seguy, L. and Perret, S. (1996) Association de cultures Maraîchères et du Géranium Rosat à une couverture de kikuyu (Pennisetum clandestinum) maîtrisée avec le Fluazifop-P-butyl. In ANPP-Quatrième colloque sur les substances de croissance, partenaires économiques des productions végétales, Paris (France), 6 February. p. 8.

Moore, H.E. (1955) 'Pelargoniums in cultivation. (I).' *Baileya*, 3(1), 5–25 and 40–46.

Naragund, V.R. and Divakar, N.G. (1983) 'Varietal evaluation in scented geranium (*Pelargonium graveolens* L'Hérit.).' *Indian Perfumer*, 27(1), 19–21.

Narayana, M.R., Prakasa Rao, E.V.S., Rajeswara Rao, B.R. and Sastry, K.P. (1986a) 'Geranium cultivation in India: potentials and prospects.' *Pafai J.*, 8(4), 25–29.

Narayana, M.R., Prakasa Rao, E.V.S., Rajeswara Rao, B.R. and Sastry, K.P. (1986b) New agronomic practices for growing geranium in south India. In Plantation Crops: Opportunities and Constraints, Oxford-IBH Publ. Co., 1, 339–342.

Naves, Y.R. (1934) 'Les essences de géranium, production, composition, caractères analytiques.' *Les Parfums de France*, (137), 168–180.

Naves, Y.R. (1944) '126. Etudes sur les matières végétales volatiles: XXXI- sur l'isolement des huiles essentielles dissoutes dans les eaux de distillation.' *Helvetica Chimica Acta*, 27, 1103–1108.

Naves, Y.R. (1951) 'The composition of essential oil of *Pelargonium tomentosum* Jacquin.' Perfumery and Essential Oil Record, 42, 113 and 121.

Naves, Y.R. (1957) 'On citronellol (rhodinol) and geraniol in geranium and rose oils.' Perfumery & Essential Oil Record, 118–119.

Naves, Y.R., Brus, G. and Allard, J. (1935) 'Effet Raman et chimie: contribution à l'étude de l'isomérie citronellol-rhodinol au moyen de la spectrographie Raman.' *Comptes Rendus des Séances de l'Académie des Sciences (Paris)* Tome, 200, 1112.

Naves, Y.R., Lamparsky, D. and Ochsner, P. (1961) 'Etudes sur les matières végétales volatiles CLXXIV(1): présence de tétrahydropyrannes dans l'huile essentielle de géranium.' *Bull. Soc. Chim. de France*, 645–647.

Naves, Y.R. and Ochsner, P. (1962) '47. Etudes sur les matières végétales volatiles: CLXXX- sur les cis-[méthyl-2-propène]-yl-2-méthyl-4-tétrahydropyrannes.' *Helvetica Chimica Acta*, 45(47), 397–399.

Naves, Y.R., Oschner, P., Thomas, A.F. and Lamparsky, D. (1963) 'Etudes sur les matières végétales volatiles: CLXXXVI (I) – présence d'acétonyl-2 méthyl-4 tétrahydro-pyranne dans l'huile essentielle de Géranium.' Bulletin de la Société Chimique de France, 1608–1611.

Naves, Y.R. and Tullen, P. (1961) '229. Etudes sur les matières végétales volatiles: CLXXIX – Synthèse des cis- et trans-[methyl-2-propene-1]-yl-2-methyl-4-terrahydropyrannes.' *Helvetica Chimica Acta*, 44(229), 1867–1872.

Obaladze, L.V. and Kuchuloriya, L.T. (1986) 'The geranium variety Yubileinaya.' Maslichnye Kul'tury, (2), 34.

Paillat, J.M. (1987) Mécanisation de la récolte du géranium: mise au point d'une machine adaptée. CEEMAT-REUNION, Saint-Denis (Reunion Island). pp. 1–4.

Payet, J. (1981) Caractéristiques morphologiques, anatomiques et cytologiques des Pelargonium à feuilles odorantes. Réunion Cytologie et Cytogénétique INRA-DIJON, Dijon (France), pp. 24–28.

Perret, S., Michellon, R., Boyer, J. and Tassin, J. (1996) 'Soil rehabilitation and erosion control through agro-ecological practices on Reunion Island (French overseas territory, Indian Ocean).' *Agric. Ecosyst. Environ.*, 59(3), 149–157.

Pesnelle, P., Corbier, B. and Teisseire, P. (1969) Comparison between Bourbon and African Geranium oils at the level of their sesquiterpenic components. 1st International Perfumery Congress, Paris (France). pp. 45–49.

Pesnelle, P., Corbier, B. and Teisseire, P. (1971a) 'Comparaison entre les essences de Géranium Bourbon et de Géranium Afrique au niveau de leurs constituants sesquiterpéniques.' *Recherches*, (18), 45–52.

Pesnelle, P., Corbier, B. and Teisseire, P. (1971b) 'Essences de Géranium.' *Parf. Cosm. Sav. France*, 1(12), 637–640.

Peyron, L. (1962a) 'Contribution à l'étude de l'essence de Géranium des eaux obtenue dans l'hydrodistillation du Géranium Rosat du Maroc.' *C.R. Acad. Sci. Paris*, 255(22), 191–192.

Peyron, L. (1962b) 'Contribution à l'étude de l'essence de géranium rosat.' Parfumerie Cosmétique Savons , 5(6), 1–4.

Pieribattesti, J.-C. (1982) Contribution à l'étude de quelques huiles essentielles de la Réunion. In Organic Chemistry, Université d'Aix-Marseille, Aix-Marseille (France), 242.

Prakasa Rao, E.V.S., Ganesha Rao, R.S. and Ramesh, S. (1995) 'Seasonal variation in oil content and its composition in two chemotypes of scented geranium (*Pelargonium* sp.).' *J. Essent. Oil Res.*, 7(2), 159–163.

Prakasa Rao, E.V.S., Singh, M. and Ganesh Rao, R.S. (1986) 'Effect of nitrogen fertilizer on geranium (*Pelargonium graveolens* L'Hérit ex. Ait.), cowpea and blackgram grown in sole cropping and intercropping systems.' *Intern. J. Tropical Agric.*, 4(4), 341–345.

Prakasa Rao, E.V.S., Singh, M. and Ganesha Rao, R.S. (1988) 'Effect of plant spacings and nitrogen levels on herb and essential oil yields and nutrient uptake in geranium (*Pelargonium graveolens* L'Herit. ex Ait.).' *Intern. J. Trop. Agric.*, 6(1–2), 95–101.

Prakasa Rao, E.V.S., Singh, M., Ganesha Rao, R.S. and Rajeswara Rao, B.R. (1984a) 'Intercropping studies in geranium (*Pelargonium graveolens* L'Hér. ex Ait.).' *J. Agric. Sci.*, (Cambridge) 102, 499–500.

Prakasa Rao, E.V.S., Singh, M., Narayana, M.R. and Chandrasekhara, G.C. (1984b) 'Micronutrient studies in geranium (*Pelargonium graveolens* L'Hérit.) and davana (Artemisia pallens Wall.).' *Indian Perfumer*, 28(2), 88–90.

Quilici, S., Vercambre, B. and Bonnemort, C. (1992) Les insectes ravageurs. In: C.A. Hauts (ed.) Le géranium rosat à la Réunion. Saint-Denis (Reunion Island), *Graphica*, 79–90.

Rajeswara Rao, B.R. and Bhattacharya, A.K. (1997) 'Yield and chemical composition of the essential oil of rose-scented geranium (*Pelargonium* sp.) grown in the presence and absence of weeds.' *Flav. Frag. J.*, 12, 201–204.

Rajeswara Rao, B.R., Bhattacharya, A.K., Kaul, P.N. and Chand, S. (1993) 'Behaviour of rose geranium (*Pelargonium* species) cuttings in response to IBA treatment.' *Indian Perfumer*, 37, 194–196.

Rajeswara Rao, B.R., Bhattacharya, A.K., Kaul, P.N. and Ramesh, S. (1992) 'The essential oil profiles of rose scented geranium (*Pelargonium* spp.) biomass dried prior to distillation.' *Indian Perfumer*, 36(4), 238–240.

Rajeswara Rao, B.R., Bhattacharya, A.K., Singh, H.B. and Mallavarapu, G.R. (1999a) 'The impact of wilt disease on yield and quality of two cultivars of rose-scented geranium (*Pelargonium* species).' *J. Essent. Oil Res.*, 11(6), 769–775.

Rajeswara Rao, B.R., Kaul, P.N., Mallavarapu, G.R. and Ramesh, S. (1996) 'Effect of seasonal climatic changes on biomass yield and terpenoid composition of rose-scented geranium (*Pelargonium* species).' *Biochem. System. Ecol.*, 24(7–8), 627–635.

Rajeswara Rao, B.R., Kaul, P.N., Mallavarapu, G.R. and Singh, R. (1999b) 'Comminution of plant material and its effect on the quality of rose-scented geranium (*Pelargonium* species) oil.' *J. Essent. Oil Res.*, 11(5), 589–592.

Rajeswara Rao, B.R., Sastry, K.P., Prakasa Rao, E.V.S. and Ramesh, S.I. (1990) 'Variation in yields and quality of geranium (*Pelargonium graveolens* L'Hérit. ex Ait.) under varied climatic and fertility conditions.' *J. Essent. Oil Res.*, 2(2), 73–79.

Ravid, U. and Putievsky, E. (1983) *Pelargonium graveolens* L.: influence de la date de récolte et de la nature des feuilles sur le taux de l'huile essentielle et ses principaux constituants. In 4th Symposium I.S.H.S.

Ravid, U., Putievsky, E., Katzir, I., Ikan, R. and Weinstein, V. (1992) 'Determination of the enantiomeric composition of citronellol in essential oils by chiral GC analysis on a modified gamma-cyclodextrin phase.' *Flav. Frag. J.*, 7(4), 235–238.

Robbins, S.R.J. (1984) *Geranium oil: market trends and prospect.* In IFEAT World Council Meeting, Cairo (Egypt). pp. 8p.

Rojahn, W. and Klein, E. (1977) 'La citronellyle-diethylamine, un constituant nouveau de l'essence de géranium Réunion (*Pelargonium graveolens*).' Dragoco Report 24(6), 150–152.

Romanuk, M., Herout, V., Sorm, F., Naves, Y.R., Tullen, P., Bates, R.B. and Sigel, C.W. (1964) 'Sur les terpènes: CLXI – structure chimique de la pélargone, cétone sesquiterpénique de l'essence de Géranium Bourbon.' *Collection Czechoslov. Chem. Commun.*, 29, 1048–1058.

Sabina Aiyanna, S., Farooqi, A.A., Shivashankar, K. and Khan, M.M. (1998) 'Effect of fertilization methods on biomass, oil yield and economics in geranium (*Pelargonium* sp.) in India.' *J. Essent. Oil Res.*, 10(1), 51–56.

Sanderovich, D., Putievsky, E. and Ravid, U. (1983) 'Irrigation of aromatic plants: *Pelargonium graveolens*.' *Hassadeh*, (63), 1980–1983.

Sinaretty, N. and Trémel, L. (1992) Les maladies. In: C.A. Hauts (ed.), *Le géranium rosat à la Réunion*. Saint-Denis (Reunion Island), *Graphica*, 69-78.

Singh, M. (1999) 'Effect of soil moisture regime, nitrogen and modified urea materials on yield and quality of geranium (*Pelargonium graveolens*) grown on alfisols.' *J. Agric. Sci.*, 133, 203–207.

Singh, M., Chandrashekhara, R.S., Ganesha Rao, R.S. and Prakasa Rao, E.V.S. (1996) 'Effect of irrigation and levels of nitrogen on herb and oil yield of geranium (*Pelargonium* sp.) under semiarid tropical India.' *J. Essent. Oil Res.*, 8(6), 653–656.

Southwell, I.A., Stiff, I.A. and Curtis, A. (1995) 'An Australian Geranium oil.' *Perf. Flav.*, 20(4), 11–14.

Srinivasan, P.S., Sambandamurthy, S., Veerannah, L. and Kandaswamy, O.S. (1979) 'A note on the effect of herbicides on the yield and oil content in geranium (*Pelargonium graveolens* L.).' *South Indian Horticulture*, 27(3), 138–139.

Sun Handong, Ding Jingkai, Ding Lisheng, Yi Yuanfen, and Wu Jiangyun (1985) 'The chemical constituents of Geranium oil.' *Acta Botanica Yunnanica*, 7(2), 233–237.

Tamai, T., Tokumasu, S. and Shinohara, K. (1958) 'Studies on the breeding of *Pelargonium* species used for the essential oil production: I – artificially induced tetraploid plant in *Pelargonium roseum*.' *Jpn. J. Breeding*, 7(3), 131–140.

Tamai, T., Tokumasu, S. and Yamada, K. (1963) 'Studies on the breeding of *Pelargonium* species used for the essential oil production: II – artificially induced tetraploid plant in *Pelargonium denticulatum*.' *Jpn. J. Breeding*, 13(3), 143–148.

Teisseire, M.M.P. and Corbier, B. (1964) Recherches sur l'oxyde de rose et sur des substances hétérocycliques voisines. In 3rd. Int. Cong. Essential Oils, Plodiv (Bulgaria). pp. 68–79.

Ter-Heide, R., De Valois, P.J., Wobben, H.J. and Timmer, R. (1975) 'Analysis of the acid fraction of Réunion Geranium oil (*Pelargonium graveolens* L'Hér. ex Ait.).' *J. Agr. Food Chem.*, 23(1), 57–60.

Tiberghien, B. (1989) Etude sur la distillation du géranium. Analyse des pratiques de distillation et des effets de la récolte du géranium sur la distillation. ISTOM, Le Havre (France), pp. 1–66.

Timmer, R., Heide, R., De Valois, P.J. and Wobben, H.J. (1971) 'Qualitative analysis of the most volatile neutral components of Réunion Geranium oil (*Pelargonium roseum* Bourbon).' *J. Agric. Food Chem.*, 19(6), 1066–1068.

Tokumasu, S. (1970) 'Comparison of anther and pollen development between male-sterile tetraploids in Pelargonium roseum.' *Jpn. J. Breeding*, 20(4), 211–218.

Trémel, L. (1992) La lutte contre les mauvaises herbes. In: C.A. Hauts (ed.), *Le géranium rosat à la Réunion*. Saint-Denis (Reunion Island), *Graphica*, 63–67.

van der Walt, J.J.A. (1985) 'A taxonomic revision of the type section of *Pelargonium* L'Hérit. (Geraniaceae).' *Bothalia*, 15(3 & 4), 345–385.

van der Walt, J.J.A. and Demarne, F.-E. (1988) '*Pelargonium graveolens* and *P. radens*: a comparison of their morphology and essential oils.' *S. Afr. J. Bot.*, 54(6), 617–622.

Vérin, P., Bianchini, J.-P., Raharivelomanana, P., Claude-Fontaine, A., Brugel, P. and Cambon, A.(1998) 'Réalisation d'une banque de données d'esters en spectrométrie de masse.' Rivista Italiana EPPOS (Numero Speciale), 592–595.

Vernin, G. and Metzger, J. (1983) 'Etude des huiles essentielles par CG-SM-banque Specma: essences de Géranium.' Parfums, Cosmétiques, *Arômes*, (52), 51–61.

Vidalie, H. (1980) Influence de l'âge du pied-mère et des conditions climatiques sur le rendement en boutures du Pelargonium x hortorum. In *Le Pelargonium (acquisitions nouvelles)*. Angers (France), E.N.I.T.H., pp. 61–65.

Viljoen, A.M., van der Walt, J.J.A., Swart, J.P.J. and Demarne, F.-E.(1995) 'A study of the variation in the essential oil of *Pelargonium capitatum* (L.) L'Hérit. (Geraniaceae). Part II. The chemotype of P. capitatum.' *J. Essent. Oil Res.*, 7(6), 605–611.

Weiss, E.A. (1967) 'Effects of nitrogen, phosphate and time of cutting on green material and oil yields from geranium (Pelargonium hybrid) in west Kenya.' *Expl. Agric.*, 3(2), 99–103.

Wolff, R.E., Ma, J.C.N. and Lukas, G. (1963) 'Sur la structure chimique des pélargones A et B, cétones sesquiterpéniques isolées de l'huile essentielle de Géranium Bourbon.' *C.R. Acad. Sci.*, Paris, 1784–1786.

Wüst, M., Beck, T., Dietrich, A. and Mosandl, A. (1996) 'On the biogenesis of rose oxide in *Pelargonium graveolens* L'Héritier and *Pelargonium radens* H.E. Moore.' *Enantiomer*, 1, 167–176.

Wüst, M. and Mosandl, A. (1998) On the biogenesis of rose oxide in Pelargonium graveolens (Geraniaceae). In 29th International Symposium on Essential Oils, Frankfurt am Main (Germany), September 6–9.

Wüst, M. and Mosandl, A. (1999) 'Important chiral monoterpenoid ethers in flavours and essential oils – enantioselective analysis and biogenesis.' *Eur. Food Res. Tech.*, 209(1), 3–11.

Wüst, M., Reindl, J., Fuchs, S., Beck, T., and Mosandl, A. (1999) 'Structure elucidation, enantioselective analysis, and biogenesis of nerol oxide in Pelargonium species.' *J. Agric. Food Chem.*, 47(8), 3145–3150.

Wüst, M., Rexroth, A., Beck, T., and Mosandl, A. (1998a) 'Mechanistic aspects of the biogenesis of rose oxide in *Pelargonium graveolens* L'Héritier.' *Chirality*, 10(3), 229–237.

Wüst, M., Rexroth, A., Beck, T., and Mosandl, A. (1998) Structure elucidation of *cis*- and *trans*-*rose* oxyde ketone and its enantioselective analysis in Geranium oils. In 29th International Symposium on Essential Oils, Frankfurt am Main (Germany), September 6–9.

Yan Dongwei, Zhang Zhengju, and Ouyang Ning (1994) 'Analysis of the chemical compositions of essential oils from scented leaves of *Pelargonium* hybrids acclimated in Yunnan province.' *Fruits*, 49(1), 22.

Yoshida, T., Ikawa, S. and Morisada, S. (1968) 'Seasonal variation of both the percentage yield of essential oil and the chemical composition in *Pelargonium* species.' *Proc. Crop Sci. Soc. of Jpn.*, 37(4), 565–569.

Zobenko, L.P. and Arinshtein, A.I. (1989) 'Results and problems of breeding essential oil crops.' Selektsiya i Semenovodstvo, Moscou(1), 10–12.

19 Cultivation and distillation of Geranium oil from *Pelargonium* species in India

B.R. Rajeswara Rao

INTRODUCTION

Rose-scented geraniums are grown in various parts of India, from the plains to high altitudes, and in spite of some differences in the major components of the oils, the oils are accepted on the domestic market in India.

LOCATIONS

The experimental crops have been grown in the South Indian State of Andhra Pradesh in Hyderabad; in the State of Karnataka in Bangalore; in the State of Tamil Nadu in the Pulney Hills, the Nilgiri Hills and in the North Indian State of Uttar Pradesh in Gaza, Kanatal, Lucknow, Pantnagar, Ranichauri. The altitude varies from: 120 m in Lucknow, 540 m in Hyderabad, 900 m in Bangalore, to 1500–2400 m in Pulney Hills and Nilgiri Hills.The climate is equally variable from semi-arid, sub-tropical in Lucknow to cool, sub-tropical in the Hills.The temperature variation is commensurate with the locality and can rise as high as 40 °C and as low as 5 °C; there is a wide range of temperatures in each area itself.

GENOTYPES

Two main genotypes are commercially cultivated in India: the Algerian or Tunisian type and the Bourbon or Reunion type, whilst a third, Kelkar or Egyptian is grown in limited areas. These are shown in Figures 19.1–19.3, and illustrate the similarity between the leaves of the three cultivars. The only way to distinguish between the cutivars is through their odour and therefore their chemical composition. The main distinguishing sesquiterpenes for Bourbon and Egyptian oils, 6,9-guaiadiene and 10-epi-γ-eudesmol respectively are sometimes both present in the Bourbon, Algerian and Egyptian types of oil grown in India; the Egyptian type however has a higher ratio of 10-epi-γ-eudesmol to 6,9-guaiadiene and the Bourbon a higher ratio of 6,9-guaiadiene to 10-epi-γ-eudesmol. (Kaul *et al.*, 1997; Ram *et al.*, 1997). The Kelkar oil abounds in geraniol, from 34 to 44 per cent with a low citronellol content, from 8 to 11 per cent. The physico-chemical characteristics of two Indian Geranium oil are shown in Table 19.1. To avoid confusion with the International names, the two main types have been

Figure 19.1 Close-up of cultivar Algerian. Courtesy: Dr K.P. Sastry, CIMAP Field Station, Kodaikanal, India.

Figure 19.2 Close-up of cultivar Bourbon. Courtesy: Dr K.P. Sastry, CIMAP Field Station, Kodaikanal, India.

Figure 19.3 Close-up of cultivar Kelkar. Courtesy: Dr K.P. Sastry, CIMAP Field Station, Kodaikanal, India.

Table 19.1 Physico-chemical characteristics of Geranium oil 'Reunion type' and 'Indian type'

Physico-chemical Characteristics	*Reunion*	*Indian*
Relative density at 30 °C	0.879–0.891	0.884–0.899
Refractive index at 30 °C	1.460–1.464	1.464–1.474
Optical rotation at 30 °C	−11 to −13	−7 to −11
Acid value maximum	5	10
Ester value	65–75	50–76
Ester value after acetylation	217–227	205–230
Carbonyl value expressed as iso-menthone % by weight, maximum	13	16

Source: (Bureau of Indian Standards Specification, IS: 587–1965)

renamed: Hemanti (Algerian or Tunisian), Bipuli (Bourbon or Reunion) and Kunti (Kelkar or Egyptian).

The composition of the two main genotypes varies somewhat when the plants are grown in different sites (Rajeswara Rao *et al.*, 1990; Singh *et al.*, 1996), but they are more acceptable on the market than the Kelkar.

Ambiguity prevails over the botanical nomenclature adopted in India for all these genotypes, which are referred to either as *Pelargonium* species or as *Pelargonium graveolens* L'Herit. ex Aiton (Rajeswara Rao, 1999).

Attempts have been made to develop new strains using Gamma irradiation (Cobalt 60, dose: 1 KR) (Angadi and Vasantha Kumar, 1995), leaf cuttings (Kulkarni *et al.*,

1998), tissue culture (Anonymous, 1997, 1999) and by crossing genotypes in the normal way.

New Algerian-type strains, have resulted in e.g. a higher content of geraniol than normal, equal to that of the Bourbon (strains IIHR-2, PG-7, Rn Pb). Another clone (64) is identical to that of the Australian genotype (Kulkarni *et al.*, 1997). However, some strains contain substantial quantities of iso-menthone.

PROPAGATION, GROWTH CONDITIONS, DISEASE AND AGE OF PLANTS FOR HARVESTING

Plants are propagated from stem cuttings, which are either terminal i.e. soft stem, or hard stem cuttings: both produce similar essential oil (EO). Changes in the EO composition has been shown to occur for a number of reasons: fully expanded old leaves, petioles and tender stems contain small amounts of EO, whilst the young, expanding leaves have a greater density of oil glands than the old. Planting dates make a substantial difference to the oils eventually produced; there are different changes in the concentration of citronellol or geraniol in different cultivars, depending on the month they are planted out. The effect of plant growth regulators have been studied and found to contribute to the quality of EO produced: Triacontanol and Mixatalol, NAA, Ethrel and Mepiquat chloride had some enhancing effect on linalool, iso-menthone, citronellol and geraniol (Rajeswara Rao, 1999).

Diseases like wilt, caused by fungi or Little leaf disease caused by phytoplasma, cause not only plant losses but also loss of quality of the EO produced (Rajeswara Rao, 1999). This is possibly through a direct effect of the organism or an indirect effect through reduced photosynthesis etc. Geraniums grown in hilly areas are often infested with nematodes: *Criconemoides* sp., *Helicotylenchus dihystera*, *Meloidogyne incognita*, *Meloidogyne hapla*, *Pratylenchus* sp., *Scutellonema conicephalum* and *Xiphenema* sp. damage the crop and can inflict losses of up to 76 per cent (Anita and Vadivelu, 1997). A number of other diseases occur including: root rot (*Fusarium* sp., *Pythium* sp., *Phytophora* sp.,), tip rot or tip burn (*Gleosporium* sp.), stem rot (*Rhizoctonia solani*); rust (*Puccinia* sp.), root galls (*Agrobacterium tumifaciens*). Control measures using various pesticides are used to control these.

Changes in temperature, humidity, rainfall, have a great effect on EO production, especially regarding their composition. Hot months favour accumulation of citronellol, whilst cool months with that of geraniol and its esters. Harvesting at different times of the day even on the same day influences the EO composition. The same was true of the height of the plants when harvested.

In general, the crop is harvested at flowering at high altitudes and during full growth elsewhere. In hilly areas, the crop is ready for harvesting with well-developed foliage within 7–8 months of planting and in plains, 4–6 months after planting (Doraswamy and Sundaram, 1982; Rajeswara Rao, 1999) Subsequent harvests are at 3–5 month intervals, depending on weather conditions and crop management. Plants are harvested using a sickle on sunny days and not when raining. In the hills, three to four harvests are made per year and the plants give a good yield for 8–10 years after wards, but in South Indian plains, plants last only for 2–4 years; in North Central Indian plains, the geranium plants are cultivated solely as an annual crop (Ram *et al.*, 1997). Only the top 15–20 cm length of shoot is harvested, leaving enough biomass for re-growth, in areas where multiple harvests are obtained.

DISTILLATION

The leaves, stems and flowers collected are allowed to wilt or wither to reduce their volume, as this allows a greater biomass to be distilled at any one time and cuts down on the cost of distillation etc. Wilting even for a day increased the per cent citronellol and geraniol in the EO in an Algerian cultivar. Usually, the harvested biomass is wilted for a few hours, but in the hot climates of the plains, there is a loss in the yield of EO after just few hours. Comminution enhances oil yield and also has a significant effect on the composition (Rajeswara Rao, 1999). The method of distillation again has a considerable influence. Hydrodistillation or steam distillation will affect different cultivars in a distinct manner.

The oil is then filtered through cotton, muslin or filter paper to exclude extraneous matter and then treated with anhydrous sodium chloride or sulphate, allowed to stand for a few hours for absorption of free and dissolved water, and then refiltered, filled up to the brim in airtight containers, capped tightly and stored in a cool, dry place. Quality can remain high for many years.

Occasionally, the oil is coloured red due to reacting with the metal (mild steel) of the distillation vessel. Tartaric acid usually restores the proper colour. If the red colour persists, then the oil is re-distilled. For shorter periods of storage, glass or hard plastic cans and galvanised iron drums are used; longer storage requires aluminium drums or amber-coloured bottles.

REFERENCES

Angadi, S.P. and Vasantha Kumar, Y. (1995) Geranium. In: K.L. Chadha, and R. Gupta (eds) *Advances in Horticulture*, vol. 11, Malhotra Publishing House, New Delhi, India, pp. 667–687.

Anita, B. and Vadivelu, S. (1997, 1999) Management of root knot nematode, *Meloidogyne hapla* on scented geranium, *Pelargonium graveolens* L. *Indian J. Nematol.*, 27, 123–125.

Anonymous (1997, 1999) *Annual Reports*, 1996–1997, 1998–1999, Central Institute of Medicinal and Aromatic Plants, Lucknow, India, pp. 57–60; pp. 10–87.

Doraswamy, K. and Sundaram, M. (1982) Geranium cultivation in South India, In: C.K. Atal, and B.M. Kapur (eds) *Cultivation and Utilization of Aromatic Plants*, Regional Res. Lab., Jammu Tawi, pp. 573–577.

Kaul, P.N., Rajeswara Rao, B.R., Bhattacharya, A.K., Mallavarapu, G.R. and Ramesh, S. (1997) Changes in the chemical composition of rose-scented geranium (*Pelargonium* sp.) oil during storage. *J. Essent. Oil Res.*, 9, 115–117.

Kulkarni, R.N., Baskaran, K., Ramesh, S. and Kumar, S. (1997) Intra-clonal variation for essential oil content and composition in plants derived from leaf cutings of rose-scented geranium (*Pelargonium* sp.) *Ind. Crops Products*, 6, 107–112.

Kulkarni, R.N., Mallavarapu, G.R., Baskaran, K., Ramesh, S. and Kumar, S. (1998) Composition of the essential oils of two iso-menthone-rich variants of geranium (*Pelargonium* sp.) *Flav. Fragr. J.*, 13, 389–392.

Mallavarapu, G.R., Prakasa Rao, E.V.S., Ramesh, S. and Narayana, M.R. (1993) Chemical and agronomical investigations of a new chemotype of geranium, *J. Essent. Oil Res.*, 5, 433–438.

Rajeswara Rao, B.R. (1999) Rose-scented geranium (*Pelargonium* species): Indian perspective. Paper: National Seminar on the Research and Development in Aromatic Plants: *Trends in Biology, Uses, Production and Marketing of Essential Oils*, Central Institute of Medicinal and Aromatic Plants, Lucknow, India, July 30–31.

Ram, M., Singh, R., Naqvi, A.A. and Kumar, S. (1997) Effect of planting time on the yield and quality of essential oil in geranium *Pelargonium graveolens*. *J. Hort. Sci.*, 72, 807–810.

Rajeswara Rao, B.R., Sastry, K.P., Prakasa Rao, E.V.S. and Ramesh, S. (1990) Variation in yields and quality of geranium (*Pelargonium graveolens* L, Her. ex Aiton) under varied climatic and fertility conditions, *J. Essent. Oil Res.*, 2, 73–79.

Singh, A.K., Bisht, P.S., Singh, K., Kumar, D. and Naqvi, A.A. (1996) Introduction of geranium *Pelargonium graveolens* in Uttar Pradesh hills in India, *J. Med. Arom. Plant Sci.*, 18, 22–25.

20 Micropropagation and biotechnological approaches to tissue culture of *Pelargonium* species and production of essential oils of scenteds

Barry V. Charlwood and Maria Lis-Balchin

MICROPROPAGATION OF COMMERCIAL GERANIUMS

Micropropagation of commercial glasshouse geraniums was reviewed by Cassells (1992a) who pointed out that propagation of *Pelargonium* for commercial purposes has always been a problem due to poor seed production and disease. Also, in many commercial *Pelargonium*, there is cytological diversity, exemplified by the fact that different cultivars in the same group can exhibit euploidy, auto and allopolyploidy and aneuploidy. Due also to the selection of chimeral groups, the crop is largely vegetatively propagated from stock plants; this involves considerable cost due to labour, space and disease control. Of the numerous diseases of *Pelargonium* including fungal, bacterial and viral vascular pathogens, the most serious are bacterial blight caused *by Xanthomonas pelargonii* and tomato spot virus. Both reduce the vigour of stock plants, and often cause their early death.

MERISTEM CULTURE AND THE PRODUCTION OF VIRUS-FREE PLANTS

The introduction of meristem culture resulted in the production of virus-free plants (Pillae and Hildebrandt, 1968) and *Xanthomonas*-free plants (Hamdorf, 1976; Theiler, 1977; Reuther, 1983). Numerous plants could be produced from meristem tip culture by cloning, and this also facilitated the production of culture-virus-indexed (CVI) plants without the need for heat treatment (Cassells, 1982, 1986, 1988).

Protocol

The micropropagation protocol for commercial geraniums favoured by Cassells (1992a) consisted of Stage 1 which involved genetic selection, whereby chimeras, beneficially-infected varieties (Cassells *et al.*, 1982) and unstable genotypes were rejected (including some beautiful chimeras e.g. Mr Wren and the ivy-leaf L'Elegante, as well as beneficially-infected varieties like picotee-flowered varieties of ivy-leaf Mexicana and the unstable Speckles). The next stage was screening for disease and contaminants which produced axenic stock; this involves using enzyme-linked immunosorbant assay (ELISA) or DNA probes for known pathogens and culture indexing using a range of common bacterial and fungal media (Schaad, 1979) and identification of contaminants (Cassells, 1991, 1992b). Then the cloning strategy followed, which involved meristem or nodal culture, adventitious regeneration and somatic embryogenesis: this would be monitored for genetic stability. The final stage involved rooting procedure and transfer to the environment: either *in vivo* rooting in clumps or individual transplanting, then weaning by misting or fogging, or *in vitro* self-rooting or induced rooting followed by biopackaging.

Virus-screening

Meristem culture does not eliminate all known viruses, therefore screening the donor plant is imperative (Cassells, 1992 a,b); the problem also arises when using a nodal or explant culture. Virus contamination may not be expressed *in vitro*, as a high salt concentration favours masking; clonal loss can therefore result after establishment or subculturing in low-salt media (Cassells *et al.*, 1982). Symptomless donors are usually selected and these are then culture-indexed for *Xanthomonas* etc. and tested for tomato ringspot using serology. If tests are positive for viruses, then heat therapy is applied; if bacterial contamination is found the plants are grown under hard conditions or treated with antibiotics. Whitefly and aphids should be eliminated from growth areas as cross-contamination is possible.

Conditions favouring meristem culture

Meristem culture is favoured by dark conditions of the donor plant, culture at low temperature, and dark incubation of cultures and auxin apparently influences meristem establishment (Menard *et al.*, 1985; Cassells, 1992a).

Establishment of culture was achieved in about 2 weeks, by using MS (Murashige and Skoog, 1962) basal medium, supplemented by 0.825 g/L ammonium nitrate, 0.15 g/L sodium dihydrogen phosphate 1 g/L casein hydrolysate, 30 g/L sucrose, 2 mg/L IAA, 1 mg/L GA_3, 4 mg/L kinetin, 50 mg/L adenine sulphate, 100 mg/L mesoinositol, with 6 g/L agar at pH 5.8. Subculturing was done every 4–6 weeks, depending on variety. Shoots were allowed to develop and then rooting was achieved *in vitro* by placing the implant with 1 cm length shoots onto half-strength MS basal medium containing 15 g/L sucrose, 0.01 mg/L kinetin and 0.1 mg/L IAA with 6 g/L agar pH 5.8. Rooting *in vivo* was allowed by placing the shoot culture onto seed compost in a special chamber (Cassells, 1992a).

Other methods were used by different workers: Menard *et al.* (1985) used 500 mg/L polyvinylpyrrolidone in their culture establishing media, which also included 250 mg/L of casein hydrolysate, 250 mg/L yeast extract and 30 g/L sucrose and 8 g/L agar pH 5.8. Zeatin and triiodobenzoic acid was used in the differentialtion medium and half-strength MS with 0.1 mg/L NAA for rooting.

IN VITRO PROPAGATION OF REPRESENTATIVE SPECIES OF ALL SECTIONS OF *PELARGONIUM* (*GERANIACEAE*)

As there is a wide diversity of *Pelargonium* species, but a shortage of specimens available for ornamental purposes, due to difficulties in finding their location (e.g. elusive geophytes, spending most of the year underground), their inherent scarcity and also difficulties in exporting from South Africa, and also the difficulty of mass propagation from the few stock plants available, the possibility of micropropagating members of the genus was examined (Lis-Balchin, 1996). Few different *Pelargonium* species had been micropropagated apart from some scented-leaf ones (Skirvin and Janick, 1976) as the main concern was to mass-produce the main crop 'geraniums' and in particular to keep them virus-free (Debergh and Maene, 1977; Menard *et al.*, 1985; Horn, 1988; Cassells, 1992a,b). Most micropropagation was from meristems except for Pillai and Hildebrandt (1968), Stefaniak and Zenkteler (1982) who used petioles.

Micropropagation conditions used

Petioles were used as these were available in suitable numbers on all species, in contrast to meristems. Sections of petioles from a wide range of representative species of all sections of *Pelargonium* were taken from mature plants in summer, grown in a greenhouse under normal light conditions at a minimum temperature of 20 °C in soil-free compost with added grit. Plants in England were identified according to van der Walt (1977), van der Walt and Vorster (1981, 1988) or by van der Walt and Voster on site in South Africa.

The petioles were sterilised for 10 min in 10 per cent Domestos (sodium hypochlorite with added wetting agents) mixed with a few drops/100 mL of household washing-up liquid. For hirsute plants, a pre-wash of a 10 sec dip in methylated spirits was made. The explant was then rinsed three times in sterile water and 0.5 to 1 cm lengths were cut and inserted into the prepared MS solid medium in glass pots 10 cm tall and 2 cm wide with white plastic screw tops containing the appropriate hormones in the same direction as in the original plant (from stem base up to tip). The pots were incubated under fluorescent lamps at 25 °C in an 18 h photocycle of light to 6 h darkness. The callus or developing plantlets were subcultured every 6 weeks. Plantlets with shoots and roots were transferred to soil-less compost after washing off the agar mixture thoroughly under running water.

The best hormonal conditions were determined in preliminary experiments using 13 different mixtures in the basic MS agar mixture containing 0.8 per cent agar and 3 per cent sucrose. (Table 20.1). At least five cuttings were used for each determination. A range of different callus-inducing media were substituted for possible shoot-inducing media as soon as the callus was established. The root-inducing medium (k), or other media were substituted when the roots did not form for 2 weeks after the shoots were visibly evident and also green.

Micropropagation results for different species

Callus formation

The mean time taken for the initiation of callus in the different species (using the most active media composition as shown in Table 20.2) was found to vary for different species and was not correlated with any particular section. The shortest average time for callus formation was 13 days for e.g. *P. patulum* and *P. ranunculophyllum*: however, some species e.g. *P. abrotanifolium* and *P. trifidum* took 58 days. An individual species within a particular section could take 19 days (*P. exstipulatum*) to 58 days (*P. abrotanifolium*). Callus

Table 20.1 Comparison of the hormone mixtures used to initiate callus formation, shoots and roots in different species

Section	*Species*	*Callus*	*Shoot*	*Root*
Campylia (Sweet) D.C.	*P. oenothera*	a,b,c,f,j	e	k
Ciconium (Sweet) Harvey	*P. acetosum*	a,h,j	a,h	k
	P. acraem	a,b,c,d,f	e	k
	P. caylae	a,b,c,d,f	a,d,e,f	k
	P. inquinans	a,b,c,d,f	a,d,e,f	k
	P. 'monstrum'	a,b,c,d,I,j	a,d,e,I,j	k
	P. ranunculophyllum	a,b,c,f,h,j	a,e	k
	P. 'stenopatalum'	c,d,f	c,d,f	k
Cortusina D.C.	*P. dichondrifolium*	–ve	–ve	–ve
	P. echinatum	a,b,c,I,j	e,j	k
	P. mollicomum	a,b,c,f,I,j	h	k
	P. odoratissimum	a,b,c	a,c,h,j	k
	P. reniforme	a,c,d	a,c,d	k
	P. sidoides	a,b,c,d	a,c,d	k
Dibrachya (Sweet) Harvey	*P. peltatum*	a,j	a,j,e	k
Glaucophyllum Harvey	*P. glaucum*	a,b,c,d,f	a	k
Harvey	*P. patulum*	a,b,c,f,j	–ve	–ve
Hoarea D.C.	*P. appendiculatum*	a,b,c,f,j	e,j	j,k
	P. punctatum	a,b,c,j	a,c,e,h,I,j	I,k
	P. rapaceum	a,b,c,f,j	e	k
Isopetalum (Sweet) D.C.	*P. cotyledonis*	–ve	–ve	–ve
Jenkinsonia (Sweet) Harvey	*P. antidysentericum*	a,j	–ve	–ve
	P. endlicheraianum	c,j	–ve	–ve
	P. tetragonum	I,j	g,h	k
Ligularia (Sweet) Harvey	*P. abrotanifolium*	a,b,c,d,h	a,e	k
	P. barklyi	a,b,c,I,j	e,j	k
	P. dolomiticum	a,b,c,f,j	e	k
	P. exstipulatum	a,b,c	a,d	k
	P. fulgidum	h,I	h,I	k
	P. griseum	a,b,c,d	a,c,d	k
	P. hirtm	a,b,c,g	a,c,g	k
	P. hystrix	a,b,c	f,I	k
	P. pulchellum	a,b,c,f,j	e	k
	P. trifidum	a,b,c	a	k
Myrrhidium D.C.	*P. multicaule*	b,f,h	f,h	k
Otidia (Sweet) G. Don	*P. alternans*	a,b,c,f,j	a,j	k
	P. carnosum	f,g,h	f,g,h	k
	P. ceratophyllum	a,b,c,h,j	e	k
	P. dasyphyllum	I	I,j	k
Pelargonium (D.C.) Harvey	*P. capitatum*	a,b,c,f,j	a,e	c,e,k
	P. graveolens	a,b,c,f,j	e,f	c,e,k
	P. papilionaceum	a,b,c,f,j	a,e	k
	P. tomentosum	a,b,c,I,j	a,e,h	a,j,k
Peristera D.C.	*P. album*	a,b,c,d,h	a,c,d,h	k
	P. australe	a,b,c,I,j	a,I,j	a,k,l
Polyactium (Eckl. and Zeyh.) D.C.	*P. triste*	a,b,c,I,j	a,e	e,k,l

Table 20.1 (Continued)

Hormones and proportions used (mg/L):					
a	BA:NAA	5:1	f	KIN:IAA	5:5
b	BA:NAA	5:5	g	KIN:NAA	5:5
c	BA:NAA	1:5	h	KIN:IAA	5:1
d	BA:IAA	5:5	i	KIN:NAA	1:5
e	BA:IAA	10:1	j	KIN:NAA	5:1
k	NAA alone				

Notes
BA (benzyladenine); NAA (naphthylene-acetic acid; IAA (indoleacetic acid); K (kinetin).

Table 20.2 Average number of days to initiate callus in different *Pelargonium* species

Days	*Species*	*Section*
13	*P. patulum*	Glaucophyllum
	P. ranunculophyllum	Ciconium
19	*P. dolomiticum*	Ligularia
	P. rapaceum	Hoarea
	P. triste	Polyactium
	P. caylae	Ciconium
	P. exstipulatum	Ligularia
26	*P. echinatum*	Cortusina
	P. glaucum	Pelargonium
	P. papilionaceum	Pelargonium
	P. capitatum	Pelargonium
	P. hirtum	Ciconium
	P. mollicomum	Cortusins
28	*P. tomentosum*	Pelargonium
29	*P. australe*	Peristera
30	*P. appendiculatum*	Hoarea
34	*P. ceratophyllum*	Otidia
	P. graveolens	Pelargonium
	P. hystrix	Ligularia
	P. oenothera	Campylia
	P. monstrum	Ciconium
	P. album	Peristera
	P. reniforme	Cortusina
	P. 'stenopetalum'	Ciconium
	P. inquinans	Ciconium
	P. barklyi	Ligularia
37	*P. alternans*	Otidia
43	*P. pulchellum*	Ligularia
58	*P. abrotanifolium*	Ligularia
	P. trifidum	Ligularia

formation was possible in most species using BA:NAA ratios of 5:1 and 1:5; kinetin could replace BA in any auxin combination; IAA and NAA were also interchangeable. One of the main problems with the Section *Ciconium*, which also gives rise to *P.* × *hortorum* cultivars (the household geraniums), is the exudation of phenolics from the cut surfaces; this has been remarked on by Hildebrand and Harney (1988) and is best treated by more numerous washing of the cut petioles with sterile water; experiments were also conducted whereby the cut petioles were kept in sterile water for 30 min to 24 h with no greater success.

Callus formation proved impossible under all possible conditions for *P. cotyledonis*, a very high-phenolic containing species, which is endemic to the island of St. Helena and differs in many ways with the rest of the *Pelargonium* species. However, the high phenolic concentration alone was probably not the reason for this lack of success, as *P. antidysentericum*, (the underground root of which is used for treating dysentery by locals in South Africa) has a very high phenolic concentration, but proved amenable to formation of callus although further transformations were not possible.

Callus formation times for *P. australe*, *P. echinatum* and *P. tomentosum* were similar to those reported by Brown and Charlwood (1986). The majority of other species had not been micropropagated before, except for some of the geophytic and xerophytic species (De Marie, 1991), whose callus formation times were not reported, although the shoot-stage was reached.

Shoot formation

The time taken for shoots to appear (Table 20.3) was also variable for different species and again there was no observable correlation with different sections. *P. rapaceum* and *P. exstipulatum* took 4 weeks; whilst *P. tetragonum* took 12 weeks. Shoot formation in *P. echinatum* was initiated within 6 weeks, in contrast to the negative results of Brown and Charlwood (1986). The geophytes, *P. appendiculatum*, *P. punctatum* and *P. rapaceum* (*Hoarea*) formed shoots within 4–5 weeks. The xerophytic *P. hystrix* also formed shoots, though after 7 weeks, in contrast to the negative results of De Marie (1991).

Table 20.3 Average number of weeks to produce shoots in different *Pelargonium* species

Weeks	*Species*	*Section*
4	*P. rapaceum*	Hoarea
	P. exstipulatum	Ligularia
5	*P. triste*	Polyactium
	P. papilionaceum	Pelargonium
	P. dolomiticum	Ligularia
	P. australe	Peristera
	P. appendiculatum	Hoarea
6	*P. barklyi*	Ligularia
	P. echinatum	Cortusina
	P. album	Peristera
7	*P. capitatum*	Pelargonium
	P. ceratophyllum	Otidia
	P. graveolens	Pelargonium
	P. hystrix	Ligularia
	P. oenothera	Campylia
10	*P. trifidum*	Ligularia
11	*P. abrotanifolium*	Ligularia
12	*P. tetragonum*	Jenkinsonia

In the present studies, however, shoot formation did not occur in *P. patulum*, *P. antidysentericum* and *P. endlicherianum* calluses under all possible hormonal conditions, even after 4 months.

Root initiation

Root initiation was very prompt in most species when the shoot-forming callus was transferred to a hormone-free MS medium or sometimes a high auxin:cytokinin medium (Table 20.1). There were a few exceptions, including *P. tetragonum*, which formed roots in a lower kinetin: even lower NAA (0.5:0.05) than used for the rest of the species; *P. tomentosum* formed roots in a high NAA alone medium (10 ppm) or BA alone (10 ppm); *P. triste*, *P. capitatum* and *P. graveolens* formed roots on almost any type of hormone concentration or mixture used.

The ease of micropropagating the bulk of different *Pelargonium* species was in contrast to some of the earlier difficulties associated with micropropagation of *Pelargonium* cultivars (so-called Geraniums, or zonals), which were hybridised mainly from the Ciconium section.

The resulting plants appeared to be identical with their parent plants, but not enough experiments were conducted to state categorically that genetic modifications would not occur as often found when using petioles (Skirvin and Janick, 1976).

In conclusion, most of the species, from all sections, could be micropropagated successfully.

BIOTECHNOLOGICAL APPROACH

Apart from the capability to produce large amounts of biomass of clonal lines of the commercially important pelargoniums as outlined above, plant biotechnology has application in more general terms for plant improvement and for essential oil production.

SECONDARY PLANT METABOLITES

During the 1980s there emerged a significant interest in the potential use of cultured plant cells (callus and suspension) for the production of secondary compounds of importance to the pharmaceutical, food processing and cosmetic industries. Because of the wide range of monoterpene classes to be found in the many scented-leaf variants of *Pelargonium*, and the ease with which these variants may be taken into culture (Charlwood and Charlwood, 1983), this genus provided an important model system through which much valuable information was derived concerning the accumulation of lower isoprenoids in undifferentiated cultures (Charlwood and Charlwood, 1991).

Essential oil accumulation in calluses

Table 20.4 shows the partial results of one such study (Brown and Charlwood, 1986a; Brown, 1988; Charlwood *et al.*, 1989) in which calluses, derived from stem

Table 20.4 The oil content, callus characteristics and regeneration capacities of scented-leaf variants of Pelargonium[a]

Scented-leaf variant[b]	*Days to form callus*[c,d]	*Callus morphology*[d,e]	*Oil content* (μg/g *fresh weight*)[f]	*Weeks to formShoots*[d,g]
P. 'Miss Australia'	8	White/pink, friable	31.0	8
P. nervosum Sweet	7	White, friable	21.7	4
P. 'Mabel Grey'	18	Pigmented, hard	15.8	14
P. australe Willd.	13	White, friable	13.5	3
P. 'Lillian Pottinger'	18	Pigmented, friable	9.2	46
P. quercifolium (a)	13	Pigmented, friable	8.8	11
P. tomentosum (a)	10	Green, friable/hard	4.2	14
P. quercifolium (b)	13	Pigmented, friable	2.7	13
P. radula L'Her	7	Pigmented, friable/hard	0.81	18
P. tomentosum (b)	10	Green, friable/hard	0.06	3
P. 'Royal Oak'	11	White, friable	0.06	26
P. fragrans Willd.	7	Pigmented, friable	0.06	15
P. tomentosum (c)	8	Green, friable/hard	0.006	3

Notes

a Data derived from Brown (1988).

b Letters in parenthesis indicate different cell lines of the same variant.

c *Callus initiation medium*: Charlwood and Charlwood (1983) medium (pH 5.5) containing 30 g/L sucrose, 10 g/L agar, 0.2 mg/L kinetin and 1 mg/L 2,4-D.

d Incubations were at 26 °C under either continuous light (3 μE/m^{2}s) or a 16 h light/8 h dark photoperiod.

e *Callus maintenance medium*: Murashige and Skoog (1962) medium (pH 5.5) containing 30 g/L sucrose, 10 g/L agar, 5 mg/L BAP and 1 mg/L NAA.

f Average oil content of intact plant was 620 μg/g fresh weight.

g *Shoot regeneration medium*: Murashige and Skoog (1962) medium (pH 5.5) containing 30 g/L sucrose, 10 g/L agar, 0.5 mg/l BAP and 0.05 mg/L NAA.

material of around 30 different scented-leaf pelargoniums, were studied with respect to morphology, growth rate, ability to regenerate shoots and capacity for accumulation of essential oil. The monoterpene accumulation in nearly all of the callus lines investigated was very low, typically only about 0.1 per cent of that accumulated by the parent plant, although in rare cases accumulations of up to 5 per cent of the oil of the parent plant could be detected (e.g. *P. nervosum* and *P.* 'Miss Australia'). Throughout this study, no clear correlation between the level of oil accumulation and either the growth characteristics or the morphological nature of the callus material ever emerged.

Typical low levels of oil accumulation were further maintained when finely divided suspension cultures were produced from the parent callus lines. Thus for *P.* × *fragrans*, a callus line which accumulated between 0.06–0.09 μg oil/g fresh weight gave rise to a suspension culture which accumulated 1.3 μg oil/g fresh weight when grown under subdued light, although this accumulation could be increased to 11.5 μg oil/g fresh weight when the culture was incubated under photoperiod conditions (Brown and Charlwood, 1986b). Despite exhaustive attempts to augment this low level accumulation in undifferentiated cells by, for example, supplementation of the medium with plant growth regulators (PGRs), biotic and abiotic inhibitors, organic acids etc. (Charlwood *et al.*, 1988), the maximum accumulation that could ever be achieved was only 35 μg oil/g fresh weight, and this was obtained using a highly aggregated suspension culture.

Differentiation of calluses

Calluses of most scented-leaf pelargoniums which had been sub-cultured at two-weekly intervals on maintenance medium (see Table 20.4) for periods of up to 1 year could be induced to differentiate through transfer to Murashige and Skoog (1962) solid medium (MS) containing low levels of BAP and NAA (0.5 and 0.05 mg/L, respectively). Typically, shoot formation occurred within 4–14 weeks following transfer to regeneration medium (Table 20.1; Brown and Charlwood, 1986a), although for some varieties (e.g. *P.* 'Lillian Pottinger' and *P.* 'Rober's Lemon Rose') prolonged culture of up to 1 year on regeneration medium was required. A few scented-leaf pelargoniums (e.g. *P. echinatum* and *P. glutinosum*) appeared not to be competent for regeneration under these conditions.

ESSENTIAL OIL ACCUMULATION IN DIFFERENTIATED/UNDIFFERENTIATED PELARGONIUMS

Relationship between storage site (trichomes) and essential oil accumulation oil

Accumulations in shoot cultures were an order of magnitude greater than those determined in undifferentiated cultures, although qualitatively the oils differed from those found in the respective parent plants, being composed almost exclusively of monoterpene hydrocarbons with only small amounts of oxygenated species. Interestingly, stable, submerged shoot-proliferation cultures could also be formed from a number of pelargoniums, in particular for *P.* × *fragrans*. *P. tomentosum* (Charlwood and Moustou, 1988) and *P. graveolens* (Katagi *et al.*, 1986), and oil accumulation in this type of culture approached 50 per cent of that associated with the intact parent plant grown under greenhouse conditions. The reduced proportion of oxygenated monoterpenes within such shoot proliferation cultures was still observed although, unusually, for *P. tomentosum* the major components of the oil were menthone and isomenthone (in roughly equal amounts) corresponding identically with those of the parent plant. In these cultures, the density of the glandular hairs (the storage sites for the EOs) present on the surface of the shoot tissue varied both with the concentrations of PGRs present in the medium and following treatment with various herbicides (i.e. Metflurazon, Amitrol, desmethyl-norflurazon, 5C cycocel and AMO 1618): a direct correlation between glandular hair density and the accumulation of oil could be established (Charlwood *et al.*, 1989).

It would thus appear that the capacity of scented-leaf *Pelargonium* cultures to accumulate essential oils is closely associated with the availability of suitable storage sites (Charlwood, 1993). Clearly if a cell culture is to be able to accumulate product, then the product itself must not be deleterious to the producing cell, and the rate of synthesis of the product must be greater than its rate of breakdown.

Toxicity of essential oils produced to the cells

Mono- and sesqui-terpenes are actually toxic to plant cells causing an inhibition of photosynthesis and respiration and giving rise to a significant reduction in the number of mitochondria and golgi bodies. The treatment of suspension cultures of

P. × *fragrans* with a range of monoterpenes, all of which naturally occur in scented-leaf pelargoniums, resulted in a 90 per cent loss of cell viability at concentrations of additive in excess of 100 mg oil/L medium (Brown *et al.*, 1987). Surprisingly, cultured cells showed no increased resistance to those monoterpenes present in the oil accumulated by the individual parent plant. Autotoxicity of the product to the cell line thus sets an upper limit on the amount of monoterpene that a culture could be expected to accumulate under conditions where the product remains in contact with the producing cells.

Metabolism of terpenes produced

However, there is a further possible cause of product loss from undifferentiated cell cultures, and this is associated with the breakdown of the terpenes so formed. There have been many reports (Charlwood, 1993) concerning the rapid metabolism of monoterpenes which had been added to plant cells in culture, and numerous 'salvage' enzymes have been identified through which the carbon skeletons of such compounds can be returned to the pool of acetyl CoA (Berger *et al.*, 1990). In some instances the rate of breakdown of product can exceed the rate of synthesis by several orders of magnitude in which case, of course, no product can ever accumulate. Clearly this is not the case for cultures of scented-leaf pelargoniums as some product, albeit in small amount, is often observed. Theoretically then it should be possible to increase oil accumulation by simply removing the product immediately formed by the cultured cells, and sequestering the oil at some remote location.

Immobilised cells

This can most readily be performed by immobilising suspension cells through entrapment within the pores of reticulate polyurethane foam and passing medium continuously over the immobilised cells in order to wash the product away. Using such a system, a yield of oil equivalent to 110 mg oil/g fresh weight could be obtained from a suspension culture of *P.* × *fragrans* that originally accumulated 100-fold less (Charlwood and Charlwood, 1991). Alternatively, an artificial storage site, in the form of a non-toxic lipophilic oil, may be added to the finely divided suspension culture as a second phase in order preferentially to absorb any monoterpene product formed. When suspension cultures of *P.* × *fragrans* were treated in this way a minimum of a 10-fold increase in essential oil accumulation was observed (Charlwood and Brown, 1988), but under optimal incubation conditions an increase in oil accumulation of some 500-fold was attainable.

Retention of individual characteristics

Although, in general, cultures of scented-leaf pelargoniums seem not to retain their ability to accumulate the monoterpenes associated with the parent plant, this capacity returns following morphological regeneration. Thus when calliclones of a number of scented-leaf pelargoniums were grown under greenhouse conditions for 3 months after rooting, their essential oils were similar both qualitatively and quantitatively to those of the parent plants (Brown, 1998). It should be noted that environmental conditions play a significant role in determining the yield and composition of the essential oil of pelargoniums, and hence it is essential that both the parent plant and the resulting calliclones be subjected to identical growing environments if comparisons of oil product are to be made. Skirvin and Janick (1976) carried out similar studies with 166 first-generation calliclones of the variant 'Rober's Lemon Rose' and found that only one line showed an essential oil significantly different from that of the parent plant even though a number of lines showed aberrant leaf morphologies. The frequency of somaclonal variation of a specific trait is generally estimated to be between 0.2 and 3 per cent.

Production of variation in plants

Passage through tissue culture can hence provide an element of variation in the regenerated plants, thus increasing the range of genetic diversity currently available in the natural germplasm. Such a strategy has been employed in the search for commercial pelargoniums showing resistance to bacterial blight caused by *Xanthomonas campestris* pv. pelargonii (Dunbar and Stephens, 1989). In this study, calluses with shoot primordia were induced either from shoot tips and hypocotyls of germinated seeds, or from leaves excised from sterile plants, incubated on MS medium (pH 5.8) containing 20 g/L sucrose, 9 g/L agar and 2 mg/L each of either IAA and *trans*-zeatin for seed-derived material, or NAA and BAP for leaf explants. Shoots were developed during 30 days on similar medium containing no auxin and one-tenth of the cytokinin concentration, and well-developed shoots were eventually rooted on Hoagland's solution containing 7 g/L agar. Thirty-day-old calliclones were assayed for resistance to blight by rubbing the upper surface of their leaves with a sterile cotton swab that had been moistened with bacterial suspension, and the level of infection was monitored 2–3 weeks later. It was found that regenerated plantlets of *P. grandiflorum*, a *P.* × *domesticum* c.v., *P. hispidum*, *P. betulinum*, *P. scabrum* and *P. multicaule* were much more resistant to blight than were a *P.* × *hortorum* c.v. and *P. denticulatum* (Dunbar and Stephens, 1989).

Use of *Agrobacterium rhizogenes* for production of variety

The selection of calliclones with appropriate characteristics has also been attempted in order to improve the ornamental quality of scented-leaf pelargoniums. However, an alternative strategy has been used by Pellegrineschi and Davolio-Mariani (1996) who treated microcuttings of *P.* × *fragrans*, *P. odoratissimum* and *P. quercifolium* with a suspension culture of *Agrobacterium rhizogenes* (strain HRi) for 30 min and then co-cultivated the infected explants on MS medium (with one-fifth macronutrients) containing agar for 2 days, followed by further incubation on the same medium supplemented with 200 mg/l cefotaxime (to eliminate the contaminating bacteria) for up to 40 days. Hairy roots (which first appeared after 10 days from the basal ends of the microcuttings) were removed and tips were subcultured onto the same medium to produce spontaneous shoots within 2 months. Such transformed shoots were rooted and grown-on in the greenhouse to give plants that showed 2–3 times the number of branches compared to non-transformed plants. The leaves of the transformed plants were darker green than those of the controls, and were more numerous per plant (although the leaf number per branch was unchanged).

The same group (Pellegrineschi *et al.*, 1994) also transformed sterilised petioles of *P.* 'Lemon Geranium' with *A. rhizogenes* (strains A4RSII, LBA9402 and 15834) using a similar technique and, following spontaneous shoot formation, obtained regenerated transformed plants which had shorter internodal distances with increased leaf and branch formation giving a more rounded appearance. The transformants also showed accelerated rooting of cuttings yielding a shorter and more highly branched root system, whilst the leaves themselves did not yellow as rapidly as the controls.

Use of *Agrobacterium rhizogenes* for improving production of essential oil

Interestingly, the oil content of the transformants had also changed in that they accumulated up to 4.4 times more geraniol, 2.8 times more linalool, and 13 times more 1,8-cineole compared with their non-transformed counterparts, although the content of citronellol decreased slightly (Pellegrineschi *et al.*, 1994). The authors suggest that hairy-root transformation may have application in producing pelargoniums with a favourable 'globosus' aspect to the plant canopy as well as increased oil yield (by up to 10 fold in view of the increased leaf production).

Disadvantages of using *A. rhizogenes*

The transformation of plant material using *A. rhizogenes* is a relatively facile technology since each transformation event occurring at a single cell gives rise to a separate, putatively-transformed root, the tip of which may be excised and cultured to produce a non-chimeric line. However, the major disadvantages of this technique are that transformed roots of many plant species are not able to produce shoots spontaneously, and regenerated plants that have been transformed using the Ri plasmid may show altered leaf morphology, inhibition of flowering, and low fertility.

Use of *A. tumefaciens* for gene transfer

The more common method for gene transfer involves infection with *A. tumefaciens*, typically employing a disarmed strain in which the natural transfer-DNA (t-DNA) has been removed from the Ti plasmid so as not to produce crown galls upon plant infection. The gene to be transferred to the target plant is typically inserted into a small plasmid (the so-called binary vector) which is designed to permit DNA transfer to the plant and subsequent expression both of the gene required and of a selectable marker gene (normally one which confers resistance to a phytotoxin in transformed plant cells). This second plasmid is then incorporated into the disarmed *A. tumefaciens* using the process of triparental mating. Upon infection, no transformed cells can be chosen based on morphological attributes (as they can following infection with *A. rhizogenes*), but such cells are selected by growth of the infected explant on medium containing levels of the phytotoxin which are just sufficient to kill those cells which do not express the marker gene.

A number of *A. tumefaciens*-based transformation systems have been developed for use with pelargoniums. One of the earliest (Robichon *et al.*, 1995) involved infection of cut segments of cotyledons and hypocotyls, which had been removed aseptically from

8-day-old seedlings of *P.* × *hortorum*, with disarmed *A. tumefaciens* (strain EHA 101) containing a binary vector carrying hygromycin and kanamycin resistance genes and a β-glucuronidase reporter gene. Following a 20 min infection with an *Agrobacterium* suspension, the treated explants were incubated on a shoot induction medium (containing cefotaxime) for 15 days before being transferred to a regeneration medium which also contained 20 μg/ml of the selective agent hygromycin. It appears that *P.* × *hortorum* is naturally resistant to kanamycin at levels up to 400 μg/ml and hence this antibiotic could not be employed as a selective agent. The authors claimed that the transformation protocol was 20 per cent efficient, and that the resulting transformed plants possessed one or two copies of the t-DNA although most copies were not full length.

An alternative transformation strategy was described by Boase *et al.* (1996, 1998), involving infection of cut segments of leaf explants of *in vitro* grown plantlets of *P.* × *domesticum* Dubonnet with *A. tumefaciens* (strain LBA 4404 with a kanamycin resistance gene and, in some experiments, a plasmid bearing a phytochrome A gene from oat) followed by co-cultivation for 2 days in the presence of 19.6 mg/L acetosyringone. Regeneration of transformed cells took place on a selection medium containing 50 mg/L kanamycin, as well as timentin (to eliminate the remaining *Agrobacterium*). Regenerated shoots were rooted on a medium containing 200 mg/L kanamycin to ensure that there were no escapes (regenerants that were non-transformed but naturally resistant to lower levels of the phytotoxin). From 150 explants that were inoculated, the authors reported obtaining 58 kanamycin resistant shoots from which 29 rooted plantlets were recovered, 24 of which expressed the phytochrome A gene – a transformation efficiency of 16 per cent.

In order to avoid the formation of chimeric transformants, KrishnaRaj *et al.* (1997) developed a transformation strategy for *P.* 'Frensham' via somatic embryogenesis. Surface sterilised leaf petioles were soaked in a suspension of *A. tumefaciens* for 10 min and co-cultivated for 2 days on an embryo induction medium (MS containing 30 g/L sucrose, 8 g/L agar, 3.4 mg/L BAP and 0.9 mg/L NAA). After this time the tissues were transferred to new induction medium supplemented with 500 mg/L cefataxime and 100 mg/L kanamycin (for selection of transformants): 4–5 weeks later the cultures were moved to an embryo development medium (MS containing 30 g/L sucrose, 8 g/L agar, 2.8 mg/L NAA and 500 mg/L cefataxime). The authors claim that on average each segment (1 cm long) of leaf petiole produced 45 embryos within 4 weeks, and more than 80 per cent of these embryos converted into plantlets: the transformation efficiency was around 1 1 per cent.

Future genetic modifications

Genetic modification, using the techniques outlined above, may be employed to extend the genetic variation presently available for traditional plant breeding, and may be applied to the alteration of flowering characteristics, leaf colour and zonation, disease resistance, and oil quality and yield. One of the most pressing problems associated with the commercial production of pelargoniums is associated with the susceptibility of the plants to a variety of pathogens. In greenhouse grown plants, one particular problem is gray mould (or *Botrytis* blight) caused by *Botrytis cinerea* Pers.:Fr which attacks plants growing under wet, humid conditions and is presently combated by the use of chemical fungicides. However, some plants, including radish, barley and onion, produce a number of small, cysteine-rich antimicrobial proteins (AMPs) which have recently been shown to play significant roles in plant defence. The AMP from onion (*Ace*-AMP1) has a wide spectrum of antimicrobial activity *in vitro* and has been shown to be active even in the presence of cations at physiological ionic strength. Bi *et al.* (1999) transformed *P.* 'Frensham' with *A. tumefaciens* containing a binary vector which carried the signal peptide, the mature protein and the carboxyl-terminal propeptide domains of *Ace*-AMP1 cDNA driven by a CaMV 35S promoter with duplicated enhancers, using the somatic embryogenesis strategy of KrishnaRaj *et al.* (1997). Seven transformed plants were obtained following selection on kanamycin and all showed expression of the *Ace*-AMP1 protein. Using an assay method which involved infecting a 10 mm leaf disc with spores of *B. cinerea* and incubating the plant tissue under humid conditions on agar containing paraquat (to kill the plant tissue and hence speed up the sporulation of the fungus), the three transformants which were further tested all showed significantly lower sporulation densities after 6 days. Furthermore, there was a significant correlation between resistance and the level of *Ace*-AMP1 protein in the transformant.

Since culture-derived (somaclonal) variation has had only a small impact on increasing the genetic diversity available for breeding ornamental plants, it seems that direct gene transfer strategies (using *Agrobacterium* vectors or through biolistic techniques) are likely to be the method of choice for the introduction of desired traits into *Pelargonium* lines in the next decade. Despite the current low level of public acceptance of this technology, a number of laboratories are presently developing new variants of pelargoniums altered with respect to leaf and petal colour, oil quality and quantity and enhanced disease resistance.

REFERENCES

Berger, R.G., Akkan, Z. and Drawert, F. (1990) Catabolism of geraniol by cell suspension cultures of *Citrus limon*. *Biochim. Biophys. Acta*, 1055, 234–239.

Bi, Y.-M., Cammue, B.P.A., Goodwin, P.H., KrishnaRaj, S. and Saxena, P.K. (1999) Resistance to *Botrytis cinerea* in scented geranium transformed with a gene encoding the anti-microbial protein *Ace*-AMP1. *Plant Cell Rep.*, 18, 835–840.

Boase, M.R., Deroles, S.C., Winefield, C.S., Butcher, S.M., Borst, N.K. and Butler, R.C. (1996) Genetic transformation of regal pelargonium (*Pelargonium × domesticum* 'Dubonnet') by *Agrobacterium tumefaciens*. *Plant Sci.*, 121, 47–61.

Boase, M.R., Bradley, J.M. and Borst, N.K. (1998) An improved method for transformation of regal pelargonium (*Pelargonium × domesticum* 'Dubonnet') by *Agrobacterium tumefaciens*. *Plant Sci.*, 139, 59–69.

Brown, J.T. (1988) *The Production and Accumulation of Mono- and Sesqui-terpenes by Plant Cells in Culture*. PhD Thesis., University of London.

Brown, J.T. and Charlwood, B.V. (1986a) The control of callus formation and differentiation in scented pelargoniums. *J. Plant Physiol.*, 123, 409–417.

Brown, J.T. and Charlwood, B.V. (1986b) The accumulation of essential oils by tissue cultures of *Pelargonium fragrans* (Willd.). *FEBS Lett.*, 204, 117–120.

Brown, J.T., Hegarty, P.K. and Charlwood, B.V. (1987) The toxicity of monoterpenes to plant cell cultures. *Plant Sci.*, 48, 195–201.

Cassells, A.C. (1991) Problems in tissue culture: Culture contamination. In: Micropropagation of horticultural Crops, eds. R.H. Zimmerman, and P.C., Debergh, Kluwer, Dordrecht, pp. 31–44.

Cassells, A.C. (1992a) Micropropagation of commercial *Pelargonium* species and hybrids (Glasshouse Geraniums)., In: Biotechnology in Agriculture and Forestry, vol. 20. High Tech and Micropropagation IV (ed.) Bajaj, YPS., Springer-Verlag, Berlin. pp. 286–306.

Cassells, A.C. (1992b) Screening for pathogens and contaminating micro-organisms in micropropagation. In: *Techniques for Rapid Detection and Diagnosis in Plant Pathology*. (Eds) J.M. Duncan, and C. Torrance, Blackwell, Oxford. pp. 179–192.

Cassells, A.C., Minas, G. and Bailiss, K.W. (1982) Pelargonium Net vein agent and Pelargonium Petal Streak as beneficial infections of commercial *Pelargonium. Scientia Hort.*, 17, 89–96.

Charlwood, B.V. (1993) Recent advances in the production of aroma compounds in plant culture systems. In: T. van Beek and H. Breteler (eds), *Phytochemistry in Agriculture*, Oxford Science Publications, Oxford, pp. 322–345.

Charlwood, B.V. and Brown, J.T. (1988) The accumulation of mono- and sesqui-terpenes in plant cell cultures. *Biochem. Soc. Trans.*, 16, 61–63.

Charlwood, B.V. and Charlwood, K.A. (1983) The biosynthesis of mono- and sesqui-terpenes in tissue culture. *Biochem. Soc. Trans.*, 11, 592–593.

Charlwood, B.V. and Charlwood, K.A. (1991) *Pelargonium* spp. (Geranium): *In vitro* culture and the production of aromatic compounds. In: Y.P.S. Bajaj (ed.), *Biotechnology in Agriculture and Forestry Vol. 15: Medicinal and Aromatic Plants III*, Springer-Verlag, Berlin, pp. 339–352.

Charlwood, B.V. and Moustou, C. (1988) Essential oil accumulation in shoot-proliferation cultures of *Pelargonium* spp. In: R.J. Robins and M.J.C. Rhodes (eds), *Manipulating Secondary Metabolism in Culture*, Cambridge University Press, Cambridge, pp. 187–194.

Charlwood, B.V., Brown, J.T., Moustou, C., Morris, G.S. and Charlwood, K.A. (1988) The accumulation of isoprenoid flavour compounds in plant cell cultures. In: P. Schreier (ed.), *Bioflavour, 87*, De Gruyter, Berlin, pp. 303–314.

Charlwood, B.V., Moustou, C., Brown, J.T., Hegarty, P.K. and Charlwood, K.A. (1989) The regulation of accumulation of lower isoprenoids in plant cell cultures. In: W.G.W. Kurz (ed.), *Primary and Secondary Metabolism in Plant Cell Cultures*, Springer-Verlag, Berlin, pp. 73–84.

Debergh, P. and Maene, L. (1977) Rapid clonal propagation of pathogen-free Pelargonium plants, starting from shoot tips and apical meristems. *Acta Hortic.*, 78, 449–454.

De Marie, E.T. (1991) Studies of *in vitro* propagation of geophytic and xerophytic *Pelargonium* species and hybrids. *PhD Thesis*, Cornell University, USA.

Dunbar, K.B. and Stephens, C.T. (1989) An *in vitro* screen for detecting resistance in *Pelargonium* somaclones to bacterial blight of geranium. *Plant Disease*, 73, 910–912.

Hildebrand, V. and Harney, P. (1998) Factors affecting the release of phenolic exudates from explants of *Pelargonium × hortorum Bailey*, 'Sprinter Scarlet'. *J. Hort. Sci*, 63, 651–657.

Hamdorf, G. (1976) Propagation of Pelargonium varieties by stem-tip culture. *Acta Hort.*, 59, 143–151.

Horn, W. (1988) Micropropagation of *Pelargonium × domesticum* (*P. grandiflorum* hybrids). *Acta Hortic.*, 226, 53–58

Katagi, H., Takahashi, E., Nakao, K. and Inui, M. (1986) Shoot-forming cultures of *Pelargonium graveolens* by jar fermentation. *Nippon Nogeikagaku Kaishi*, 60, 15–18.

KrishnaRaj, S., Bi, Y.-M. and Saxena, P.K. (1997) Somatic embryogenesis and *Agrobacterium*-mediated transformation system for scented geraniums (*Pelargonium* sp. 'Frensham'). *Planta*, 201, 434–440.

Lis-Balchin, M. (1996) Micropropagation of *Pelargonium*. Australian *Pelargonium* Society Centenary Conference, Melborne, Australia.

Menard, D., Coumans, M. and Gaspar, T.H. (1985) Micropropagation du *Pelargonium* a partir de meristems. *Med. Fac. Landbouww. Rijksuniv. Gent.*, 50, 327–331.

Murashige, T. and Skoog, F. (1962) A revised medium for rapid growth and bioassays with tobacco tissue cultures. *Physiol. Plant.*, 15, 473–497.

Pellegrineschi, A. and Davolio-Mariani, O. (1996) *Agrobacterium rhizogenes*-mediated transformation of scented geranium. *Plant Cell Tissue Organ Cult.*, 47, 79–86.

Pellegrineschi, A., Damon, J.-P., Valtorta, N., Paillard, N. and Tepfer, D. (1994) Improvement of ornamental characters and fragrance production through genetic transformation by *Agrobacterium rhizogenes. Biotechnology*, 12, 64–68.

Pillai, S. and Hildebrandt, B. (1968) Geranium plants differentiated *in vitro* from stem tip and callus cultures. *Plant Dis. Reporter*, 52, 600–601.

Reuther, G. (1983) Propagation of disease-free *Pelargonium* cultivars by tissue culture. *Acta Hort.*, 131, 311–319.

Robichon, M.-P., Renou, J.-P. and Jalouzot, R. (1995) Genetic transformation of *Pelargonium* × *hortorum*. *Plant Cell Rep.*, 15, 63–67.

Schaad, N.W. (1979) 'Serological identification of plant pathogenic bacteria' as being in *Annual Reviews of phytopathology*, 17, 123–147.

Skirvin, R.M. and Janick, J. (1976) Tissue culture induced variation in scented *Pelargonium* spp. *J. Am. Hortic. Sci.*, 101, 281–290.

Stefaniak, B. and Zenkteler, M. (1982) Regeneration of whole plants of geraniums from petioles cultured *in vitro*. *Acta Soc. Bot. Poloniae*, 51, 167–172.

van der Walt, J.J.A. (1977) *Pelargoniums of Southern Africa*. vol. 1. Purnell, South Africa.

van der Walt, J.J.A. and Vorster, P.J. (1981) *Pelargoniums of Southern Africa*.vol. 2. National Botanic Gardens Kirstenbosch, South Africa.

van der Walt, J.J.A. and Vorster, P.J. (1988) *Pelargoniums of Southern Africa*. vol. 3. Juta, South Africa.

21 Geranium oil and its use in aromatherapy

Maria Lis-Balchin

INTRODUCTION

Aromatherapy is broadly defined as 'treatment with odours', the inhalation of which can have beneficial effects on clients through their action on the limbic system in the brain (Buchbauer, 1992; Warren and Warrenburg, 1993; Lis-Balchin, 1997). However, in England it involves the application of a very diluted essential oil (EO) or mixture of EO(s) (1–2 per cent) in a carrier oil like almond oil, which is massaged into the skin; either on hands, feet, head or the total body. Aromatherapy can also mean the addition of drops of EO to the bath or a basin of hot water, or the volatilization of the EO(s) using various burners. It usually involves counselling about diet, exercise, lifestyle etc. by the aromatherapist, who may have absolutely no qualifications to offer such advice.

In France, Germany and other parts of Europe, aromatherapy takes on a different meaning as it involves the internal usage of EO(s) as medicines, and is practised by medically qualified doctors (Lis-Balchin, 1997). This includes oral, rectal and vaginal introduction of EO(s) into the body, treatment of wounds, as well as massage using more concentrated EO(s). This 'clinical aromatherapy' has also spread to England and the USA, where totally unqualified people practise internal usage of various EO(s) after making their diagnosis of the client's medical condition. Other herbs, novel medicinal plant extracts, hydrolats, herbal oils, phytols, infusions, etc. are often included in the aromatherapy treatment (Buckle, 1997), although the herbal knowledge of the aromatherapists may be lacking (Lis-Balchin, 1999).

Looking through aromatherapy books (Tisserand, 1985; Westwood, 1991; Worwood, 1991), the definition and application of aromatherapy is found to be a mixture of the esoteric (if not paranormal aspects of plant essences and their energetics) and the medical and scientific aspects of the EO(s) as chemicals.

It remains to be seen whether aromatherapy has any actual medicinal benefits, other than stress-alleviating, through massage, and whether these are attributable only to massage with the true EO(s), especially as there is a wide difference in the actual percentage chemical composition of EO(s) obtained from different geographical sources and also different samples from plants grown in various countries where differences in hybridization has occurred and even the same plants grown under different climatic conditions etc. show differences. These differences can be further accentuated in commercial oils due to blending, deterpenation, addition of other essential oils fractions or synthetic components and also often dilution with solvents.

AROMATHERAPY BASICS

Aromatherapy applications include:

1 A diffuser, which can be powered by electricity, giving out a fine mist of the essential oil (EO).
2 A burner, with water added to the fragrance to prevent burning of the EO: about 1–4 drops of EO is added to 10 ml water. The burner can be warmed by candles or electricity. The latter would be safer in a hospital/children's room/ and even a bedroom.
3 Ceramic or metal rings placed on an electric light bulb with a drop or two of EO. This results in a rapid burnout of the oil and also lasts for a very short time due to the rapid volatalization of the EO in the heat.
4 A warm bath with drops of EO added. This results in the slow volatalization of the EO, and not in absorption of the EO through the skin as stated in aromatherapy books, as the EO does not mix with water. Pouring in an EO mixed with milk serves no useful purpose as the EO will still not mix with water; the pre-mixing of the EO in a carrier oil, as for massage, results in a nasty oily scum around the bath.
5 A bowl of hot water with drops of EO, usually used for soaking feet or used as a bidet. Again, the EO will not mix with the water. This is useful for respiratory conditions and colds, where the EO can be breathed in when the head is over the container and a towel placed over the head. This is an old way of treatment and has been used successfully with Vicks, Obas oil, *Eucalyptus* oils for numerous years.
6 Compresses using EO drops on a wet cloth, either hot or cold, to relieve inflammation, treat wounds, etc. Again, the EO is not able to mix with the water and can be concentrated in one or two areas.
7 Massage of body, hands, feet, back or all over using 2–4 drops of EO (single EO or mixture) diluted in 10 ml carrier oil (fixed, oily) e.g. almond oil or jojoba, grapeseed, wheatgerm oils, etc.

The last is the most common method used by aromatherapists.

Another method uses 'medicinal' properties of the EO after oral intake. This is not to be condoned, unless effected under a medically qualified aromatherapist.

EO drops are 'mixed' in a tumbler of hot water or presented on a sugar cube or 'mixed' with a teaspoonful of honey and taken internally. This is not true aromatherapy, as almost all the rest of the methods are based on EO volatalization and therefore largely the effect of the EO on the central nervous system via the nose and thence the limbic system which can cause a secondary effect on other parts of the body. Direct effects on the skin can also occur e.g. antiseptic action of the EO or counterirritant effect which can cause reddening of the skin and an increased blood flow to the area and could presumably ease pain and swelling. The latter could only be effected by a few EOs e.g. thyme, clove, oregano. Many EOs can be effective antiseptics, but this may not be the outcome when used as a 1–2 per cent dilution in a carrier oil.

Massage

The massage applied is usually gentle effleurage with some petrissage (kneading) with some shiatsu, lymph drainage in some cases and sometimes more vigorous massage, according to the aromatherapist's skills. The massage should be relaxing, but also able to increase the circulation of the blood and lymphatic system 'in order to release toxins', break down tension in muscles and tone weak muscles (Price, 1993).

EO blending

The actual blending of EOs is considered an art form, but is basically simple: to 10 ml of carrier oil, in a brown bottle, 1–4 drops of the same or different EO are added and the bottle stopperred. The contents are then gently mixed without shaking the bottle too much and creating air-bubbles. Some Aromatherapists swear that different people can create a different mixture simply by their own energetic, but scientifically, it may simply be a question of slight changes in the volume of EO applied by different people.

GERANIUM OIL: MISINTERPRETATION OF ITS BOTANICAL SOURCE

The wrong genus

Although Geranium oil is one of the most widely-used EOs, there is apparently great confusion regarding its botanical source and therefore its functions. 'Geranium oil', sold as an aromatherapy oil or included in perfumes and some food products, is extracted from the scented leaves of some *Pelargonium* species and cultivars and has therefore nothing to do with the genus *Geranium*. However, this is not apparent when reviewing the origin of 'Geranium oil' in aromatherapy books and aromatherapy journals, as these often implicate *Geranium maculatum*, *G. robertianum* and other *Geranium* species. Even Body Shop referred to *Geranium maculatum* as source of their Geranium oil. This arose due to the unfortunate mistake by some of the original aromatherapy book authors, who looked up the medicinal properties of 'geranium' from the many Herbals (Culpeper, 1653; Grieve, 1937) thinking it was the same species.

There are amazing botanical concoctions e.g. 'The oil is extracted from the species *Pelargonium*, *Geranium Robert* or lemon plant' (Worwood, 1991). Tisserand (1985) informs us that: '*Pelargonium odorantissimum graveolens* is found on wasteland and was used by the ancients as a remedy for wounds and tumours'. He is undoubtedly referring to *G. robertianum* (Culpeper, 1653) as the quote is partly extracted from this source.

Lawless (1995) informs us that 'the British plant herb Robert (*Geranium robertianum*) and the American cranesbill, (*G. maculatum*) are the most widely-used types in herbal medicine today', but seemingly does not realise that this has nothing to do with *Pelargonium*.

Moreover, geranium species are usually used as a tea or alcoholic extract, which is taken orally (Culpeper, 1653; Grieve, 1937) and volatile EOs are not mentioned. The use of *G. macrorrhizum* EO is one exception, however this is produced mainly for its use in perfumery as 'Zdravetz' oil and is mainly confined to Bulgaria.

The wrong species

Frequent misnomers for the origin of 'Geranium oil' include: '*Pelargonium odorantissimum*', which is often misspelt as '*P. odorantissimum*' (Lawless, 1992; Valnet, 1982) and also another version: '*Pelargonium odorantissium*' (Westwood, 1991), which is an actual species with apple-scented leaves (van der Walt, 1977).

Other misnomers include: *P. asperum*, *P. roseum* Willd. and *P. graveolens*. The latter species may have contributed to the parentage of some Geranium oil originating in Africa, but the species itself has a more distinctive peppermint aroma (Demarne and van der Walt, 1989; Lis-Balchin, 1991). The cultivar known as *P.* cv. 'Rosé' which gives rise to the commercial 'Geranium oil, Bourbon' is, most probably, a hybrid *between P capitatum* × *P. radens* (Demarne and van der Walt, 1989).

Geranium oil contains mainly citronellol and geraniol and their esters and differs completely from that of a true Geranium oil e.g. *Geranium robertianum* oil (Pedro *et al.*, 1992) or that of *G. macrorrhizum* (Ognyanov, 1985).

A further misconception is that of the name 'Geranium Rose', which has been interpreted by some aromatherapists as meaning that geranium was distilled over rose (Price, 1993). This would be rather difficult as geranium leaves would smother the rose petals!

MISCONCEPTIONS ABOUT THE FUNCTION OF 'GERANIUM EXTRACTS' AND 'GERANIUM OIL'

Genus *Geranium* usage and Geranium oil

The actual usages of the geranium extracts mentioned in old herbals are associated with their tannin content and other water-soluble chemicals e.g. flavonoids, in the leaves or roots. Essential oils, on the other hand, are steam-distilled volatiles and do not contain these components.

Valnet (1980) gave Geranium oil's major attributes as 'its vulnerary powers (according to Ancients) and its power to mend fractures and eliminate cancers'; he then provided the following list of properties: 'internal use as astringent, tonic, antiseptic, antidiabetic, anticancer; external use as cicatrising agent, antiseptic, analgesic, parasiticide, insect repellent for mosquitos and gnats'. The indications for use vary with usage i.e. internal use: 'adrenal cortex deficiency, gastroenteritis, uterine haemorrhage, sterility, jaundice, urinary stones, gastric ulcer, cancer'; external use is indicated by: 'engorgement of breasts, sore, burns, cancers, tonsillitis, ophthalmia, facial neuralgia, gastric and lumbar pain, oedema of legs, herpes, shingles, scurf, lice, etc'. The directions for oral use are given as for 'Herb Robert', including infusions of the fresh or dried leaves!

The same botanical mistakes have been made by Tisserand (1985) who quotes directly from Culpeper's herbal (1653) about Herb Robert. Tisserand (1985) also lists properties of Geranium oil as: 'analgesic, antidepressant, antiseptic, cicatrisant, diuretic, haemostatic, sedative, stimulant of adrenal cortex, tonic and vulnerary'. Its uses are for: 'aphthae, burns, cancer (uterine), depression, dermatitis, diabetes, diarrhoea, eczema, engorgement of breasts, gastralgia, glossitis, haemorrhage, jaundice, kidney stones, nervous tension, neuralgia, pediculosis, ringworm, shingles, skin care, sterility, stomatitis, throat infections, ulvers (ulcers?), and wounds'.

Further attributes of Geranium oil

The myth is perpetuated by other authors e.g. Worwood (1991) who says that Geranium oil is advocated for: 'depression, menstrual problems, diarrhoea, diabetes, sores, neuralgia, bleeding, circulatory problems, eczema, sore throats, nervous tension, kidney stones'. The following home uses are advocated by Lawless (1992): 'acne, bruises, broken capillaries, burns, congested skin, cuts, dermatitis, eczema, haemorrhoids, lice, oily complexion, mature skin, mosquito repellent, ringworm, ulcers, wounds, cellulitis, engorgement of breasts, oedema, poor circulation, tonsillitis, adrenocortical glands and menopausal problems, nervous tension and stress-related conditions'. Other attributes include 'having a stimulating effect on the lymphatic system and a tendency to balance extremes on both the physical and emotional levels' (Westwood, 1991). The same author lists the following as indications for use: 'abrasiveness, excessive attachment, lack of balance, emotional extremes, lack of harmony, lack of moderation, mood swings, overpowering, excessive talkativeness, lack of tolerance and workaholic'. The Geranium oil has also 'yin' qualities and 'its ruling planet is Venus' (Tisserand, 1985).

It seems, therefore, that Geranium oil is effective for just about every malady. Some of the remarks in aromatherapy books are even more incredulous e.g. 'It (Geranium oil) is reputed to help in cases of uterine and breast cancer – and if nothing else, would certainly help the patient to relax and cope with the pain' (Worwood, 1991).

But, where is the proof for all these different functions?

FACTUAL EVIDENCE FOR GERANIUM OIL EFFECTS

The only factual evidence for *Pelargonium* activity must be taken from the folk-medicinal usage of the plants in their native country (southern Africa), as they were not available in Europe in any quantity till the eighteenth century and even then were used solely as ornamentals.

Folk-medicinal usage of *Pelargonium* species

Pelargonium species were used in South Africa by the local population and also by the Boers (Pappe, 1868; Watt and Brandwjik, 1962). Several *Pelargonium* species were used for antidysenteric purposes, and the more tannin-containing root was used for syphilis. Wooden articles made for sale were sprinkled with a decoction of some scented *Pelargonium* species to ensure a quick sale. Some of the folk-medicinal properties of *Pelargonium* species include:

Antidysenteric/antidiarrhoea/anticolic action

P. antidysentericum root
P. bowkeri leaf
P. cucullatum root and leaves
P. flabellifolium (syn. *P. luridum*) root
P. pulveratum root
P. reniforme root
P. sidaefolium
P. transvaalense root
P. triste root

Antihelmintic

P. triste root

Antiseptic

P. peltatum

Neuralgia

P. ramosissimum (syn. *P. tragacanthoides*)

Haemorrhoids

P. pulveratum leaves

Nephritis

P. cucullatum

Fever

P. luridum root
P. alchemilloides
P. transvaalense root

Wounds/abcesses

P. alchemilloides
P. cucullatum leaves
P. reniforme leaves

Syphilis/gonorrhoea

P. sidaefolium

Menstrual flow initiator

fumarioides (syn. *P. minimum*)
P. grossularioides
P. reniforme

Astringent

P. antidysentericum
P. peltatum leaf
P. reniforme root
P. luridum root

Colds

P. ramosissimum (syn *P. tragacanthoides*)

Abortifacient

P. grossularioides

Liver complaints in animals

P. reniforme

Emmanogogue

P. grossularioides

Unfortunately, all these attributes have been due to:

(a) different *Pelargonium* species to those used in the production of Geranium oil.
(b) they implicate mostly the water-soluble components, as they involved teas and infusions and not volatile EO.
(c) the plant extracts are mainly used internally and not massaged into the skin.

The pharmacological activity of the *water-soluble extracts* of the two genera (*Geranium* and *Pelargonium*) are, however, not very different, as *Geranium incanum* tea was used by indigenous South Africans as an antihelmintic, but the Europeans used it as an infusion for venereal diseases; the Southern Sotho tribes in South Africa used *Geranium canescens* as a remedy for colic, diarrhoea, dysenteries and fevers (Pappe, 1868; Watt and Brandwjik, 1962). But this still does not account for the volatile EO functions of *Pelargonium*-derived 'Geranium oil' quoted in most aromatherapy books, as virtually no scientific literature was available at the time.

Pharmacological action of *Pelargonium* oils

The lipophylic essential oils of *Pelargonium* species have mainly a spasmolytic effect on smooth muscle, except for *P. grossularioides*, which was used as an abortifacient in Southern African folk medicine (Watt and Breyer-Brandwijk, 1962) and has been shown to have a spasmogenic action on smooth and uterine muscle *in vitro* (Lis-Balchin and Hart, 1994).

Pelargonium EOs from leaves of the many different species and cultivars (other than those grown to produce commercial Geranium oil) have very different odours and chemical compositions, but most of the floral-smelling ones act through cyclic AMP as the secondary messenger; others with odours which are more pine or menthol-like have a different mode of action (Lis-Balchin and Hart, 1998). There is therefore some correlation between their mode of action and their odour and chemical composition. Furthermore, there was a distinct correlation between the following three criteria:

1 the chemical composition;
2 the predicted effect of the oil on smooth muscle *in vitro* (relaxation, contraction or both); and
3 the actual effect on man (either relaxation, stimulation or a mixture of both) as predicted by aromatherapists (Lis-Balchin and Hart, 1997a).

This indicated that effects shown on the isolated guinea-pig ileum would mimic the effect on the whole body.

Exploitation of the medicinal properties of *Pelargonium*

One of the more successful modern medicines derived from the original folk medicinal usage is Umckaloabo, which is extracted from *P. reniforme* and *P. sidoides* tubers (Kolodziej *et al.*, 1995; Kolodziej and Kaiser, 1997) and used to treat respiratory conditions. However, this involves internal oral usage of the water-soluble (rather than volatile) components of the drug and is therefore not true 'aroma' therapy.

Scientific proof of antimicrobial EO efficacy *in vitro*

Many plant EOs are extremely potent antimicrobials *in vitro* and can have a substantial effect on many bacteria (Maruzella and Henry, 1958; Deans and Ritchie, 1987) e.g. thyme or oregano oils, which inhibited 25 out of 25 different bacteria (Lis-Balchin, 1995; Lis-Balchin *et al.*, 1996, 1998). Geranium oil was one of the most potent antimicrobials after these two oils, but there was a large variation in the activity of different commercial geranium oils from different sources (Bourbon, Egyptian, Chinese) or supplier (Lis-Balchin *et al.*, 1996). The actual chemical composition of the oils was also variable and there was no correlation between the composition and the bioactivity, suggesting that there was some degree of adulteration.

CLINICAL AROMATHERAPY TRIALS

Virtually no trials have been conducted using Geranium oil, as lavender oil has almost always been the choice EO.

Use in childbirth

Studies of the use of aromatherapy in childbirth (Burns and Blaney, 1994) were not very conclusive, mainly because a large number of different EOs were used at different times and in different ways and there was a bias towards the use of a few oils e.g. lavender and clary sage, the latter having a probably undeserved (and unproven) status of being oestrogenic.

High dose chemotherapy

A study at the London clinic into the use of EOs for the treatment of chemotherapy-induced side effects in a group of patients undergoing high dose chemotherapy, with stem cell rescue for breast cancer (Gravett *et al.*, 1995) was reported at an Aromatherapy conference but no scientific publications followed. Groups of patients were not randomly allocated, and no-double blinding was attempted.

Treatment for mucositis (damage to mouth lining due to chemotherapy), where mouth washes often give burning sensation was changed to: one drop Tea tree: one drop bergamot: one drop geranium in half a tumbler of boiled warm water 5 times a day. Gargling with swallowing was allowed! This is therefore internal medicine, rather than aromatherapy. No statistically significant difference was found, but the actual experiment can be severely criticised, as EOs are *not* soluble in water and the un-dispersed globules of the oil could have caused even further damage to the mucosa. Another group suffering from diarrhoea, which was normally treated with codeine phosphate were treated with Buscopan and 'aromatherapy', consisting of geranium 15 drops, German chamomile 10 drops, patchouli 1 drop, turmeric phytol 10 drops mixed in 50 ml of sweet almond oil (i.e. a 5 per cent dilution) which was initially applied by abdominal massage twice a day, but due to nausea and diarrhoea 'oral administration in small doses in an alcoholic vehicle, usually sherry' was substituted. The justification for using geranium was 'that Geranium is a traditional remedy for gastroenteritis and stomatitis', which refers to the true geranium and its water-soluble 'tea' and not a lipophylic, volatile EO from a different genus! No significant differences between groups, was again found, but the one good outcome was that patients taking this mixture had a fragrant diarrhoea.

Lack of statistically-significant difference between massage with and without essential oils

There are few clinical studies and none of them show a statistically-significant difference between massage with and without EOs (Vickers, 1996); there are no studies using Geranium oil specifically. Studies involving lavender oil have shown no difference between three treatments involving aromatherapy massage, massage alone and giving the patients 30 min of rest in an intensive care unit (Dunn *et al.*, 1995). Another study using massage, with and without various EOs, on children with atopic eczema showed no differences after several weeks of treatment, however, a continuation of the study using EO massage showed a possible sensitization effect, as the symptoms became worse after subsequent massages following a three-month period of rest (Anderson *et al.*, 2000).

Use in depression

Studies in Italy and France, have offered very little in the way of scientific evidence on the efficacy of EOs and perfumes on patients and have been reported mainly in Trade Journals (Rovesti and Colombo, 1973). There was apparently some success in the treatment of depressed patients, but there is little data supplied as to the precise diagnosis of the patients, their symptoms, which of these symptoms were relieved, the number of patients involved and the statistical significance or otherwise of the results. Under such circumstances, the evidence is at best anecdotal. Most of the recent work has not been published in peer-reviewed journals and has consisted of single case studies of various treatments e.g. Franchomme and Penoel (1990), which largely involve internal use of EOs.

Use as anti-ageing products

Some EOs are strong antioxidant agents e.g. thyme and clove oils, and have recently been shown to counteract ageing in animals, as measured by the changes in the lipid composition of tissue membranes (Dorman *et al.*, 1995). However, the antioxidant values for 18 commercial geranium samples studied (Lis-Balchin *et al.*, 1996) showed inconsistencies and suggested that antioxidants may have been added to some of the oils and not to others.

Psychological and physiological effects of Geranium oil

Many fragrances have been shown to have an effect on mood and in general, pleasant odours generate happy memories, more positive feelings and a general sense of well-being (Warren and Warrenburg, 1993). Much of this type of research has been conducted by perfumery companies (Jellinek, 1956) to boost sales and some EOs have also been used in hospitals in the USA to create a more happy and positive atmosphere and also in offices and factories in Japan to enhance productivity.

Many EO vapours have been shown to depress contingent negative variation (CNV) brain waves in human volunteers (Table 21.1) and these are considered to be sedative. Others increase CNV and are considered stimulant. The effects of inhaling different EOs on the CNV is compared to the effect on mouse motility and the direct effect of the EO on smooth muscle *in vitro* (Table 21.1). Although there is a great difference in the application of the oils and the measurement of their effect, there is surprisingly, frequent agreement. However, in the case of Geranium oil, both a sedative and stimulant effect is shown for CNV studies, unlike that for lavender and sandalwood which show sedative effects throughout. However, even valerian, a well-known sedative showed some stimulant effects in CNV studies. This suggests that either there is a different effect through concentration or it depends on the individual's liking/disliking of a given smell.

There is some evidence that certain EOs can lower blood pressure, if it is elevated e.g. nutmeg (Warren and Warrenburg, 1993), but Geranium oil has not been tested for this specific function.

Table 21.1 Sedative and stimulant EOs

Essential oil	*Sedative*	*Stimulant*
Geranium	2,3,5	2,3
Rose	5	2,3
Jasmine	5	1,2,4
Lavender	1,2,4,5	
Sandalwood	1,2,3,4,5	
Valerian	1,3,5	3

Sources: Kubota *et al.* (1992): CNV studies in man; Torii *et al.* (1988): CNV studies in man; Manley (1993): CNV studies in man; Buchbauer *et al.* (1991, 1993): Jager *et al.* (1992): motility of mice; Lis-Balchin *et al.* (1997a,b): smooth muscle *in vitro*.

Direct effect on tissues after skin absorption?

There is no scientific evidence, as yet, regarding the direct action of EOs, applied through massaging of the skin, on specific internal organs. This is despite some evidence that certain EO components can be absorbed either through the skin or lungs (Buchbauer *et al.*, 1992; Jager *et al.*, 1992). Furthermore, although many EOs are very active on many different animal tissues *in vitro* (Lis-Balchin and Hart, 1997b), we have no idea as yet whether their activity in minute amounts (as used in aromatherapy massage) can benefit the patient through direct action on target organs or tissues (Vickers, 1996) rather than through the odour pathway leading into the mid-brain's 'limbic system' and thence through the normal sympathetic and parasympathetic pathways. There is also no proof that synergism occurs when mixtures of EOs are used (Lis-Balchin *et al.*, 1998).

Future clinical application of aromatherapy

What could be achieved by using aromatherapy as an adjunct to clinical medicine especially in hospitals and general practice? So far there have been many 'successes' in various areas, notably hospices. There are no miracle cures, but an alleviation of suffering and possibly pain, mainly through relaxation due to gentle massage, a nice odour like Geranium oil, and the presence of someone who cares and listens to the patient. This is probably also the case in geriatric wards, in general wards, in the treatment of severely physically and mentally-challenged children and adults etc. There is a need for this kind of healing contact and the added power of odour, and aromatherapy fits this niche.

Nurses and other healthcare professionals have the wish to learn and train in the use of aromatherapy in favour of all the other alternative therapies (Trevelyan, 1996). The medical profession is also turning towards any branch of alternative medicine, which is useful in the treatment of patients whose symptoms are largely based on stress and who do not respond to conventional medicine.

Possible toxicity of Geranium oil

From the toxicological aspect, there is, however, the danger of causing dermatitis in sensitive people (Rudzki *et al.*, 1976). Geranium oil has only been implicated rarely and all references are due to contact dermatitis and sensitization. Most of the references are to one of the main components geraniol (Lovell, 1993). Patch tests to geraniol proved negative but dermatitis to perfumes containing Geranium oil has been shown in a few cases (Klarmann, 1958). Sensitization to geraniol using a maximization test proved negative (Opdyke, 1975). Latest reports from Japanese studies, using patients with ordinary cosmetic dermatitis and pigmented cosmetic dermatitis, who showed a positive allergic responses to a wide range of fragrances (Nakayama, 1998), gave a list of Class A fragrances which were termed common cosmetic sensitizers and primary sensitizers. This Class included Geranium oil, geraniol, sandalwood oil, artificial sandalwood, musk ambrette, jasmine absolute, hydroxycitronellal, Ylang ylang oil, cinnamic alcohol, cinnamaldehyde, eugenol, balsam of Peru and lavender oil.

There is also the danger of airborne contact allergic dermatitis through overuse of EOs and their continued storage in the home (Schaller and Korting, 1995). There may also be danger in the overuse of EOs during pregnancy and childbirth. Studies during childbirth in particular should take into account the baby's health, as there is always the danger of over-sedation of the infant and the subsequent lack of the breathing reflex after birth.

CONCLUSION

The numerous aromatherapeutic uses for Geranium oil are yet to be scientifically validated, although there is every reason to accept the scientific evidence that inhalation of a pleasant aroma and its action through the limbic system has a relaxing effect, as has massage; theoretically, the two used together could relieve many stress-related conditions like dermatitis, asthma, intestinal problems and headaches could be alleviated.

REFERENCES

Anderson, C., Lis-Balchin, M. and Kirk-Smith, M. (2000) Evaluation of massage with essential oils on Childhood Atopic Eczema. *Phytother. Res.*, 14, 452–456.

Buckle, J. (1997) *Clinical Aromatherapy in Nursing*. Arnold, London.

Buchbauer, G., Jirovetz, L., Jager, W., Dietrich, H., Plank, C. and Karamat, E. (1991) Aromatherapy: evidence for the sedative effects of the essential oils of lavender after inhalation. *Z. Naturforsch.*, 46, 1067–1072.

Buchbauer, G. (1992) Biological effects of fragrances and essential oils. *Perf. Flav.*, 18, 19–24.

Buchbauer, G., Jager, W., Jirovetz, L., Ilmberger, J. and Dietrich, H. (1993) Therapeutic properties of essential oils and fragrances. In: R. Teramishu, R.G. Buttery and H. Sugisawa (eds), *Bioactive Volatile Compounds from Plants*, 159–165. ACS symposium series 525. Washington, DC: American Chemical Society.

Burns, E. and Blaney, C. (1994) Using Aromatherapy in childbirth. *Nurs. Times*, 90, 54–58.

Culpeper, N. (1653) *The English Physitian Enlarged*, George Sawbridge, London.

Deans, S.G. and Ritchie, G. (1987) The antibacterial properties of Plant Essential Oils. *Int. J. Food Microbiol.*, 5, 165–180.

Demarne, F. and van der Walt, J.J.A. (1989) Origin of the rose-scented *Pelargonium* grown on Reunion Island. *S. Afr. J. Bot.*, 55, 184–191.

Dorman, H.J.D., youdim, K.A., Deans, S.G. and Lis-Balchin, M. (1995) Antioxidant-rich plant volatile oils: *in vitro* assessment of activity. 26th Int. Symp. Essential Oils, Hamburg, Germany, Sept., 10–13.

Dunn, C., Sleep, J. and Collett, D. (1995) Sensing an improvement: an experimental study to evaluate the use of aromatherapy, massage and periods of rest in an intensive care unit. *J. Adv. Nursing*, 21, 34–40.

Franchomme, P. and Penoel, D. (1990) *Aromatherapie exactement*. Paris. Roger Jollois.

Gravett, P.J., Finn, M. and Hallasey, S. (1995) An investigation of the use of essential oils for the treatment of chemotherapy-induced side-effects in a group of patients undergoing high dose chemotherapy, with stem cell rescue for breast cancer. Paper given at Aromatherapy Conference, AROMA '95, UK.

Grieve, M. (1937) *A Modern Herbal*. Reprinted 1992. Tiger Books International, London.

Jager, W., Buchbauer, G., Jirovetz, L. and Fritzer, M. (1992) Percutaneous absorption of lavender oil from a massage oil. *J. Soc. Cosmet. Chem.*, 43, 49–54.

Jellinek, P. (1956) *Die Psychologischen Grndlagen der Parfumerie*. Alfred Hutig Verlag, Heidelberg.

Klarmann, E.G. (1958) Perfume dermatitis. *Ann. Allergy*, 16, 425–434.

Kolodziej, H. and Kaiser, O. (1997) *Pelargonium sidoides* DC. Neuste Erkenntnisse zum Verstandnis des Phytotherapeutikums Umckaloabo. *Z. Phytother.*, 19, 141–151.

Kolodziej, H., Kaiser, O. and Gutman, M. (1995) Arzneilich verwendete Pelargonien aus Sudafrika. *Dtsch. Apotheker Ztg.*, 135, 853–864.

Kubota, M., Ikemoto, T., Komaki, R. and Inui, M. (1992) Odor and emotion-effects of essential oils on contingent negative variation. *Proc. 12th Int. Congress on Flavours, Fragrances and Essential oils*, Vienna, Austria, Oct. 4–8. pp. 456–461.

Lawless, J. (1992) *The Encyclopaedia of Essential Oils*. Element, Dorset.

Lis-Balchin, M. (1991) Essential oil profiles and their possible use in hybridization of some common scented geraniums. *J. Essent. Oil Res.*, 3, 99–195.

Lis-Balchin, M. and Hart, S. (1994) A pharmacological appraisal of the folk medicinal usage of *Pelargonium grossularioides* and *Erodium cicutarium* (Geraniaceae). *J. Herbs. Spices. Med. Plants*, 2, 41–48.

Lis-Balchin, M. (1995) *The Chemistry and Bioactivity of Essential Oils*. Amberwood Publishing Ltd. Surrey.

Lis-Balchin, M., Deans, S.G. and Hart, S. (1996) Bioactivity of commercial Geranium oil from different sources. *J. Essent. Oil Res.*, 8, 281–290.

Lis-Balchin, M. (1997) Essential oils and 'Aromatherapy': their modern role in healing. *J. Roy. Soc. Health*, 117, 324–329.

Lis-Balchin, M. and Hart, S. (1997a) Correlation of the chemical profiles of essential oil mixes with their relaxant or stimulant properties in man and smooth muscle preparations *in vitro*. *Proc. 27th Int. Symp. Ess. Oils*, Vienna, Austria, 8–11 Sept. 1996. Ch. Franz, A. Mathe and G. Buchbauer (eds), Allured Pub. Corp., Carol Stream, Ill. pp. 24–28.

Lis-Balchin, M. and Hart, S. (1997b) Pharmacological effect of essential oils on the uterus compared to that on other tissue types. *Proc. 27th Int. Symp. Ess. Oils*, Vienna, Austria, 8–11 Sept. 1996. C.H. Franz, A. Mathe and G. Buchbauer (eds), Allured Pub. Corp., Carol Stream, Ill. pp. 29–32.

Lis-Balchin, M., Deans, S.G. and Hart, S. (1997a) A study of the changes in the bioactivity of essential oils used singly and as mixtures in aromatherapy. *J. Alt. Complement. Med.*, 3, 249–255.

Lis-Balchin, M., Hart, S., Deans, S.D. and Eaglesham, E. (1997b) Comparison of the pharmacological and antimicrobial action of commercial plant essential oils. *J. Herbs. Spices. Med. Plants*, 4, 69–86.

Lis-Balchin, M. and Hart, S. (1998) Studies on the mode of action of scented-leaf *Pelargonium* (Geraniaceae). *Phytother. Res.*, 12, 215–217.

Lis-Balchin, M., Deans, S.G. and Eaglesham, E. (1998) Relationship between the bioactivity and chemical composition of commercial plant essential oils. *Flav. Fragr. J.*, 13, 98–104.

Lis-Balchin, M. (1999) Possible Health and Safety Problems in the use of Novel Plant essential oils and extracts in Aromatherapy. *J. Roy. Soc. Prom. Health*, 119, 240–243.

Lovell, C.R. (1993) *Plants and the skin*, Blackwell Scientific. Publ., Oxford.

Manley, C.H. (1993) Psychophysiological effect of odor. *Crit. Rev. Food Sci. Nutr.*, 33, 57–62.

Maruzella, J.C. and Henry, A. (1958) The *in vitro* antibacterial activity of essential oils and oil combinations. *J. Am. Pharmaceut. Assoc.*, 47, 294–296.

Nakayama, H. (1998) Fragrance Hypersensitivity and its control. In: P.J. Frosch, J.D. Johansen and I.R. White (eds), *Fragrances: Beneficial and Adverse Affects*, Springer Verlag, Berlin. pp. 83–91.

Ognyanov, I. (1985) Bulgarian Zdravetz oil. *Perf. Flav.*, 10(6), 38–44.

Opdyke, D.L.T. (1975) Monographs on fragrance raw materials. *Food Cosmet. Toxicol.*, 13, 451.

Pappe, L. (1868) *Florae Capensis Medicae*, Prodromus; 3rd ed. Cape Town.

Pedro, L.G., Pais, M.S.S. and Scheffer, J.J.C. (1992) Composition of the Essential oil of *Geranium robertianum* L. *Flav. Fragr. J.*, 7, 223–226.

Price, S. (1993) *Aromatherapy Workbook*, Thorsons, London.

Rovesti, P. and Colombo, E. (1973) Aromatherapy and aerosols. *Soap, Perfumery and Cosmetics*, 46, 475–477.

Rudzki, E., Grzywa, Z. and Bruo, W.S. (1976) Sensitivity to 35 essential oils. *Contact Derm.*, 2, 196–200.

Schaller, M. and Korting, H.C. (1995) Allergic airborne contact dermatitis from essential oils used in aromatherapy. *Clin. Exp. Dermatol.*, 20, 143–145.

Tisserand, R. (1985) *The Art of Aromatherapy*, Revised ed. C.W. Daniel Co. Ltd, Saffron Walden, pp. 231, 232.

Torii, S., Fukuda, H., Kanemoto, H., Miyanchio, R., Hamauzu, Y. and Kawasaki, M. (1988) Contingent negative variation and the psychological effects of odor. In: S. Toller and G.H. Dodds (eds), *Perfumery: The Psychology and Biology of Fragrance*. Chapman & Hall, New York.

Trevelyan, J. (1996) A true complement. *Nurs. Times*, 92, 42–43.

Valnet, J. (1982) *The Practice of Aromatherapy*. C.W. Daniels Co. Ltd, Saffron Walden.

Vickers, A. (1996) *Massage and Aromatherapy. A Guide for Health Professionals*. Chapman & Hall, London.

Warren, C. and Warrenburg, S. (1993). Mood Benefits of *Fragrance. Perf. Flavorist*, 18, 9–15.

Watt, J.M. and Breyer-Brandwijk, M.G. (1962) *The Medicinal Plants of Southern Africa*. Livingstone Ltd, Edinburgh.

Westwood, C. (1991) *Aromatherapy. Stress Management: A Guide for Home Use*. Amberwood Publishing Ltd, Dorset.

Worwood, V.A. (1991) *The Fragrant Pharmacy*. Bantam Books, London. p. 25.

22 Perfumery and cosmetic products utilising Geranium oil

Rhona Wells and Maria Lis-Balchin

ORIGINS OF COMMERCIAL 'GERANIUM OIL'

'Geranium oil', included in perfumes and cosmetic products is extracted from the scented leaves of several *Pelargonium* species and cultivars which originated from southern Africa. John Tradescant obtained the first species in 1631 for the UK and other species were introduced over the next 300 years and subsequently hybridized.

The origin of 'Geranium oil' is very confused as it is frequently referred to as *Pelargonium odoratissimum*. This is an apple-scented species and the misnomer probably arose from a particular *P. graveolens* variety which was very odouriferous. The name *P. odoratissimum* was then used by early writers of Essential Oil books (Guenther, 1951) as well as the trade distributors.

Other names used are *P. asperum* and *P. graveolens*. The latter may be a parent of some Geranium oils originating in Africa, but the main source of the oil is from a cultivar known as P. cv. 'Rosé' which gives rise to the commercial 'Geranium oil, Bourbon' and originated from hybridizations in England, in the eighteenth century; the cultivar was then exported to the South of France and Reunion and also lately to China. The 'Rosé' cultivar is most probably, a hybrid between *P. capitatum* × *P. radens* (Demarne and van der Walt, 1989).

The first plants grown for the French perfumery industry were planted in Algeria in 1847; in the 1880s extensive plantings were set out in Reunion Island. Nowadays the main sites are China, Egypt and Reunion, although the latter is decreasing its production. Geranium plants were also extensively grown in the provence region of France, but due to the high costs of production, they are no longer grown there.

THE POOR MAN'S ROSE OIL

Geranium oil and concoctions using Geranium oil components have long been used in making artificial rose oil. Examples of 'rose extenders' and various 'rose' formulations abound, based on either the essential oil of geranium and its fractions (Table 22.1) or synthetic components (Table 22.2). Other Rose bases and enhancers are provided by Curtis and Williams (1994).

Geranium oil mixes well with artificial musk, vanillin, bergamot oils, patchouli, clove and heliotropin. Geranium oil Bourbon is frequently adulterated and the real or preferred oil can only be detected by the perfumery Noses and stringent chemical analysis. Geranium oil is frequently made entirely from synthetic components.

Table 22.1 Rose extender

Constituent	*Part by weight*
Geraniol ex Palmarosa	1.5
Oil of Grasse geranium	0.5
Phenylethyl alcohol	2.5
Linalool ex bois de rose	1.2
Rhodinol ex African geranium	1.2
l-Citronellol	3.0
Nerol	0.07
Farnesol	0.03
Total	10.00

Table 22.2 Rose perfume

Constituent	*Part by weight*
Rhodinol	10
Phenylethyl alcohol	20
Cinnamic alcohol	6
Geraniol	20
Citronellol	10
Linalool	6
Hydroxycitronellol	5
Linalyl acetate	4
Eugenol	4
Rose de Grasse absolute	4
Geranium (African) oil	10
Phenylethyl acetate	1
Total	100

Rhodinol ex Geranium is used with hydroxycitronellol, linalool, geraniol, dimethyl benzyl carbide, cinnamic alcohol, phenyl ethyl alcohol, geranyl and linalyl esters in modern perfumery and cosmetic products. This rhodinol is often adulterated with synthetic rhodinol, fractions of citronella or palmarosa oils and other synthetic compounds.

Some of the geranium oils produced are not suitable for perfumery due to a earthy, potato-like, sulphide top notes often found. This depends partly on the method of production, especially the use of old, iron stills, and more so on the actual cultivar of *Pelargonium* used and the conditions under which it grows. Modern, massive, stainless steel stills ensure a better and more consistent product.

Geranium oil is still used in high quality perfumes and to a lesser extent the food processing industry. Its herbal character also lends itself to some toiletries where a delicate herbal note is required to reinforce the natural concept. It is also well loved by aromatherapists, who use it as a sedative essential oil.

FAMOUS PERFUMES CONTAINING GERANIUM OIL

Geranium oil displays green herbal, fresh yet earthy characterisitics and is frequently used in masculine fragrances especially as a heart note. Geranium is seldom found as a top note,

Table 22.3 Perfumes containing Geranium in conjunction with Lavender

Compound	*Parts*
Fougere-type perfume	
Lavender oil, French	14
Bergamot oil, FCF	8
Coumarin	12
Rose base	5
Jasmine base	4
Oakmoss absolute	6
Patchouli oil, light	2
Vetivert oil, Bourbon	10
Geranium oil, Bourbon	2
Iso-amyl salicylate	3
Source: Curtis and Williams, 1994.	
Modern Lavender water	
Lavender oil, French	45
Bergamot oil, FCF	25
Lemon oil, Sicilian	6
Neroli oil, reconstituted	4
Musk ketone	3
Sweet orange oil	3
Geranium oil, Reunion	4
Benzoin resinoid	4
Source: Curtis and Williams, 1996.	

as it tends to be long-lasting and add body to a fragrance. It is frequently used in conjunction with lavender in the true men's lavender scents such as Moustache (Rochas) or Pino Silvestre, and also the classical fougère blends (Table 22.3), where the top notes are primarily lavender linked and the heart lends itself well to the dry floral aspects of geranium. It also adds floral aspects to green fragrances such as Grey Flannel and Monsieur Lanvin.

Although geranium is more predominant in men's classical fragrances, it also appears in women's fragrances, in green florals, as in a heart note in Ivoire, Balmain (1980) as well as featuring in heart notes of classical chypres such as Cabochard, Gres (1958). The original Chypre was that of Coty (1917), a bouquet of orange, geranium, spices and oakmoss and a fragrance so individual that it inspired a whole family of chypres. Giorgio, Armani (1981) is a combination of mandarin and geranium to give touch of freshness to the top note.

Jean-Francois Laporte designed perfumes after the countries he had visited e.g. 'geranium' was a perfume inspired by his visit to Egypt, reminding him of the heady fragrance of sun-drenched geraniums stacked on a wooden cart (Barille and Laroze, 1995).

Paris, Yves St. Laurent (1983) contained geranium as one of the top notes, with mimosa, bergamot, mayflower, hawthorn and juniper. The middle notes were from Damascus rose, may-rose and violet and the base notes were provided by sandalwood, iris and amber (Edwards, 1997).

REFERENCES

Barille, E. and Laroze, C. (1995) *The Book of Perfume*, Flammarion, Paris.

Curtis, T. and Williams, D.G. (1994) *Introduction to Perfumery*, Part, III, Ellis Horwood, London.

Demarne, F. and van der Walt, J.J.A. (1989) Origin of the rose-scented Pelargonium grown on Reunion Island, *S. Afr. J. Bot.*, 55, 184–191.

Edwards, M. (1997) *Perfume Legends*, HM Editions, Paris.

Guenther, E. (1951) The Essential oils, vol. 3, van Nostrand Co., New York.

Other general references for perfumery

Arctander, S. (1960) *Perfume and Flavor Materials of Natural Origin*, Elizabeth, N.J., USA.

Haarman and Reimer (1989) *The Book of Perfume*, 5 vols., R. Gloss & Co., Germany.

Irvine, S. (1995) *Perfume*, Aurum Book, Haldane Mason Ltd., London.

Lamparsky, D. ed. (1991) *Perfumes. Art, Science and Technology*, Elsevier Science, New York.

Lefkowith, C.M. (1994) *The Art of Perfume*, Thames and Hudson, London.

Morris, E.T. (1984) *Fragrance. The Story of Perfume from Cleopatra to Chanel*, Charles Scribner's Sons, New York.

Pavia, F. (1995) *The World of Perfume*, Knickerbocker Press, New York.

Piesse, S. (1890) *Histoire des parfums*, J.B. Baillière et fils, Paris.

Poucher, W.A. (1994) *The Production, Manufacture and Application of Perfumes*. Chapman & Hall, New York.

Trueman, J. (1975) *The Romantic Story of Scent*, Aldus/Jupiter.

Wells, F.V. and Billot, M. (1988) *Perfumery Technology*, Ellis Harwood Ltd., Chichester, 2nd Ed.

23 New research: possible uses of various *Pelargonium* leaf oils and extracts as food preservatives

Maria Lis-Balchin

INTRODUCTION

Conventional food preservatives are falling into disrepute as more data accumulates regarding the toxicity of the chemicals used. Butylated hydroxytoluene (BHT) and butylated hydroxyanisole (BHA) are particularly suspect, but there are no available substitutes as they have such a wide usage in foods. Many herbs, containing essential oils (EOs), have been used as preservatives in foods for centuries, but the most biocidal of these are very odourous e.g. thyme, oreganum, clove and therefore have a restricted value.

Pelargonium EOs obtained from different species, and having a wide spectrum of chemical compositions, have shown considerable potential as antimicrobial agents (Lis-Balchin *et al.*, 1995); *Pelargonium* solvent extracts have shown a similar property (Lis-Balchin *et al.*, 1998a), including the more hydrophylic extracts (Lis-Balchin and Deans, 1996). Commercial 'geranium' oil, obtained from different sources and commercial outlets, showed considerable variation in bioactivity against 25 different bacterial species, 20 different variants of *Listeria monocytogenes* and three fungi, which was not correlated with the chemical composition or the stated country of origin (Lis-Balchin *et al.*, 1996a). The bioactivity against microorganisms was very potent for the main synthetic components citronellol and geraniol, but low for authentic Geranium oil, suggesting that differences in adulteration were responsible for the main differences in bioactivity. These results were also in line with former studies of commercial EOs (Lis-Balchin *et al.*, 1998b), where differences in bioactivity between EOs, which were similarly-labelled, proved to have different bioactivities.

The possibility of detecting adulteration was further supported by bioactivity measurements using two enantiomers of limonene (Lis-Balchin *et al.*, 1996b) and α-pinene (Lis-Balchin *et al.*, 1999), which suggested that the individual bioactivity of each enantiomer varied against different parameters, and that adulteration of commercial oils with synthetic components could possibly be proven.

ANTIBACTERIAL ACTIVITY OF ESSENTIAL OILS AND SOLVENT EXTRACTS OF *PELARGONIUM* SPECIES AND CULTIVARS

The bioactivity of 18 *Pelargonium* species and cultivars (extracted with solvents and by steam distillation) was studied against four bacteria, namely: *Staphylococcus aureus*

Table 23.1 Antimicrobial activity of plant extracts against four bacteria

Plant	Zones of inhibition, mm			
	Staphylococcus aureus	*Proteus vulgaris*	*Bacillus cereus*	*Staphylococcus epidermidis*
'Attar of Roses'				
SD	12	11	17	8
MeOH	9	8	–	13
'Radula'				
SD	8	7	10	8
MeOH	13	13	13	18
'Sweet Mimosa'				
SD	7	10	n/a	–
MeOH	14	17	12	18
P. tomentosum				
SD	6	–	n/a	–
MeOH	10	10	9	10
'Chocolate tomentosum'				
SD	8	8	n/a	8
MeOH	8	12	9	10
'Lemon Fancy'				
SD	9	13	18	12
MeOH	8	11	10	8
'Crispum variegatum'				
SD	–	–	8	7
MeOH	14	10	7	–
P. × *fragrans*				
SD	10	9	n/a	8
MeOH	28	23	16	19
Geranium oil (Commores)				
SD undiluted	10	10	9	9
SD dil ×10	8	7	6	8
SD dil ×100	–	–	6	–

Notes
SD = steam distilled extract; MeOH = methanolic extract.

(ATCC 9144); *S. epidermidis* (ATCC 12228); *Proteus vulgaris* (ATCC 13315) and *Bacillus cereus* (NCIMB 6349).

The solvent extracts were made using fresh *Pelargonium* leaves, which were sequentially, but not exhaustively extracted with petroleum spirits followed by methanol. The solvent extracts were all reduced in a rotary evaporator (at temperatures under 40 °C). Essential oils were obtained using a Clavenger-type apparatus, from fresh leaves after 2 h of distillation.

Bioactivity and chemical composition

The bioactivity of 'Attar of Roses' was similar to the published activities of commercial geranium oils, which themselves showed a considerable span (Figure 23.1). This was not surprising as the source of commercial Geranium oil is from a cultivar very similar to that of the latter. The bioactivity of different *Pelargonium* EOs were largely correlated with their chemical composition: the rose-like geranium oils were potent antibacterials,

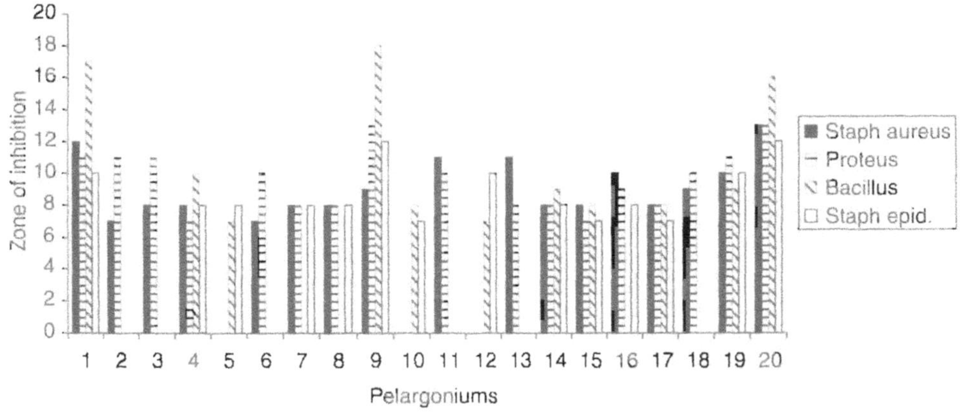

Figure 23.1 Antibacterial effect of Pelargonium extracts (steam distilled). 1 = 'Attar of Roses', 2 = 'Lady Plymouth', 3 = 'Pink Little Gem', 4 = 'Radula', 5 = 'Rober's Lemon Rose', 6 = 'Sweet Mimosa', 7 = *P. tomentosum*, 8 = 'Chocolate tomentosum', 9 = 'Lemon Fancy', 10 = *P. crispum* 'variegatum', 11 = 'Clorinda', 12 = 'Copthorne', 13 = 'Oak cv.', 14 = 'Village Hill Oak', 15 = *P. denticultum*, 16 = *P.* × *fragrans*, 17 = *P. odoratissimum*, 18 = 'Orsett', 19 = Geranium oil (commercial), 20 = Cinnamon oil (commercial)

Table 23.2 Chemical composition of *Pelargonium* oils

Pelargonium	*Main components*
'Attar of Roses'	Citronellol, geraniol
'Rober's lemon Rose'	Citronellol, citronellyl formate, isomenthone, linalool, sesquiterpenenes;
P. denticulatum	*p*-Cymene, hexenyl butyrate, limonene, sesquiterpenes;
P. × *fragrans*	α-Pinene, methyl eugenol, fenchone, limonene;
'Clorinda'	β-Pinene, α-phellandrene, *p*-cymene, sesquiterpenes;
'Lemon Fancy'	Neral, geranial, sesquiterpenes;
Crispum variegatum	Neral, geranial, sesquiterpenes;
Geranium oil (Commercial)	Citronellol, geraniol, citronellyl formate, isomenthone.

as were their main components: citronellol and geraniol, confirming previous results (Lis-Balchin *et al.*, 1996a,b).

The minty and the commercial Vicks ointment-like species showed very different low antibacterial effects. *P. tomentosum* which is largely composed of the two components menthone and isomenthone, showed the lowest bactericidal activity; however, the 'Chocolate tomentosum' or ' Chocolate Peppermint' which is a cultivar of *P. tomentosum*, was more active, the activity being a reflection of the high concentration of monoterpenes and sesquiterpenes in its composition.

'Lemon Fancy', containing neral and geraniol i.e. citral, as its main components, had the most potent activity against *B. cereus* and *P. vulgaris*, a reflection of synthetic citral itself. However, 'Crispum variegatum' with a similar composition, did not have the same activity as 'Lemon Fancy', suggesting that other components also play a part.

The camphoraceus, 'Vicks-like' group: 'Clorinda', 'Copthorne', 'Oak-cultivars', 'Orsett', *P. denticulatum*. *P.* × *fragrans*, and the apple-scented *P. odoratissimum*, had a similar antibacterial action. However, their chemical composition was very different.

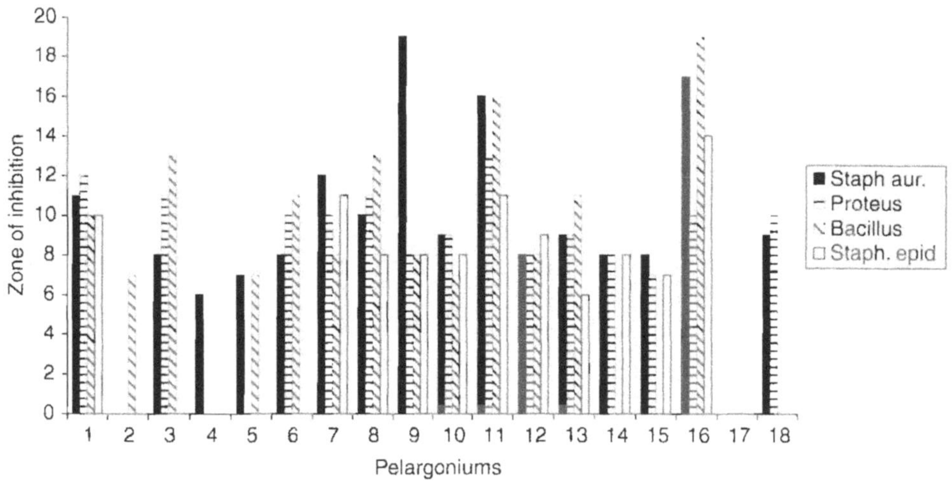

Figure 23.2 Antibacterial effect of Pelargonium extracts (petroleum ether). 1 = 'Attar of Roses', 2 = 'Lady Plymouth', 3 = 'Pink Little Gem', 4 = 'Radula', 5 = 'Rober's Lemon Rose', 6 = 'Sweet Mimosa', 7 = *P. tomentosum*, 8 = 'Chocolate tomentosum', 9 = 'Lemon Fancy', 10 = *P. crispum* 'variegatum', 11 = 'Clorinda', 12 = 'Copthorne', 13 = 'Oak cv.', 14'Village Hill Oak', 15 = *P. denticultum*, 16 = *P.* × *fragrans*, 17 = *P. odoratissimum*, 18 = 'Orsett'.

Pelargonium oils with a high level of limonene, a component of citrus oils, e.g. 'Sweet Mimosa', 'Copthorne', and the 'Oak cultivars', had a low biological activity, similar to that of citrus oils like lemon (Lis-Balchin *et al.*, 1998b).

Hydrophobic extracts

The petroleum spirit extracts, which resemble the steam distilled samples, except for the additional extraction of various other components, shared many similarities with the latter and had activities in the same range, but due to the dilution factors involved, the antibacterial activities were in fact much higher (Figure 23.2).

Hydrophylic extracts

The hydrophylic extracts in the series, proved to have more potent antibacterial activity than the EOs (Figure 23.3): this suggests that the flavonoids, tannins and other phenolics are the effective antimicrobial agents in the whole plant (Lis-Balchin and Deans, 1996; Lis-Balchin *et al.*, 1998c).

ESSENTIAL OILS OF *PELARGONIUM* SPECIES AND CULTIVARS: USE AS PRESERVATIVES IN FOOD PROCESSING

Using a Quiche filling as a model food system

Using a quiche filling as a model food system, the antimicrobial activity of *Pelargonium* EOs was investigated against *Salmonella enteriditis* (PT4 WT 132344),

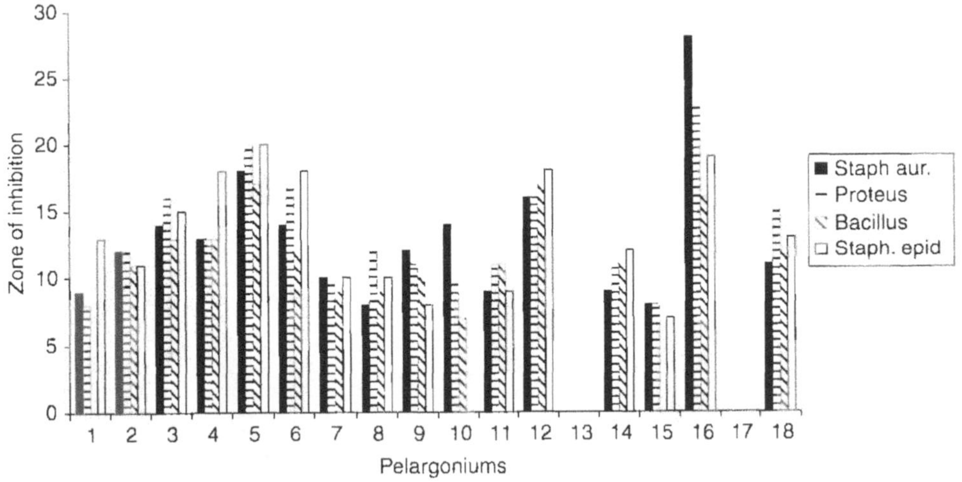

Figure 23.3 Antibacterial effect of Pelargonium extracts (methanol). 1 = 'Attar of Roses', 2 = 'Lady Plymouth', 3 = 'Pink Little Gem', 4 = 'Radula', 5 = 'Rober's Lemon Rose', 6 = 'Sweet Mimosa', 7 = *P. tomentosum*, 8 = 'Chocolate tomentosum', 9 = 'Lemon Fancy', 10 = *P. crispum* 'variegatum', 11 = 'Clorinda', 12 = 'Copthorne', 13 = 'Oak cv.', 14 = 'Village Hill Oak', 15 = *P. denticultum*, 16 = *P.* × *fragrans*, 17 = *P. odoratissimum*, 18 = 'Orsett'.

Listeria innocua (NCTC 10528), *Saccharomyces luduigii* (a cider isolate), and *Zygosaccharomyces bailii* (NCTC 1766). The EOs were in concentrations ranging from 250 to 500 ppm (Lis-Balchin *et al.*, 1998b). The quiche filling was composed of two medium eggs, mixed with 200 ml milk and 750 g of mixed frozen vegetables which were previously boiled till soft and mashed up; the mixture was autoclaved prior to inoculation, and EOs were added at 250, 500 and 1000 ppm to different portions.

At 250 ppm, 'Attar of Roses', *P. filicifolium*, 'Paton's Unique', 'Sweet Mimosa', and *P. glutinosum* showed considerable antimicrobial activity, compared with commercial cinnamon, clove, thyme and coriander (Figure 23.4). The most active against *Z. bailii* was thyme oil and 'Sweet Mimosa'; the activity of EOs was lower against the bacteria, with 'Paton's Unique' having the most potent activity against *S. enteriditis*. The log cfu/g^{-1} reduction was around 0.1 (for *Salmonella enteriditis* and *Listeria innocua*) to 2.6 (for *Salmonella luduigii* and *Zygosaccharomyces bailii*).

At 500 ppm, thyme oil was most inhibitory against *S. luduigii*, followed by 'Madam Nonin' and *P. glutinosum* and was considerably more active than cinnamon, clove, and coriander oils (Figures 23.5a,b). Thyme, 'Madam Nonin', 'Paton's Unique' and 'Sweet Mimosa' were all equally active against *Z. bailii*. Coriander, followed by *P. glutinosum*, 'Paton's Unique', Geranium oil (commercial) and 'Sweet Mimosa' were strongly active against *S. enteriditis* and *L. innocua*. 'Clorinda, *P. filicifolium* and 'Lady Plymouth, were almost equally good. The log cfu/g^{-1} reduction was around 1.4 (for *Salmonella enteriditis*) to 4.3 (for *Salmonella luduigii*).

The comparatively low dosages of *Pelargonium* EOs showing antimicrobial activity, used in this study, was more beneficial than higher doses of mint oil used by Tassou *et al.* (1995), who also used a lower inoculum in their model food. This strongly suggests that

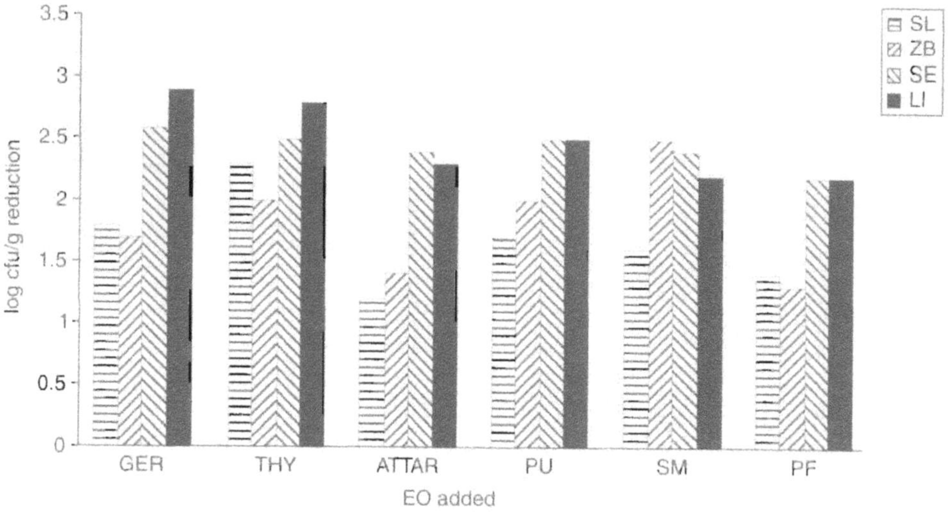

Figure 23.4 Antimicrobial activiy of essential oils (at 250 ppm) against *Saccharomyces ludwigii* (SL), *Zygosaccharomyces bailii* (ZB), *Salmonella enteriditis* (SE) and *Listeria innocua* (LI) respectively. GER = Commercial Geranium oil, THY = Commercial thyme oil, ATTAR = 'Attar of Roses', PU = 'Paton's Unique', SM = 'Sweet Mimosa', *PF* = *P. filicifolium*.

Table 23.3 *Pelargonium* essential oils and hydrophilic extracts useful as antimicrobial agents

P. 'filicifolium'
P. × *fragrans*
P. exstipulatum
P. 'Lady Plymouth'
P. scabrum
P. tomentosum
P. 'Chocolate Peppermint'
P. vitifolium
P. glutinosum
P. odoratissimum
P. graveolens
P. 'Rose'/'Attar of Roses'
P. quercifolium
P. cucullatum
P. citronellum
P. papilionaceum
P. 'Clorinda'
P. 'Purple Unique'

Pelargonium EOs have a great potential as antimicrobial agents in foods, as they are non-toxic and have a more pleasant and not over-bearing odour potential than the strong antimicrobial EOs. The probability of their commercial usage in foods, when mixed with other natural antimicrobials (Lis-Balchin and Deans, 1997), is under investigation.

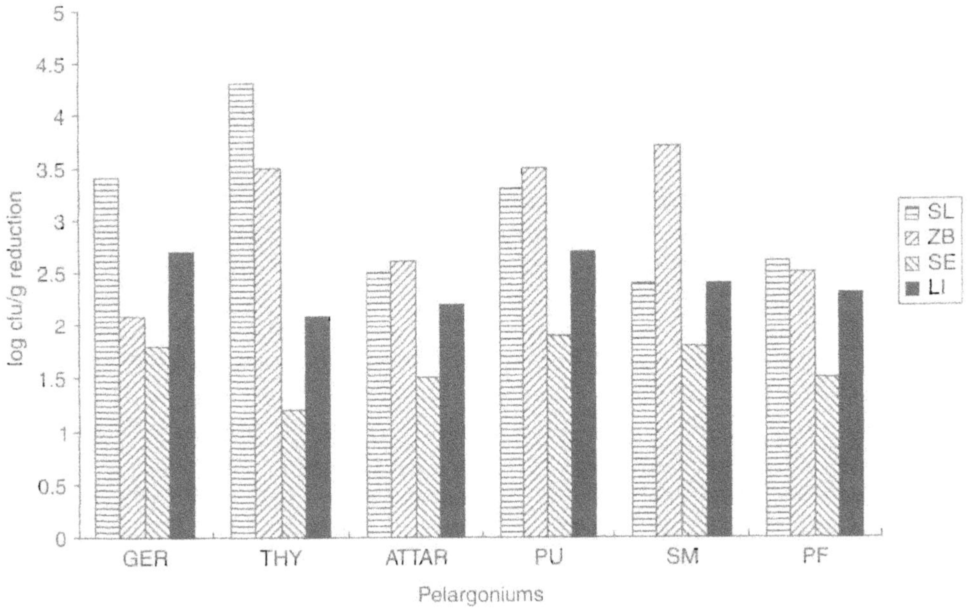

Figure 23.5a Antimicrobial activiy of essential oils (at 500 ppm) against *Saccharomyces ludwigii* (SL), *Zygosaccharomyces bailii* (ZB), *Salmonella enteriditis* (SE) and *Listeria innocua* (LI) respectively. GER = Commercial Geranium oil, THY = Commercial thyme oil, ATTAR = 'Attar of Roses', PU = 'Paton's Unique', SM = 'Sweet Mimosa', PF = *P. filicifolium*.

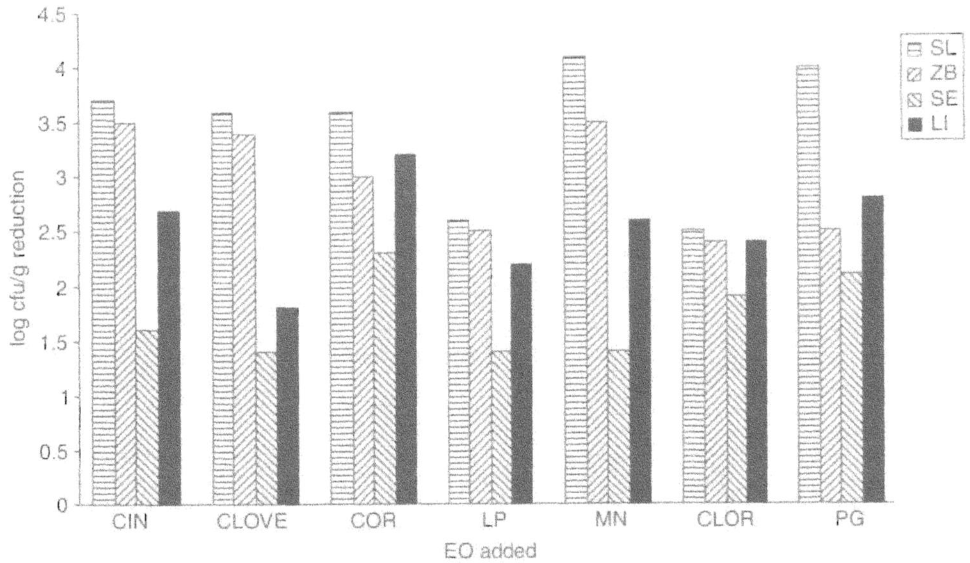

Figure 23.5b Antimicrobial activiy of essential oils (at 500 ppm) against *Saccharomyces ludwigii* (SL), *Zygosaccharomyces bailii* (ZB), *Salmonella enteriditis* (SE) and *Listeria innocua* (LI) respectively. CIN = Commercial cinnamon oil, CLOVE = Commercial clove oil, COR = Commercial coriander oil, LP = 'Lady Plymouth', MN = 'Madam Nonin', CLOR = 'Clorinda', PG = *P. glutinosum*.

COMPARATIVE ANTIMICROBIAL ACTIVITY OF *PELARGONIUM* AND SELECTED PLANT ESSENTIAL OILS AND THEIR RESPECTIVE HYDROSOLS IN A MODEL FOOD SYSTEM

Steam/water distillation of EOs inevitably results in the production of a hydrosol as well as leaving the residual plant material. This hydrosol residue is often not utilized, except for a few floral products like rose-water. It seemed that there is therefore a possible loss of opportunity for salvaging the hydrosols if they proved to have some commercially feasible bioactivity like, for example, antimicrobial activity.

Preliminary studies using the more water-soluble extracts of various *Pelargonium* plants indicated that these had indeed considerable antimicrobial activity. These studies were carried out using the *in vitro* agar well test (Lis-Balchin and Deans, 1996). The idea, which in theory seemed plausible, seemed therefore worthwhile to be studied in a model food system.

Plant EOs were obtained by distillation using a glass Clavenger-type laboratory still, from fresh leaves of *Pelargonium* 'Attar of Roses'. Other, potentially very active antimicrobial plants were also tested. These comprised: dried leaves of meadowsweet (*Filipendula ulmaria*), dried cinnamon bark (*Cinnamomum verum*), dried bay leaves (*Laurus nobilis*), clove buds (*Eugenia caryophyllus*) (Lis-Balchin *et al.*, 2000). The hydrosol remaining, after the distillation of the EO in each case, was reduced by a hundred-fold using a rotary evaporator at 50 °C.

A Model Food System, consisting of 250 g Porridge oats, 200 ml whole milk and 300 g of mixed vegetables (Tesco frozen vegetables) was mixed well together with (Maximum recovery diluent, 1:1) (MRD), and then 10 g portions were inoculated with a 24 h subculture of either *Staphylococcus aureus* (gram-positive) or *Escherichia coli* (gram-negative) at 10^4 cfu/ml and incubated with or without a plant EO or its hydrosol (at 1000 ppm) for 24 h at 37 °C for *E. coli* and 30 °C for *S. aureus*. The log cfu g^{-1} reduction in the bacterial numbers (estimated using the pour-plate technique, after incubation at 25 °C × 48 h) was then compared against controls.

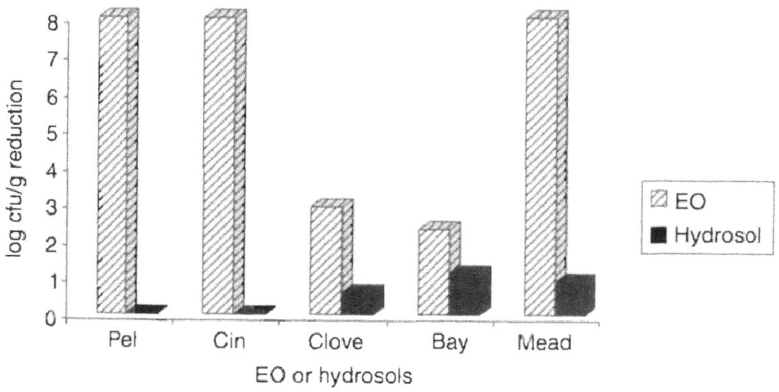

Figure 23.6 Antimicrobial action of essential oils and hydrosols against *Staphylococcus aureus* in porridge.

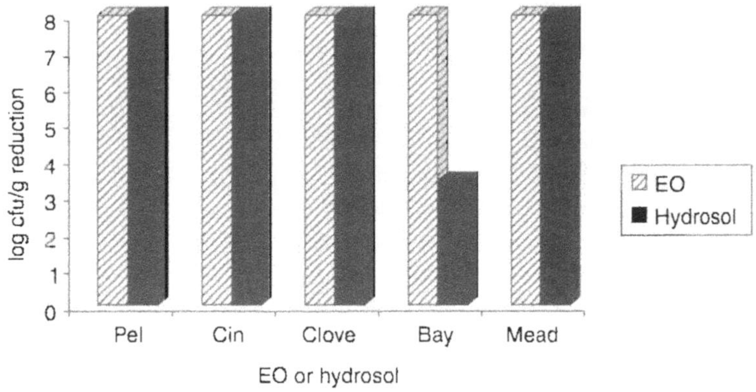

Figure 23.7 Antimicrobial action of essential oils and hydrosols against *Staphylococcus aureus* in MRD.

Activity against *Staphylococcus aureus* in the porridge system

The results indicated that meadowsweet oil (at 1000 ppm) was totally effective against both *S. aureus* and *E. coli*, but its hydrosol was only slightly effective against *S. aureus* in the porridge system (Figure 23.6); the oil was totally effective from 250 ppm. The clove, cinnamon and *Pelargonium* oils were very effective against *S. aureus*, with bay oil slightly so; however there was only slight antibacterial effect using cinnamon and bay hydrosols and the rest of the hydrosols were ineffective. The effect against this bacterium in MRD was very potent (Figure 23.7).

Activity against *E. coli* in the porridge system

Meadowsweet oil had a very strong effect against *E. coli* in porridge, and there was a decreasing effect with cinnamon, clove and *Pelargonium* oils respectively; there was no significant effect against *E. coli* by any plant hydrosol (Figure 23.8).

Activity in MRD

Antimicrobial effectiveness of all plant EOs against both the bacteria in the MRD was greatly enhanced (Figure 23.7), suggesting that the porridge-based food system was sequestering the oils or the bacteria or simply inhibiting the effect of EOs; the effectiveness of hydrosols from meadowsweet, *Pelargonium*, clove and to a lesser extent cinnamon and bay in MRD, again showed the difficulties of the practical application of potential antimicrobial agents in some foods.

The results using hydrosols, i.e. the hydrophylic extracts in actual food systems were therefore disappointing, as the results in an *in vitro* assay had shown great promise.

Possible explanation for poor antimicrobial results of hydrosols in a food system

The fact that all hydrosols were effective against *S. aureus* in the liquid MRD, and the lack of action of the hydrosols, compared to the potent action of the corresponding EOs against *E. coli*, suggested that the difference in composition of the food system and MRD and the difference in the water-solubility of the extracts was largely responsible for this difference in bioactivity.

Taking first the divergent activity of the EO and its corresponding hydrosol against *S. aureus* in the food system compared to that in MRD, one can see that hydrosols seem to be effective in an aqueous medium, but not in a medium containing fat, protein and carbohydrate. The main problematic food component seems to be the fat content. This theory was suggested by preliminary results using a bechamel sauce as the food system. This high-fat system (Table 23.4) proved to be completely detrimental in attaining any positive antimicrobial results at all using hydrosols. Reducing the fat content (Table 23.4) in the porridge system had a positive effect on the antimicrobial action of the hydrosols. The probable reason for this is that the fat-soluble EOs are probably sequestered in the fatty components whilst the bacteria are to be located mainly in the watery component and therefore unavailable to the EO. Therefore the more fat there is, the greater sequestering of the EOs occur.

The difference in fat solubility between the hydrosols and their corresponding EOs probably accounts for their difference in antimicrobial activity in the food system.

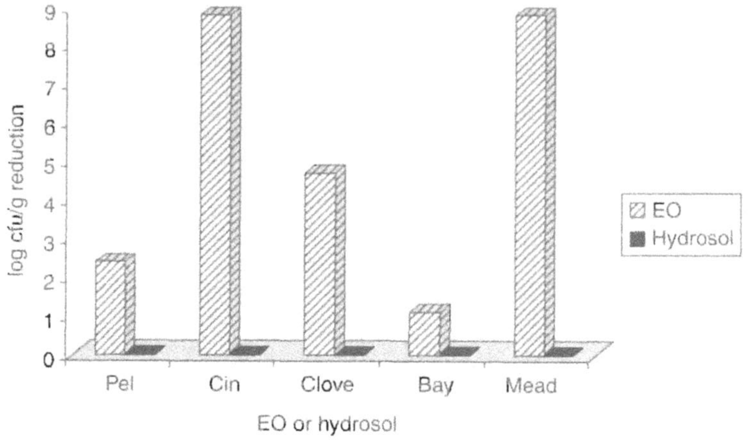

Figure 23.8 Antimicrobial action of essentials oils and hydrosols against *E. coli* in Porridge.

Table 23.4 Nutritional composition of model food systems

Food component	*Per* 10 g (g)
Porridge	
Protein	0.33
Fat	0.22
Carbohydrate	0.92
Bechamel sauce	
Protein	0.33
Fat	0.73
Carbohydrate	0.77

Notes
Composition of Bechamel sauce: 10 g butter; 10 g flour; 200 ml milk

However, the very reasons given above could not be used as an argument for the lower antimicrobial activity shown by the hydrosols compared with that of the EOs. If we consider the fact that both the bacteria and the hydrosols have a preference for the aqueous phase of the medium, this would suggest that the hydrophylic hydrosols would be more active. However, the reverse is apparent (Figure 23.7). There is therefore a much more complex explanation for this differential effect.

CONCLUSION

The potential usefulness of *Pelargonium* EOs and more hydrophylic extracts is only beginning to emerge. The main *Pelargonium* EOs so far studied and shown to be strongly antimicrobial are shown in Table 23.4. As there is a strong indication that the hydrophylic extracts are even more potent as antimicrobials, there remain more than 240 species and thousands of cultivars to study in order to ascertain which are the most active antimicrobial agents and which could be used commercially in food processing and in household products like cleaners and air-fresheners and even the great range of cosmetics.

REFERENCES

Lis-Balchin, M., Hart, S., Deans, S.G. and Eaglesham, E. (1995) Potential agrochemical and medicinal usage of essential oils of *Pelargonium* species. *Herbs. Spices Med. Plants*, **3**, 11–22.

Lis-Balchin, M. and Deans, S.G. (1996) Antimicrobial effects of hydrophylic extracts of Pelargonium species (Geraniaceae). *Letts. Appl. Microbiol.*, **23**, 205–207.

Lis-Balchin, M. and Deans, S.G. (1997) Studies on the potential usage of mixtures of plant essential oils as synergistic antibacterial agents in foods. *Phytother. Res.*, **12**, 1–4.

Lis-Balchin, M., hart, S., Deans, S.G. and Eaglesham, E. (1996a) Comparison of the pharmacological and Antimicrobial action of commercial plant Essential Oils. *J. Herbs, Spices & Med. Plants*, **4**, 69–86.

Lis-Balchin, M., Deans, S.G. and Hart, S. (1996b) Bioactivity of commercial Geranium oil from different sources. *J. Essent. Oil Res.*, **8**, 281–290.

Lis-Balchin, M., Buchbauer, G., Hirtenlehner, T. and Resch, M. (1998a) Antimicrobial activity of novel *Pelargonium* essential oils added to a quiche filling as a model food system. *Lett. Appl. Microbiol.*, **27**, 207–210.

Lis-Balchin, M., Deans, S.G. and Eaglesham, E. (1998b) Relationship between the bioactivity and chemical composition of commercial plant essential oils. *Flav. Fragr. J.*, **13**, 98–104.

Lis-Balchin, M., Buchbauer, G., Ribisch, K. and Wenger, M.-T. (1998c) Comparative antibacterial effects of novel Pelargonium essential oils and solvent extracts. *Lett. Appl. Microbiol.*, **27**, 135–141.

Lis-Balchin, M., O chocka, R.J., Deans, S.G. and Hart, S. (1999) Differences in bioactivity between the enantiomers of α-pinene. *J. Essent. Oil Res.*, **11**, 393–397.

Lis-Balchin, M., Buchbauer, G., Astrid Brandstetter, A. and Andrea Bauer, A. (2000) Comparative antimicrobial activity of *Pelargonium* and other selected plant essential oils and their respective hydrosols in a model food system. Paper presented at: Int. Ess. Oil. Symp., Hamburg, Sept. 10–13. To be published.

Tassou, C.C., Drosinos, E.H. and Nychas, G.J.E. (1995) Effects of essential oil from mint (mentha piperita) on *Salmonella enteriditis* and *Listeria monocytogenes* in modal food systems at 40 °C and 100 °C. *J. Appl. Bacterial.*, **78**, 593–600.

24 *Pelargonium reniforme* and *Pelargonium sidoides*: their botany, chemistry and medicinal use

Herbert Kolodziej

INTRODUCTION

Traditionally, the plant kingdom has been recognised as an inestimably rich source of potentially active metabolites not only as drugs, but also as unique leads that could serve as a starting point for the synthesis of new chemical analogues. This is best documented by the very great extent to which current medicines have their origins in plants. For example, artemisinin, codeine, digoxin, morphine, quinine, and taxol represent plant constituents of considerable therapeutic value, while salicin and khellin promoted the development of non-steroidal antiinflammatory agents and that of the antiasthmatic agent cromoglycinic acid, respectively. Thus, a successful *modus operandi* for finding promising agents involves the exploration of plants of traditional medical systems.

Pelargonium species indigenous to areas of southern Africa are highly valued by traditional healers and the native population for their curative properties. Among those traditional herbal medicines is umckaloabo, which originates from *Pelargonium sidoides* DC (De Candolle), and *Pelargonium reniforme* Curt (Curtis). Whereas *Pelargonium* species are well known for the beauty of their flowers, representing very popular ornamental plants in Europe, little is known of the medical practice with pelargoniums in folk medicine in areas of southern Africa.

Our interest in umckaloabo is explained by the therapeutic use amongst the local native population of southern Africa and its present utilisation in modern phytomedicine in Europe (Umckaloabo®, ISO-Arzneimittel, Ettlingen, Germany). The hitherto limited information on the chemical constituents of *P. sidoides* and *P. reniforme* and the underlying biologically active principle(s) prompted a more detailed chemical and pharmacological study on this plant medicine. This report represents a summary of the latest botanical facets, structural studies and recent pharmacological investigations of the titled *Pelargonium* species mainly carried out in our research group, and provides an overview of the clinical studies on the herbal medicinal product *Umckaloabo*® giving credence to its efficacy and safety in the claimed clinical indications.

PELARGONIUM RENIFORME CURT.

Botany

Species of *Pelargonium* are all very similar and have been much confused in the past. For example, the existence of gradual variation between *P. reniforme* and *P. sidoides* adds to general problems of taxonomic classification, as reflected in the past by numerous revisions in the framework of the Linneaen taxonomic system (van der Walt and Vorster, 1988; Dreyer *et al.*, 1992). In this context, a brief illustration of the taxonomic variations of both *Pelargonium* species may be justified. Following the first description of the Geraniaceae by Burmann (1738), L'Heritier de Brutelle (1788) outlined a number of genera including *Pelargonium* in his fundamental edition 'Geraniologia'. At the turn of the century, the botanical names *Pelargonium reniforme* Curt (Curtius, 1800) and *Geranium sidaefolium* Thunb. (Thunberg, 1794), the latter being subsequently changed into the present taxon *Pelargonium sidoides* DC. (De Candolle, 1824), have been given to two newly discovered species. However, in his revision of the Geraniaceae, Harvey and Sonder (1859) considered *P. sidoides* as a variety of *P. reniforme* and turned it to *P. reniforme* Curt. var. *sidaefolium*. Adopting de Brutelle's earlier taxonomic treatment of *Pelargonium*, these authors divided the genus into 15 sections and placed the titled *Pelargonium* species in section *Cortusina*. A sixteenth section was later recognised by Dreyer *et al.* (1992). Treatment by Knuth (1912) of the Geraniaceae resulted in the re-naming of *P. reniforme* var. *sidaefolium* at the species level to *P. sidaefolium* (Thunb.) R. Knuth. Wettstein's revision to *P. sidoides* DC., rewards De Candolle as first author of this species (Wettstein, 1935). Based on geographical, chemical, karyotypic, anatomical and morphological criteria, Dreyer *et al.* (1992) replaced the Section *Cortusina* s. l. by the Sections *Cortusina* (DC.) Harv. s. str., and *Reniformia* (Knuth) Dreyer, *comb. nova*. In the most recent treatment of the species *P. reniforme*, Dreyer *et al.* (1995) reported on the existence of the ssp. *reniforme* Curt., and ssp. *velutinum* (Eckl. and Zeyh.) Dreyer *stat. nov. et comb. nov.* The present taxonomic classification of *P. reniforme* and *P. sidoides* with reference to the essential textbook 'Strasburger – Lehrbuch der Botanik' (Sitte *et al.*, 1998) is shown in Table 24.1.

The pink-flowered *P. reniforme* (Figure 24.1) is an attractive erect shrublet of up to 1 m in height, developing from a tuberous rootstock. The zygomorphous flower heads are borne on tall slender stalks; each flower has five lanceolate petals, with two distinctive

Table 24.1 Taxonomic classification of *P. reniforme* and *P. sidoides*

Taxon	*Taxonomic category*	
Kingdom	Euca	
Subkingdom	Cormophyta	
Division	Spermatophyta	
Subdivision	Magnoliophytina	
Class	Rosopsida	
Subclass	Rosidae s. lat.	
Superorder	Rosanae	
Order	Geraniales	
Family	Geraniaceae	
Tribe	Geraniae	
Genus	*Pelargonium*	
Section	Reniformia (Knuth) Dreyer, *comb. nova*	
Species	*P. reniforme* Curt.	*P. sidoides* DC.
Subspecies	*P. reniforme* ssp. *reniforme* Curt.	*P. reniforme* ssp. *velutinum* (Eckl. and Zeyh.) Dreyer *stat. nov. et comb. nov.*

Reniform leaves ('reniforme') Flower head with distinctive stripes and dots

Morphology of *Pelargonium reniforme* ssp. *reniforme*

Figure 24.1 *P. reniforme* CURT. (*See Colour plate 14*)

stripes on the upper two petals. The reniform leaves with crenate or finely lobed margins are a characteristic feature of this species that is reflected in its botanical name '*reniforme*'. Most leaves have a velvety texture and greyish-green colour due to the presence of matted hairs (van der Walt and Vorster, 1983, 1988). The triporate pollen is ovoid and distinctly white, at least in terms of the ssp. *reniforme*; the non-availability of ssp. *velutinum* material excluded a similar microscopic inspection of its pollen. Noteworthy is that the two subspecies of *P. reniforme* not only differ morphologically, but also in area of distribution (Dreyer *et al.*, 1995). *Pelargonium reniforme* ssp. *reniforme* is confined to the region at Port Elizabeth where it is fairly widespread at low altitudes. All the material extending in the coastal districts further north and south and in inland areas is represented by the ssp. *velutinum* (Figure 24.2). The only notably morphological difference between the two subspecies lies in the shape and arrangement of leaves and in the length of petioles; ssp. *velutinum* shows reniform to predominantly cordate leaves and conspicuously longer petioles, reminiscent of those of *P. sidoides* (*vide infra*).

Chemical constituents

Although *Pelargoniums* have a long tradition, limited chemical sampling of members of this genus produced mainly common organic acids, derivatives of cinnamic acid, flavonoids, and phytosterols (Hegnauer, 1966, 1989). With the exception of the detection of the unique alkaloids (epi)elaeocarpidin in hybrids (Lis-Balchin *et al.*, 1996a,b), all recent papers deal with these types of secondary products. Owing to the persistent interest in perfumery, reflected by numerous papers on Geranium oil (*P. graveolens*, hybrids and other *Pelargonium* species) (van der Walt and Vorster, 1981; Kaiser, 1984; van der Walt and Demarne, 1988; Southwell and Stiff, 1995), essential oils (EOs) of distinct *Pelargonium* species have been the subject of detailed studies (Demarne and van der Walt, 1990, 1992, 1993a; Demarne *et al.*,

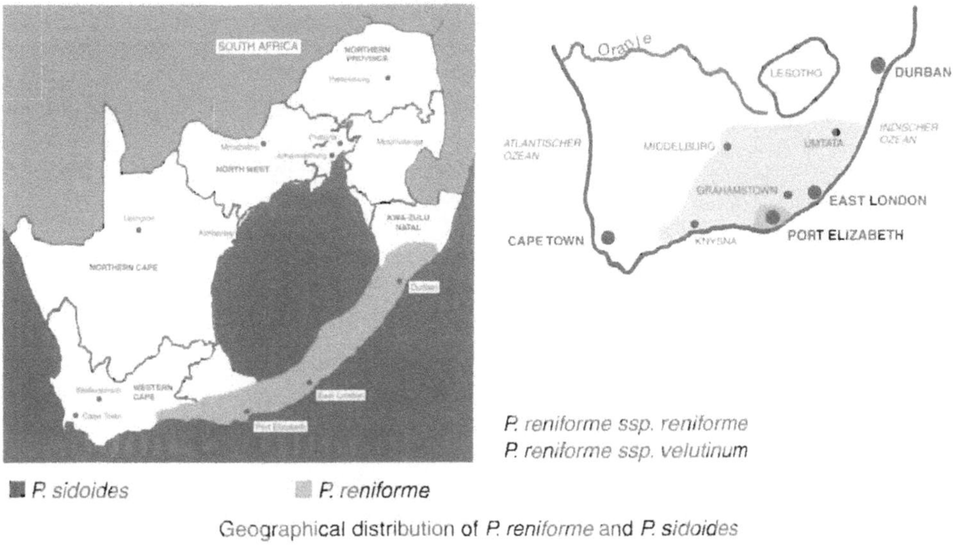

Figure 24.2 Geographical distribution of *P. reniforme* and *P. sidoides* ssp.

1993b). Reports on the metabolites of *P. reniforme* are hitherto limited and have revealed the occurrence of coumarins, tannins, flavonoids, carbohydrates and hydroxycinnamic acids (Wagner and Bladt, 1975). However, it should be stressed that chemical studies have not been carried out on distinct plant extracts of either *Pelargonium* species. Current evidence points also to erroneous identification of the plant material claimed to be *P. reniforme* in the earlier investigation. For example, the coumarin pattern and here the striking presence of 7-hydroxy-5,6-dimethoxycoumarin (umckalin) (**71**), typical of *P. sidoides*, in the previous paper on alleged *P. reniforme* conflicts with current information (*vide infra*). Independent support for this conjecture is also provided by the geographical origin of the former plant material, i.e. Grahamstown, an area where the ssp. *velutinum* commonly occurs. Thus, it would be of great interest to extend our studies to this subspecies. With botanically defined plant materials available through cultivation, studies on the individual patterns of constituents of each *Pelargonium* species were permitted for the first time. It should be noted that the plant material studied conformed with ssp. *reniforme*.

Root material

Structural examination of root metabolites of *P. reniforme* ssp. *reniforme* led to the characterisation of a total of 19 various metabolites including five simple phenolic acids (**1**, **2**, **5–7**), three hydroxycinnamic acid derivatives (**8**, **9**, **12**), six coumarins (**22–27**), four flavonoids (**28–31**), and one phytosterol (**63**) (Table 24.2). The majority of these metabolites has been encountered in relatively low yields. Noteworthy, however, is the presence of gallic acid (**1**) and its methyl ester (**2**) in fairly high amounts.

Monomeric flavan-3-ols, which apparently represent the precursors of associated condensed tannins (proanthocyanidins), were only detectable in traces. A substantial

Table 24.2 Constituents of *P. reniforme* ssp. *reniforme*

No.	*Compound*	*Roots*	*Herb*	*Flowers*
	Phenolic acids, Phenylpropanoids and Derivatives			
1	Gallic acid	+	+	+
2	Gallic acid methyl ester	+	+	
3	Gallic acid ethyl ester		+	
4	Gallic acid butyl ester		+	
5	*p*-Hydroxybenzoic acid	+		
6	Protocatechuic acid	+		
7	Vanillic acid	+		
8	Caffeic acid	+		
9	Ferulic acid	+		
10	*p*-Coumaric acid		+	
11	*p*-Coumaroyl-4-*O*-*β*-D-glucopyranoside		+	
12	*p*-Coumaraldehyde	+		
13	Shikimic acid 3-*O*-allate		+	
14	Shikimic acid 3,5-di-*O*-gallate		+	
15	*p*-Hydroxyphenylethanol		+	
16	*p*- Hydroxyphenyl acetic acid		+	
17	*p*-Hydroxybenzyl alcohol		+	
18	Glycerol-1-gallate		+	
19	Glucogallin		+	
20	(α,*β*)-3,4-Di-*O*-galloylglucopyranoside			+
21	Salidroside-6-*O*-gallate		+	
	Coumarins			
22	7-Hydroxy-6-methoxycoumarin (Scopoletin)	+	+	
23	6,7,8-Trihydroxycoumarin	+		
24	8-Hydroxy-6,7-dimethoxycoumarin (Fraxidin)	+		
25	6-Hydroxy-5,7-dimethoxycoumarin (Fraxinol)	+		
26	5,6-Dihydroxy-7-methoxycoumarin (Isofraxetin)	+		
27	8-Hydroxy-5,6,7-trimethoxycoumarin	+		
	Flavonoids			
28	Kaempferol 3-*O*-*β*-D-glucoside	+		
29	Kaempferol 3-*O*-*β*-D-galactoside	+		
30	Quercetin 3-*O*-*β*-D-glucoside	+		
31	Myricetin 3-*O*-*β*-D-glucoside	+		
32	Kaempferol 7-*O*-*β*-D-glucoside		+	+
33	Kaempferol 3-*O*-*β*-D-rutinoside			+
34	Quercetin 7-*O*-*β*-D-glucoside		+	+
35	Quercetin 3-*O*-*β*-D-rutinoside		+	+
36	Naringenin 7-*O*-*β*-D-glucoside		+	+
37	Dihydrokaempferol			+
38	Dihydroquercetin (Taxifolin)		+	+
39	Taxifolin 7-*O*-*β*-D-glucoside		+	+
40	Luteolin 7-*O*-*β*-D-glucoside		+	
41	Vitexin		+	
42	Vitexin 2″-*O*-gallate		+	
43	Orientin		+	
44	Orientin 2″-*O*-gallate		+	
45	Isovitexin		+	
46	Isovitexin 2″-*O*-gallate		+	
47	Isoorientin		+	+
48	Isoorientin 2″-*O*-gallate		+	
	Hydrolysable Tannins			
49	Brevifolin carboxyclic acid		+	+

50	Phyllantusiin E		+	
51	Phyllantusiin E *O*-methyl ester		+	
52	Strictinin		+	
53	Isostrictinin		+	+
54	Corilagin		+	+
55	Phyllantusiin C		+	+
56	Pelargoniin A		+	
57	Pelargoniin B		+	
58	Pelargoniin C		+	
59	Pelargoniin D		+	
	Miscellaneous			
60	(+)-cyclolariciresinol-2α-β-D-glucoside		+	
61	Stereoisomer of 60		+	
62	4,6-dihydroxyacetophenone 2-*O*-β-D-glucoside		+	
63	β-sitosterol	+		

proportion of the proanthocyanidins present in the roots is represented by oligomeric and polymeric forms, with catechin and gallocatechin entities as dominating extender units. A plausible interpretation is that precursors, when available at low concentration in the presence of an excess of competitive nucleophiles are immediately and quantitatively converted into high-molecular weight products. An illustrative general structure of these complex molecules is presented as depicted in formula (69).

Other characteristic constituents of *P. reniforme* included a remarkable series of simple coumarins as regards the high degree of aromatic functionalisation including hydroxyl and methoxyl groups. Such oxygenation patterns are very rarely found in the plant king-

Phenolic acids, Phenylpropanoids and Derivatives

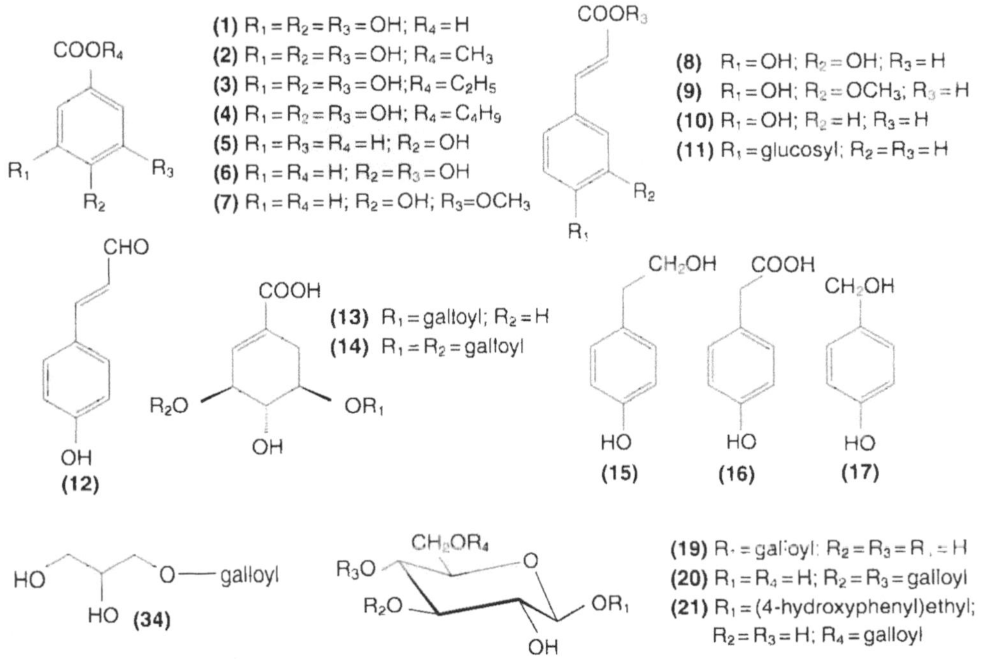

Coumarins

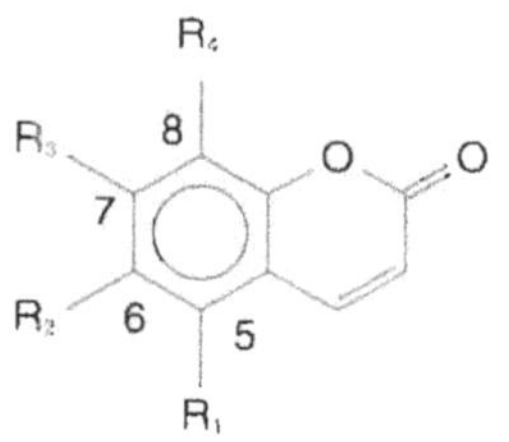

(22) $R_1=R_4=H$; $R_2=OCH_3$; $R_3=OH$
(23) $R_1=H$; $R_2=R_3=R_4=OH$
(24) $R_1=H$; $R_2=R_3=OCH3$; $R_4=O$
(25) $R_1=R_3=OCH_3$; $R_2=OH$; $R_4=H$
(26) $R_1=R_2=OH$; $R_3=OCH_3$; $R_4=H$
(27) $R_1=R_2=R_3=OCH_3$; $R_4=OH$

dom, but apparently typical for the genus *Pelargonium*. Apart from the widely distributed di-substituted scopoletin (22), all the coumarins possess tri- and tetra-substituted oxygenation patterns on the aromatic nucleus. Amongst these, 6,7,8-trihydroxycoumarin (23) and 8-hydroxy-5,6,7-trimethoxycoumarin (27) represent novel metabolites of the above class of secondary products.

Aerial parts

In contrast to the roots, the aerial parts of *P. reniforme* have not been the subject of chemical studies, which may be attributed, at least in part, to less relevant therapeutic uses. The results of our systematic examination of the metabolites present in the aerial parts of *P. reniforme* are included in Table 24.2.

Hydroxylated benzoic and cinnamic acids, gallic acid derivatives, flavonoids and tannins are the principal phenolic substances found in the aerial parts of *P. reniforme*. As regards flavonoids, the parent methanol extract afforded a complex mixture of flavonols (32–35), flavanones (36), dihydroflavonols (37–39) and flavones (40–48) (Table 24.2). Noteworthy is the presence of a series of C–C linked β-D-glucopyranosides (41–48) based on the flavones apigenin and luteolin. Although the C-6-glucopyranosylflavones, isovitexin (45) and isoorientin (47), and their C-8 regiomers, vitexin (41) and orientin (43), exhibit a wide taxonomic distribution (Jay, 1994), their occurrence in this natural source is of some chemotaxonomic significance in that C-glycosyl-flavonoids are considered to be poorly represented or generally lacking in the Geraniaceae (Jay, 1994; Hegnauer, 1989). These metabolites are accompanied by a unique series of 2″-galloyl analogues (42, 44, 46, 46), representing the first *O*-galloyl derivatives of C-glucosylflavones (Latté *et al.*, 2002).

In addition to the aforementioned flavonoids, the ethyl acetate and *n*-butanol phases of *P. reniforme* afforded a unique series of *O*-galloylated metabolites, including simple galloyl esters (2–4), shikimic acid 3-(13) and 3,5-di-gallate (14), glycerol-1-gallate (18), glucogallin (19), salidroside-6″-gallate (21), and (α,β)-3,4-di-*O*-galloylglucopyranoside (20) (Table 24.2) (Latté and Kolodziej, 1999). The majority of the identified isolates represents new or rarely found secondary products. For example, the occurrence of glycerol-1-gallate (18) has hitherto been demonstrated only in *Mallotus japonicus* (Euphorbiaceae) (Saijo *et al.*, 1990) and in a *Rheum* species (Polygonaceae) (Nonaka and Nishioka, 1983), while that of (α,β)-3,4-di-*O*-galloylglucopyranoside (20) is confined to *Macaranga tamarius* (Euphorbiaceae) (Lin *et al.*, 1990) and that of salidroside-6″-gallate (21) to *Quercus phillyraeoides* (Fagaceae) (Nonaka *et al.*, 1989) and *Q. stenophylla* (Nonaka *et al.*, 1982). On the other hand, gallic acid butyl ester (4) represents a new natural product. The remarkable broad range of *O*-galloylated compounds present in

Flavonoids

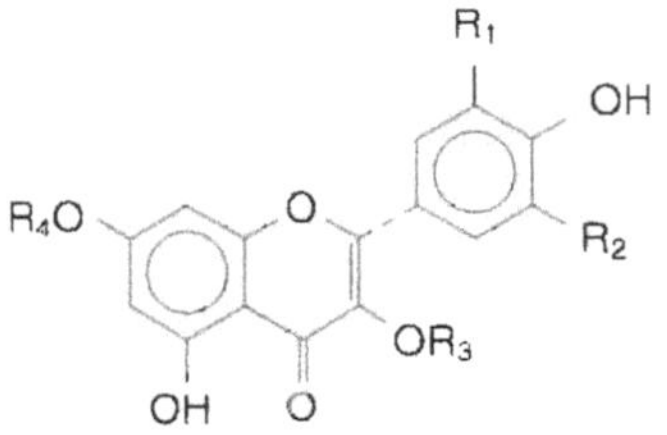

(28) $R_1 = R_2 = R_4 = H$; $R_3 =$ glucosyl
(29) $R_1 = R_2 = R_4 = H$; $R_3 =$ galactosyl
(30) $R_1 = OH$; $R_2 = R_4 = H$; $R_3 =$ glucosyl
(31) $R_1 = R_2 = OH$; $R_3 =$ glucosyl; $R_4 = H$
(32) $R_1 = R_2 = R_3 = H$; $R_4 =$ glucosyl
(33) $R_1 = R_2 = R_4 = H$; $R_3 =$ rutinosyl
(34) $R_1 = OH$; $R_2 = R_3 = H$; $R_4 =$ glucosyl
(35) $R_1 = OH$; $R_2 = R_4 = H$; $R_3 =$ rutinosyl

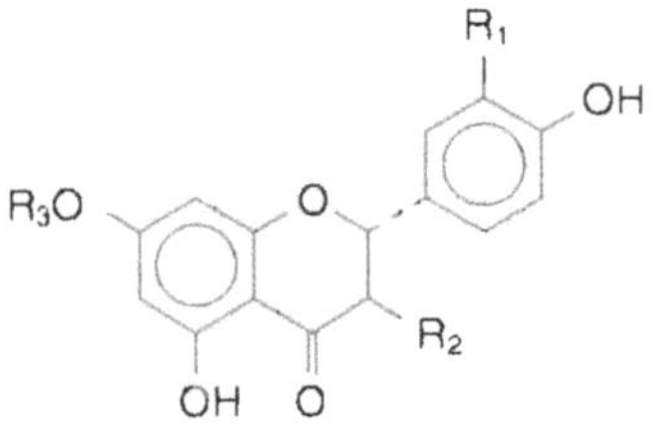

(36) $R_1 = OH$; $R_2 = H$; $R_3 =$ glucosyl
(37) $R_1 = R_3 = H$; $R_2 = OH$
(38) $R_1 = R_2 = OH$; $R_3 = H$
(39) $R_1 = R_2 = OH$; $R_3 =$ glucosyl

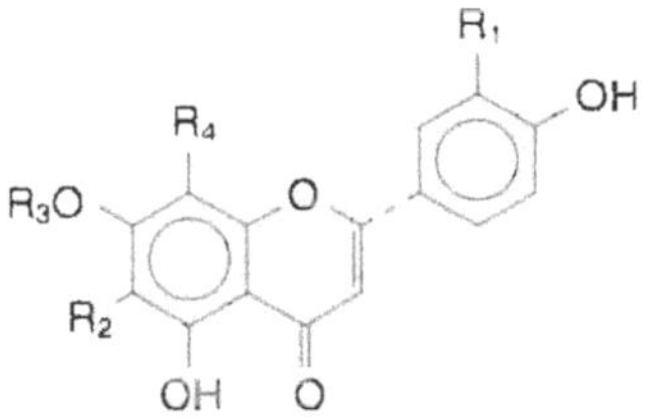

(40) $R_1 = OH$; $R_2 = R_4 = H$; $R_3 =$ glucosyl
(41) $R_1 = R_2 = R_3 = H$; $R_4 =$ glucosyl
(42) $R_1 = R_2 = R_3 = H$; $R_4 =$ (2″-galloyl)glucosyl
(43) $R_1 = OH$; $R_2 = R_3 = H$; $R_4 =$ glucosyl
(44) $R_1 = OH$; $R_2 = R_3 = H$; $R_4 =$ (2″-galloyl)glucosyl
(45) $R_1 = R_3 = R_4 = H$; $R_2 =$ glucosyl
(46) $R_1 = R_3 = R_4 = H$; $R_2 =$ (2″-galloyl)glucosyl
(47) $R_1 = OH$; $R_2 =$ glucosyl; $R_3 = R_4 = H$
(48) $R_1 = OH$; $R_2 =$ (2″-galloyl)glucosyl; $R_3 = R_4 = H$

P. renifome suggests galloylation of secondary products to be a putative chemotaxonomic marker for the genus *Pelargonium*. Although this conjecture is supported by similar observations made for the morphologically closely related species *P. sidoides*, further studies are required to place this hypothesis beyond doubts.

A major group of the metabolic pool of the aerial parts of *P. reniforme* is represented by a wealth of tannins. Here, polymeric proanthocyanidins are associated with ellagitannins, which are conspicuously absent from the root material. Their occurrence explains the traditional use of the aerial parts as wound healing agent, which may be attributed, at least in part, to their astringent action. Ellagitannins, receiving currently increasing interest, display a remarkable array of biochemical and pharmacological actions, including antiviral, antimicrobial and antitumoral properties (Okuda *et al.*, 1995).

Identified members included strictinin (52) and isostrictinin (53), composed of a central glucose core, adopting the $^4C_1$ conformation, a β-linked galloyl group at C-1 and a hexahydroydiphenoic (HHDP) moiety, formed by oxidative C–C coupling of two adjacent galloyl groups. In the case of strictinin (52) the HHDP group bridges the 4,6-position, while isostrictinin (53) represents the isomeric 2,3-coupled analogue.

Noteworthy is the co-occurrence of a series of structural variants in *P. reniforme*, similarly based on a 1-*O*-galloyl-β-D-glucopyranose precursor which itself adopts the

Hydrolyzable tannins

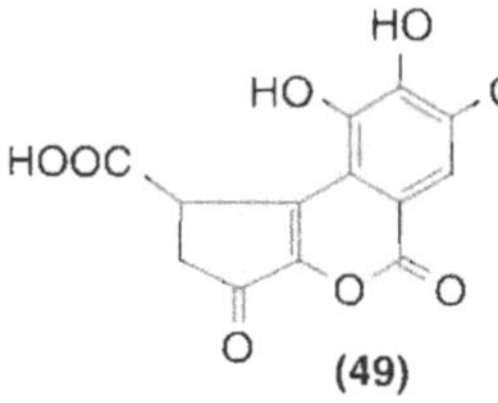

(49)

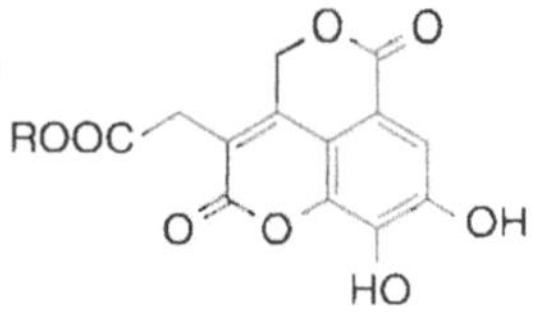

(50) R = H
(51) R = CH_3

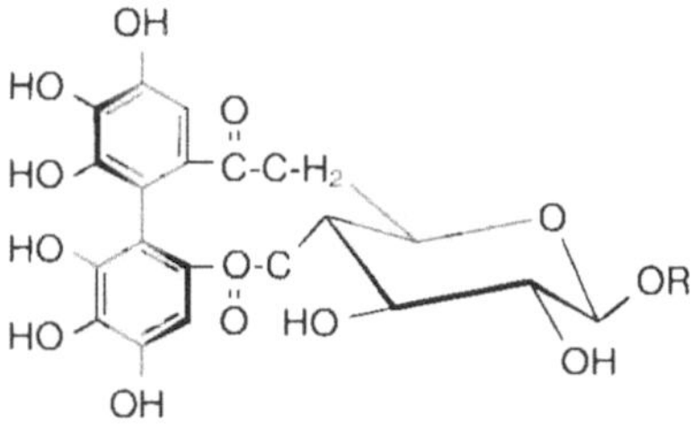

(52) R = galloyl

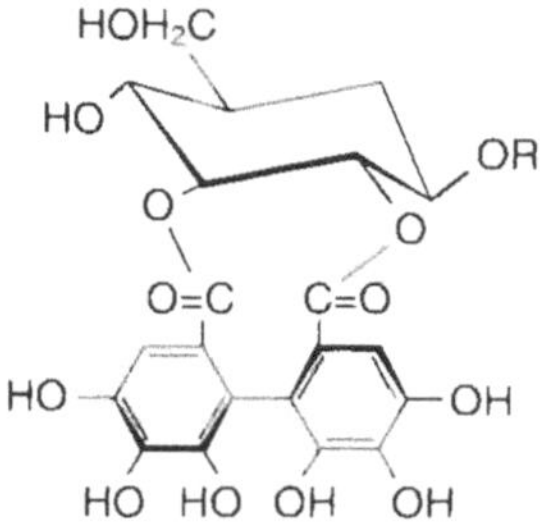

(53) R = galloyl

less favourable $^1C_4$ conformation. However, the galloyl ester groups at C-3 and C-6 are ideally aligned in a *axial* position for the feasible coupling in this energetically unfavourable chair conformation. These metabolites included corilagin (54) and a series of corilagin-based ellagitannins such as the rarely found phyllantusiin C (55) (Yoshida *et al.*, 1992b; Liu *et al.*, 1999; Amakura *et al.*, 1999) and four structurally related unique ellagitannins (56–59), designated as pelargoniins A, B, C, and D (Latté and Kolodziej, 2000). A remarkable common structural feature of these analogues is the presence of an oxidized DHHDP entity, bridging the 2,3-position. From their close structural relationship to geraniin, a biogenetic relationship may be postulated, though geraniin itself appears to be conspicuously absent from members of the genus *Pelargonium*.

Generally oxidative coupling of galloyl ester groups via the $^4C_1$ form of the glucopyranose precursor is much more widely encountered in plants than that via the alternative $^1C_4$ conformation (Haslam, 1998). According to present evidence plants appear to specialise in just one of these two distinctive forms of oxidative metabolism of the precursor (Haslam, 1988). This is the first example of the co-occurrence of ellagitannins with $^4C_1$ and $^1C_4$ glucose cores demonstrated for plants belonging to Geraniaceae. Up to now, only a few plants of the Euphorbiaceae (Saijo *et al.*, 1989a,b), Melastomataceae (Yoshida *et al.*, 1992a) and Onagraceae (Haddock *et al.*, 1982) have been shown to contain both forms of metabolites.

Our detailed chemical study on *P. reniforme* has also led to the isolation of two structurally closely related lignan glucosides (Table 24.2), the first members of this group of secondary products to be reported for the Geraniaceae and characterised by means of spectroscopic methods as (+)-isolariciresinol-2α-β-glucopyranoside (60) and an isomer of undefined stereochemistry (61) due to limited sample quantity. To date, the natural occurrence of (+)-isolariciresinol-2α-β-glucopyranoside (60) is hitherto limited to *Populus nigra* (Thieme and Benecke, 1969) and *Stemmadenia minima* (Achenbach *et al.*, 1992). Noteworthy is also the detection of 4,6-dihydroxyacetophenone

Ellagitannins based on a $^{1}C_4$-glucose precursor

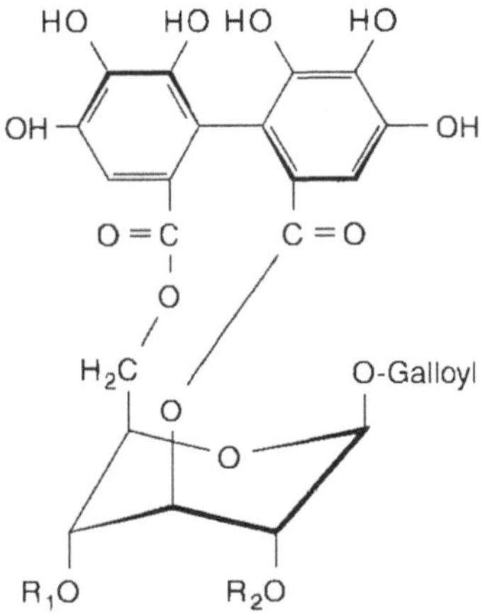

(54) $R_1 = R_2 = H$

(55) $R_1, R_2 =$

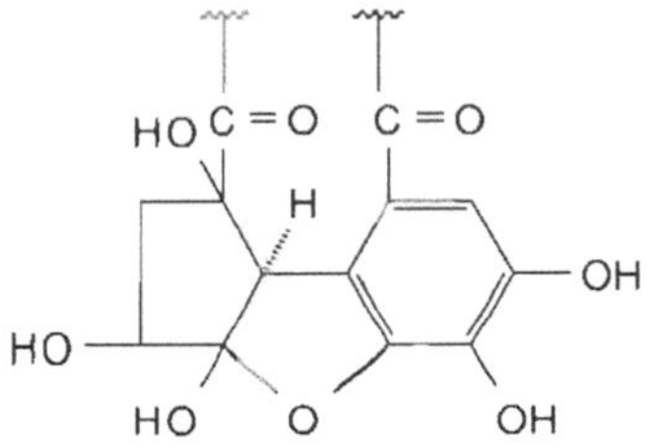

(56) $R_1, R_2 =$

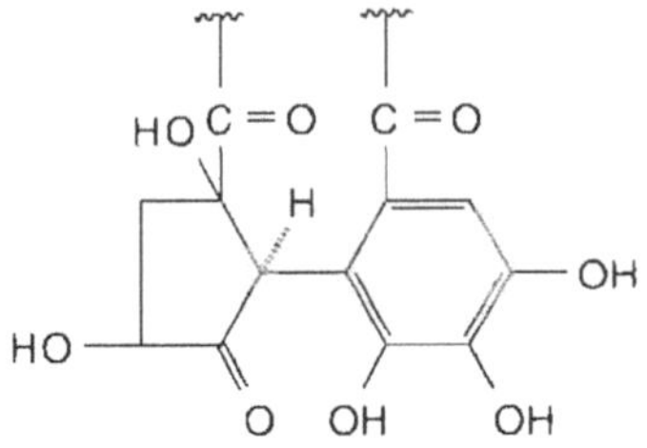

(57) $R_1, R_2 =$

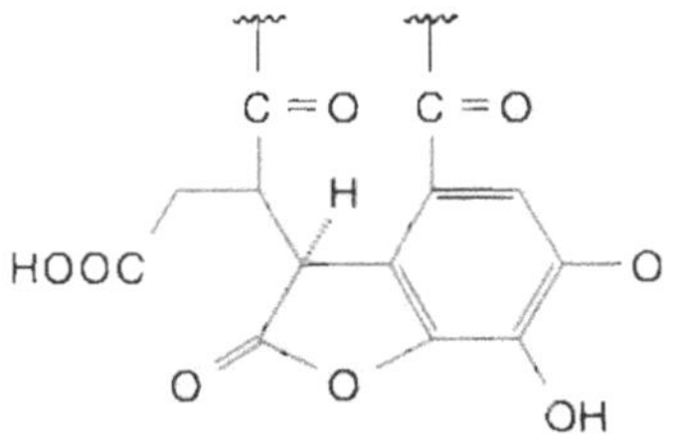

(58) $R_1, R_2 =$

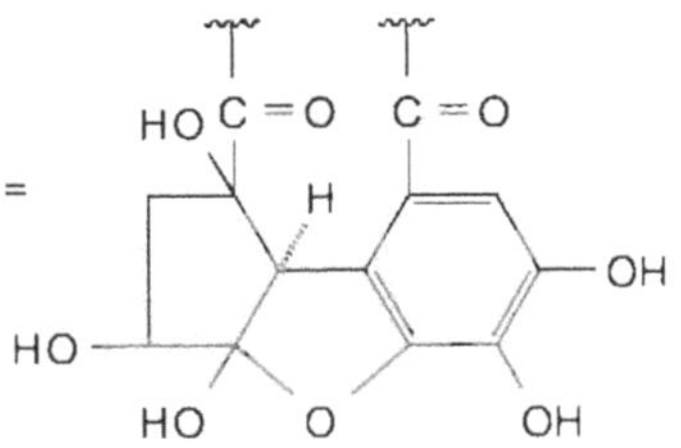

(59) $R_2 =$

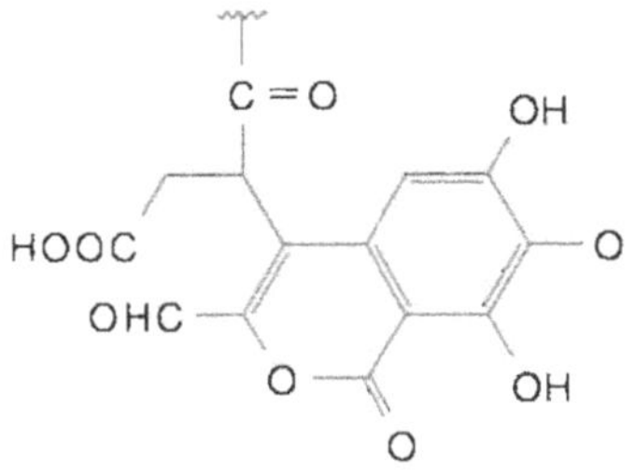

Lignans

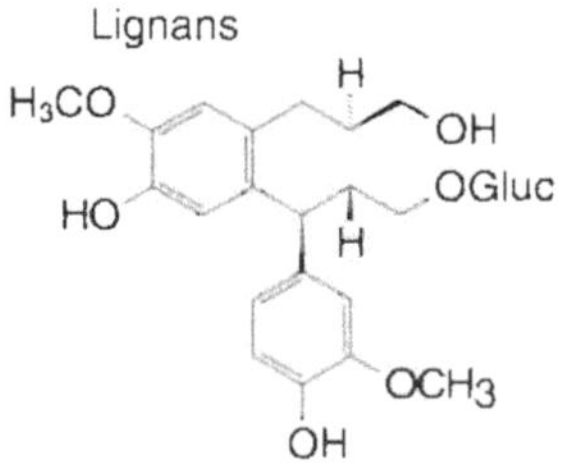

(60)

Acetophenone

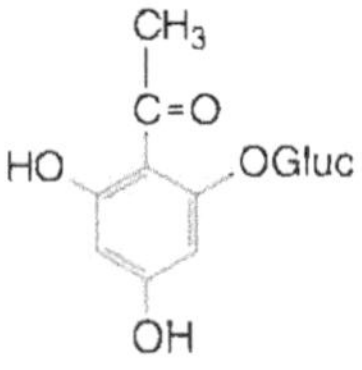

(62)

Phytosterols

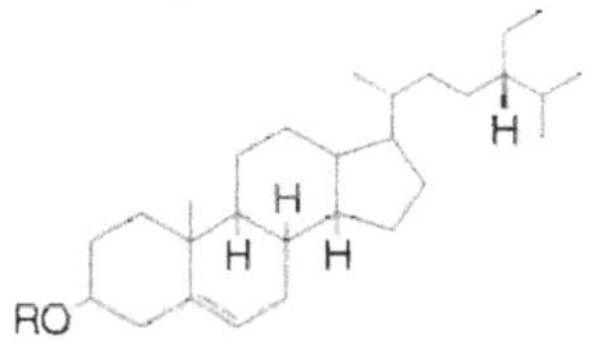

(63) R = H

(76) R = glucosyl

2-*O*-β-D-glucopyranoside (62) in that acetophenones being reported to occur only in a limited number of plant families (Singh *et al.*, 1997).

Essential Oil

Many *Pelargonium* species have a characteristic scent, based on the content of EOs, which provides for their utilisation in perfumery, cosmetic and food applications (Gildemeister and Hoffmann, 1959). Also worthy of mention is the observation that a trichome exudate, comprising a mixture of anacardic acids and tetrahydropyrans (Walters *et al.*, 1989 and 1990), is a potent chemical defence against small pest species, but also that some *Pelargonium* species are effectively repellent to insects and, hence, are widely employed in southern Africa by the native population in this respect. Although *Pelargonium* EOs, sporadically incorrectly declared as *Geranium* oils, have a long tradition, detailed studies on the chemical composition of EO of distinct *Pelargonium* species are rare (Demarne and van der Walt, 1990, 1992, 1993a; Demarne *et al.*, 1993b).

The details of the composition of the volatile fraction of *P. reniforme* are presented elsewhere (Kayser *et al.*, 1998). Therefore, only a brief summary will be presented here with an informative table of the chemical composition of the EOs of both *Pelargonium* species in terms of chemical classes (Table 24.3). Hydrodistillation of dried aerial parts produced a strongly aromatic-scented EO in 0.71 per cent yields related to the dry weight. About 230 components were detected of which 81 were unambiguously identified, accounting for about 49 per cent of the total peak area. The majority of the unknown compounds were represented by sesquiterpenes, as concluded from their mass spectral data and their retention times. The leaf oil consists of a complex mixture of different substances, with sesquiterpenes (ca. 60 per cent, inclusive of the unknowns) as the dominating components. Noteworthy is the relatively high level of sesquiterpene hydrocarbons (19.4 per cent) including α-muurolene, β- and Δ-selinene, cyclosativene and calamene. The monoterpenes comprised 4.7 per cent of the EO, with oxygenated components as the most abundant group (3.8 per cent).

Table 24.3 Composition of the essential oils of *P. reniforme* and *P. sidoides*

Class	*P. reniforme* *Number of components*	(%)	*P. sidoides* *Number of components*	(%)
Hydrocarbons/Oxyg. Terpenoids	4	0.4	12	4.1
Monoterpenes	Σ 26	Σ 4.7	Σ 41	Σ 16.3
Hydrocarbons	10	0.9	11	0.9
Alcohols	7	2.5	18	8.4
Aldehydes			1	0.2
Esters			4	3.0
Ketones	7	1.2	6	3.7
Oxides	2	0.1	1	0.1
Sesquiterpenes	Σ 30	Σ 36.9	Σ 37	Σ 32.9
Hydrocarbons	18	19.4	23	8.1
Oxyg.	12	17.5	16	24.8
Diterpenes	1	0.5	2	0.8
Phenylpropanoids			5	8.5
Fatty Acids	7	4.5	7	4.7

Detection of anacardic acids (Walters *et al.*, 1990) *via* the characteristic salicylic fragment *m/z* 138 and the presence of constituents with reputed repellent properties such as α-pinene, limonene, carvone and β-myrcene in the EO provide for a rational explanation for reported insect deterrency by *Pelargonium* species (Thorsell *et al.*, 1998). Also, the furan-type constituents identified (2-pentylfuran, perillene) and naphthalene derivatives (naphthalene, 1,2-dihydro-1,1,6-trimethylnaphthalene) are suggested to contribute significantly to plant protection against insects (Rodriguez-Saona *et al.*, 1999).

PELARGONIUM SIDOIDES DC

Botany

As the following botanical description may lead to the impression of a clear distinction between *P. reniforme* and *P. sidoides*, it is appropriate to recall the existence of gradual variations (*vide supra*). In its unadulterated form, this species (Figure 24.3) can be distinguished by the shape of the leaves and the colour of the flowers (van der Walt and Vorster, 1988). Dark red, but commonly almost black flowers are borne on long, slender stalks, spreading outwards from a woody base. The rosette-like plant has crowded, velvety, cordate, long-stalked leaves with short glandular trichomes sparsely interspersed. The triporate pollen is globose in shape and yellowish, but apparently not orange as indicated by van der Walt and Vorster (1988); accordingly, the various pollen characters do represent a diagnostic feature for the discrimination of the two *Pelargonium* species (*vide supra*). This species is predominantly found over large parts of the interior of southern Africa, but also occurs in coastal mountain ranges up to 2300 m (van der Walt and Vorster, 1983 and 1988) (Refer also to Figure 24.2).

Figure 24.3 *P. sidoides* DC. (*See Colour plate 15*)

Table 24.4 Constituents of *P. sidoides*

No.	*Compound*	*Roots*	*Herb*
	Phenols		
1	Gallic acid	+	+
2	Gallic acid methyl ester	+	+
13	Shikimic acid 3-*O*-gallate		+
19	Glucogallin		+
	Hydrolysable Tannins		
49	Brevifolin carboxyclic acid		+
54	Corilagin		+
	Flavonoids		
47	Isoorientin		+
64	Quercetin	+	+
65	Taxifolin 3-*O*-*β*-glucoside		+
	Flavan-3-ols		
66	(+)-Afzelechin	+	
67	(+)-Catechin	+	+
68	(+)-Gallocatechin	+	+
	Coumarins		
22	7-Hydroxy-6-methoxycoumarin (Scopoletin)	+	+
23	6,7,8-Trihydroxycoumarin	+	
70	6,8-Dihydroxy-7-methoxycoumarin	+	
71	7-Hydroxy-5,6-dimethoxycoumarin (Umckalin)	+	
72	7-Acetoxy-5,6-dimethoxycoumarin	+	
73	5,6,7-Trimethoxycoumarin	+	
74	6,8-Dihydroxy-5,7-dimethoxycoumarin	+	
75	5,6,7,8-Tetramethoxycoumarin (Artelin)	+	
	Phytosterols		
76	Sitosterol-3-*O*-*β*-D-glucoside	+	

Chemical constituents

A similar picture of a broad metabolic profile is also visible for *P. sidoides* based on our current data, though this study is not completed yet. Here, only the characteristic constituents of the roots and the aerial parts are briefly dealt with. Additional compounds unambiguously characterised are included in Table 24.4.

Root material

Compositional studies provided similar proof of the occurrence of oligomeric and polymeric flavan-3-ols in significant amounts. Here, the putative precursors (+)-afzelechin (66), (+)-catechin (67) and (+)-gallocatechin (68) could be successfully isolated from this source. One can conclude that the high-molecular weight tannins produced by *P. sidoides*, but also other *Pelargonium* species in its root tissues ultimately provide a significant barrier to invading microorganisms and other pests.

In addition to the high content of proanthocyanidins with their presumed ecological function, the root extract of *P. sidoides* also contained a wealth of highly oxygenated simple coumarins, unique in its composition (Table 24.4). For example, 6,8-dihydroxy-5,7-dimethoxycoumarin (74), 5,6,7-trimethoxycoumarin (73), 6,8-dihydroxy-7-

Flavan-3-ols

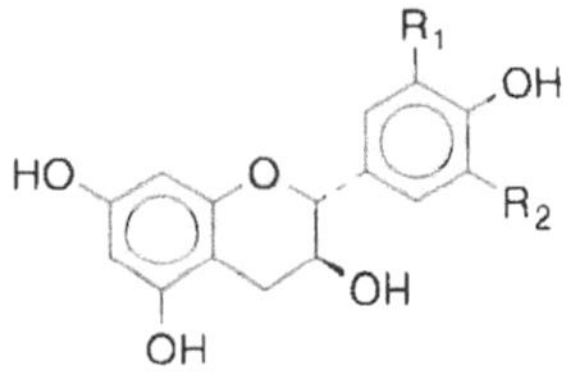

(66) $R_1 = H; R_2 = H$
(67) $R_1 = OH; R_2 = H$
(68) $R_1 = OH; R_2 = OH$

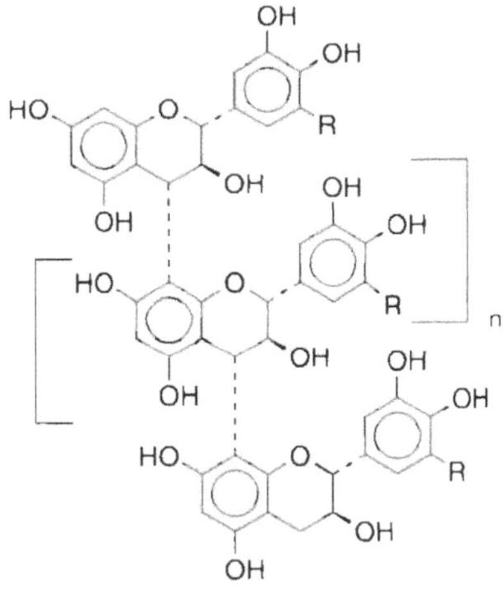

(69) R = H or OH

methoxy-coumarin (70), and 7-acetoxy-5,6-dimethoxy-coumarin (72) represent new natural products; the latter being the first natural compound known hitherto within this group possessing an acetoxy function. Comparison of the coumarin patterns in the roots of *P. sidoides* and *P. reniforme* clearly showed that they express conspicuously distinct variations (Table 24.1 and 24.2). Of the twelve identified coumarins, the two species share the ubiquitous scopoletin (22) and the unique 6,7,8-trihydroxycoumarin (23) only, the latter may therefore serve as useful chemotaxonomic marker. Also, discrimination between *P. sidoides* and *P. reniforme* may readily be verified by characteristic fingerprint chromatograms. It should also be noted that there is much divergence in concentrations, with significantly higher yields of coumarins in *P. sidoides*. With the exception of the characteristic 6,7,8-trihydroxy-coumarin (23), the remaining coumarins are present only in small amounts in *P. reniforme*.

Aerial parts

Again, polymeric proanthocyanidins were associated with members of hydrolysable tannins, as evidenced by the identification of brevifolin carboxyclic acid (49) and corilagin (54). Owing to difficulties regarding the characterisation of the exact molecular structure of extended flavan-3-ol entities, emphasis is currently given to the structural assessment of additional ellagitannins that have been detected in the herbal parts.

Additional notes concern the very limited occurrence of coumarins, in contrast to the abundance in the roots, and the presence of flavonoids, gallic acid derivatives and related phenolic compounds in the aerial parts. Here, the concurrent demonstration of the presence of isoorientin (47), when taken in conjunction with unpublished results, should be indicative not only of the existence of additional C-linked glycosylflavones in this plant source, but also of a much wider distribution of this class of secondary products in the genus *Pelargonium* than hitherto anticipated. Based on similar sampling evidence, it is ventured to say that there is a close similarity in the phenolic and flavonoid profile between the two species, already discernible from the present data in Tables 24.1 and 24.2.

Flavonoids

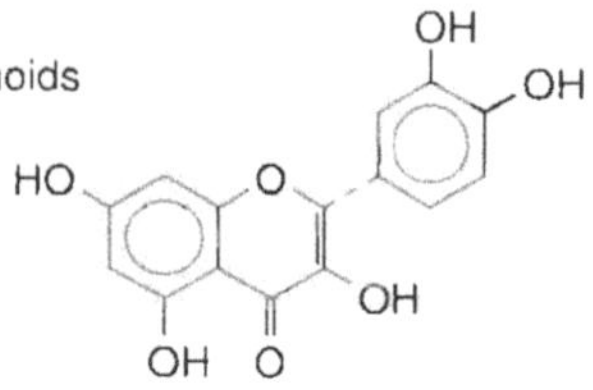

(64)

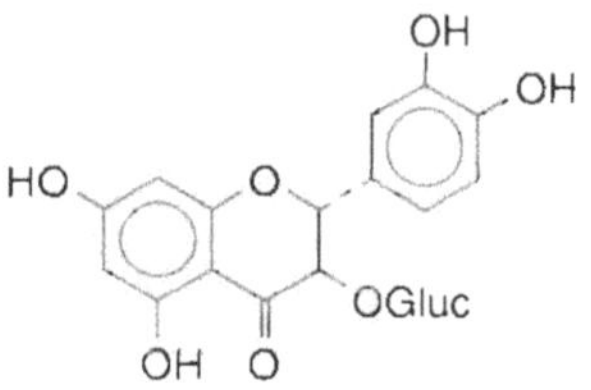

(65)

Coumarins

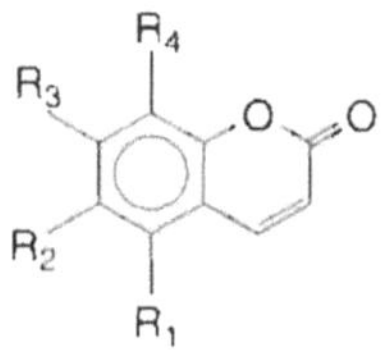

(70) $R_1 = H$; $R_2 = R_4 = OH$; $R_3 = OCH_3$
(71) $R_1 = R_2 = OCH_3$; $R_3 = OH$; $R_4 = H$
(72) $R_1 = R_2 = OCH_3$; $R_3 = OAc$; $R_4 = H$
(73) $R_1 = R_2 = R_3 = OCH_3$; $R_4 = H$
(74) $R_1 = R_3 = OCH_3$; $R_2 = R_4 = OH$;
(75) $R_1 = R_2 = R_3 = R_4 = OCH_3$

Essential Oil

The EO, obtained in 0.52 per cent on a dry-weight basis, was similarly analysed by gas chromatography–mass spectrophotomety (GC–MS), leading to the unambiguous identification of 102 components (Kayser *et al.*, 1998). Again, sesquiterpenes proved to be the dominating constituents (61 per cent, inclusive of the unknowns), with caryophyllene (2.3 per cent) and caryophyllene epoxide (13 per cent) as the most abundant compounds. In the monoterpene fraction, oxygenated members constitute the most abundant group. As can be seen in Table 24.3, the composition of the EO of *P. sidoides* differed significantly from that of *P. reniforme* by the presence of considerable amounts of phenylpropanoids (8.5 per cent), which were apparently absent or only present as trace components in the latter. Methyleugenol (4.3 per cent) and elemicin (3.6 per cent) were the most abundant phenylpropanoids, accounting for almost 94 per cent of this fraction. This divergence in the EO profiles should be of relevance for a clear differentiation of the morphologically closely related *Pelargonium* species. This finding could also be of significance for other *Pelargonium* species, but requires further studies.

It is also worth noting here that the likewise detection of, e.g. anacardic acids, furfural, geranyl acetate, linalyl acetate and phenylpropanoids, reputed for their insect deterrent potential, in the EO of *P. sidoides* provides additional evidence that *Pelargonium* species are effectively repellent to insects. For this purpose, *Pelargonium* oils are widely employed in southern Africa by the native population.

TRADITIONAL USE

The traditional drug umckaloabo originates from the root material of *P. sidoides* and *P. reniforme*. Etymologically, the name 'umckaloabo' comes from the Zulu words 'umKhulkane' for complaints associated with lung disorders and 'uHlabo' which means chest pain (Bladt, 1974), reflecting a major traditional indication. Initially, only the latter species has been suggested to form the plant source of this traditional herbal medicine (Bladt, 1974), but, according to present evidence, the origin of umckaloabo is strongly associated with the morphologically closely related species *P. sidoides* (*vide supra*) (Kolodziej and Kayser, 1998). Therefore, it is most likely that some medicinal records apply to mixtures prepared from both species. In our hands, the root material of both species comprised of pieces differing in thickness but apparently devoid of conspicuously morphological features. Noteworthy, however, are pieces with tuberous segments that are apparently characteristic of roots of *P. sidoides*, but this morphological feature has only been found for older materials (Kolodziej *et al.*, 1995b).

Umckaloabo enjoys a wide reputation by traditional healers and is highly valued by the southern African native population for its curative and palliative effects in the treatment of gastrointestinal disorders, hepatic disorders, wounds and respiratory tract disorders (Watt and Breyer-Brandwyk, 1962; Hutchings, 1996). For example, decoctions are used for the treatment of cough and chest pain including tuberculosis. Infusions of the roots of *P. sidoides* and *P. reniforme* are applied to treat gastrointestinal disorders such as diarrhoea. In addition, umckaloabo is claimed to cure hepatic disorders and menstrual complaints such as dysmenorrhea. The aerial parts of these *Pelargonium* species are employed in wound healing.

The therapeutic use of umckaloabo in traditional medicine has been noticed by European settlers in the eighteenth century. At the turn of the century Major Stevens introduced a secret medicine, Steven's Consumption Cure, in England for the treatment of tuberculosis (Helmstädter, 1996). Stevens believed that he recovered from tuberculosis by the administration of a decoction of umckaloabo prepared by a traditional healer. In 1920, the Swiss physician Sechehaye heard of Steven's cure, treated 800 patients in the following 9 years and reported successful cases to the Medical Society of Geneva, though in most cases tuberculosis had not been definitely proven. Following the well documented therapeutic use since then in the treatment of lung disorders, umckaloabo has survived and found its place in modern phytomedicine. Nowadays, ethanolic preparations using the *Pelargonium* species medicinally (*Umckaloabo*®, ISO-Arzneimittel, Ettlingen, Germany) are successfully employed to treat ENT and respiratory tract infections.

Regarding the treatment of gastrointestinal disorders such as diarrhoea in folk medicine, the fairly high concentrations of proanthocyanidins (ca. 9 per cent) occurring in this indigenous medicine may explain the traditional use. Owing to the high degree of polymerisation, absorption of these tannins from the digestive tract with an intact mucosa is highly unlikely, thereby acting as an effective astringent useful in these conditions. For example, the hitherto assumed precipitation of proteins in the epithelial surface of the gut should form a protective layer along the intestinal lumen. Also, some of the beneficial antisecretory effects which tannins exert in these conditions may well follow from their interaction with toxins produced by pathogenic bacteria in the intestine (Hänsel, 1991; Hör *et al.*, 1995).

A similar rationale explanation based on the presence of tannins may be provided for the application of the aerial parts as wound healing agent, while curative effects, if any, related to hepatic disorders may tentatively be explicable on the basis of the radical scavenging activities of the broad range of phenolic compounds. In the following sections, the therapeutic claims of umckaloabo (*P. reniforme*/*P. sidoides*) in traditional medicine and in modern phytomedicine in Europe are evaluated and discussed in greater detail.

PHARMACOLOGY

Following the therapeutic use of umckaloabo (*P. reniforme*/*P. sidoides*) in traditional medicine for the treatment of respiratory tract infections and its present utilisation in modern phytomedicine in Europe, this herbal medicine is still the subject of considerable research aimed at establishing the chemical and pharmacological basis of the activity which has been clearly shown in a number of clinical studies (*vide infra*). In principle, the claimed clinical efficacy in ENT and respiratory tract infections could be explained by antibacterial activities and/or stimulation of the non-specific immune system.

To obtain clues as to whether these botanically defined *Pelargonium* species have antimycobacterial potencies as claimed in earlier records, extracts and some typical phenol and coumarin isolates were tested for this particular biological activity in collaboration with the Tuberculosis Antimicrobial Acquisition and Coordinating Facility, Alabama, USA. At 12.5 μg/ml, the crude extract of *P. reniforme* showed some interesting inhibitory activity (96 per cent) against *Mycobacterium tuberculosis* in a primary screen, while that of *P. sidoides* was inactive. This finding does not necessarily imply lack of antimycobacterial properties of the latter species, possibly detectable at higher sample concentrations, but clearly indicates the need for establishing with certainty the precise botanical species traditionally used. Such a criterion is, among others, liable to lead to inconsistencies in reported bioactivity and clinical efficacy. For example, confusions regarding the origin of umckaloabo have led to lack of credibility in therapy (Helmstädter, 1996). With the minimum inhibitory concentration (MIC) of 100 μg/mL, determined in a broth microdilution Alamar Blue assay, the extract of *P. reniforme* was only moderately active against *M. tuberculosis*, when compared with the MIC of 0.06 μg/mL of the clinically used drug, rifampicin, as a reference compound. Also, none of the isolated constituents exhibited antimycobacterial activities under the experimental conditions.

Antibacterial activity

Due to ambiguities regarding the botanical identity of the previously studied species (*vide supra*), the reported biological activities are limited to our own studies. The qualitative evaluation of the antibacterial activity of extracts and isolated constituents was accomplished using the agar dilution method. Samples were tested for *in vitro* activity against a panel of microorganisms (Kayser and Kolodziej, 1997) including three Gram-positive (*Staphylococcus aureus*, *Steptococcus pneumoniae*, and beta-hemolytic *Streptococcus* 1451) and five Gram-negative bacteria (*Escherichia coli*, *Klebsiella pneumoniae*, *Proteus mirabilis*, *Pseudomona aeruginosa* and *Haemophilus influenzae*), pathogens that are primarily responsible for numerous respiratory tract infections. Whereas the crude *Pelargonium* root extracts were found to be moderately active against the tested bacteria with MICs of 5–7.5 mg/ml, the ethyl acetate, *n*-butanol and water phases exhibited fairly high antibacterial effects as evident from MICs of 0.6–1.2 μg/ml (Table 24.5). No significant differences were found between *P. sidoides* and *P. reniforme* extracts regarding this particular biological activity.

Regarding the antibacterial potential of isolates, the highly oxygenated coumarins, 7-hydroxy-5,6-di-methoxycoumarin (71) and 6,8-dihydroxy-5,7-di-methoxycoumarin (74) represented the most potent candidates with MICs of 200–500 μg/ml (Table 24.5). Another major contributing factor towards antibacterial activity of umckaloabo extracts proved to be gallic acid methyl ester (2) with MICs of 250–500 μg/ml against

Table 24.5 Antibacterial activity of *P. sidoides* and *P. reniforme* (MIC values in μg/ml)

Extract	*E. coli*	*Klebs. pneum.*	*Staph. aureus*	*Pseud. aerug.*	*Prot. mirab.*	*β-hem. Strept.*	*Strep. pneum.*	*Haem. influen.*
Crude extracts								
P. sidoides	5000	7500	7500	5000	5000	7500	7500	5000
P. reniforme	7500	5000	2500	5000	5000	7500	7500	5000
Fractions								
P. sidoides								
H_2O	600	600	600	600	1200	600	600	1200
EtOAc	1200	1200	2500	2500	2500	1200	2500	2500
n-BuOH	1200	1200	1200	1200	1200	1200	2500	2500
P. reniforme								
H_2O	600	600	1200	1200	1200	600	1200	1200
EtOAc	1200	2500	2500	2500	2500	2500	1200	2500
n-BuOH	1200	1200	1200	2500	2500	2500	2500	2500
Compounds								
Simple Phenols								
(1)	2000	1000	500	500	500	2000	2000	2000
(2)	500	250	250	250	250	500	1000	500
(67)	>8000	>8000	>8000	>8000	>8000	>8000	>8000	>8000
Coumarins								
(22)	500	500	500	400	400	1000	1000	1000
(71)	200	200	200	200	200	500	500	500
(73)	1000	1000	1000	1000	1000	1000	500	500
(74)	220	250	220	220	220	500	500	500
Penicillin G	25	5	5	5	6	166	16	16

all of the tested bacteria but *S. pneumoniae* (1000 μg/ml). The demonstrated antibacterial activity of umckaloabo may explain, at least in part, the documented clinical efficacy of *Pelargonium*-containing herbal medicines.

Immunmodulatory properties

Clearly, the antibacterial activity of umckaloabo extracts and their constituents is considerably inferior to commercial antibiotics like penicillin G (MICs of 5–25 μg/ml), and, hence, these results cannot satisfactorily explain the documented clinical efficacy in the claimed conditions. Respiratory tract infections are frequently primarily due to viruses including coxsackie-, parainfluenza-, influenza-, echo-, adeno- and rhino-viruses, thus leading to the assumption of an immunomodulatory activity associated with *Pelargonium*-containing herbal medicines.

To assess the immunostimulating potential of this traditional medicine, functional bioassays including an *in vitro* model for intracellular infection with *Leishmania* parasites (Kiderlen and Kaye, 1990), an extracellular *Leishmania* growth assay (Kayser *et al.*, 1997), a fibroblast-antivirus protection assay (IFN activity) (Marcucci *et al.*, 1992), and a fibroblast-lysis assay (TNF) (Wagner and Jurcic, 1991), as well as a biochemical assay for nitric oxides (iNO) were employed (Ding *et al.*, 1988). These test models serve as a useful tool for the evaluation of activation of host defense mechanisms against intracellular microorganisms and viral infections associated with the traditional use and present therapeutic use of umckaloabo.

Leishmanicidal activity

In view of the use and known therapeutic effects of *Pelargonium*-containing herbal medicines, an *in vitro* model for visceral leishmaniasis was selected in which murine macrophages are infected with the obligate intracellular protozon *Leishmania donovani*. Pronounced leishmanicidal effects against intracellular amastigote forms of *L. donovani* were observed for the parent methanol extracts, but also for the petrol ether, ethyl acetate and *n*-butanol phases obtained thereof, as reflected by EC_{50} values ranging from <0.1 to 3.3 μg/ml. The EC_{50} value of Pentostam® as a reference agent was 7.9 μg/ml under the same conditions. Subsequent bioassay-guided fractionation of the active fractions led to the characterisation of gallic acid (EC_{50} 4.4 μg/ml) and its methyl ester (EC_{50} 12.5 μg/ml) as potentially leishmanicidal active constituents, both being present in significant amounts in umckaloabo. In contrast, coumarins isolated from the same plant source proved to be inactive in concentrations up to 25 μg/ml in this *in vitro* infection model. On the other hand, neither extracts nor characteristic compounds of umckaloabo proved to be directly toxic to extracellular, promastigote forms of *L. donovani* (EC_{50} > 25 μg/ml). Similarly, sample toxicity for non-parasitised macrophages were significantly less prominent in this concentration range.

These findings suggest that the observed activity against intracellular *L. donovani* might well be due to activation of sample-treated macrophages for enhanced *Leishmania* kill, though stage-specific sensitivity of *Leishmania* amastigotes for these products can not completely be excluded.

Release of nitric oxide

Reactive nitrogen intermediates such as nitric oxide (NO) play a firmly established role as antimicrobial effector molecules produced by activated macrophages (Nathan and Hibbs, 1991; Nussler and Biliar, 1993). Using the Griess assay, all samples were able to induce NO production, but in a different amount (Figure 24.4). Compared to the

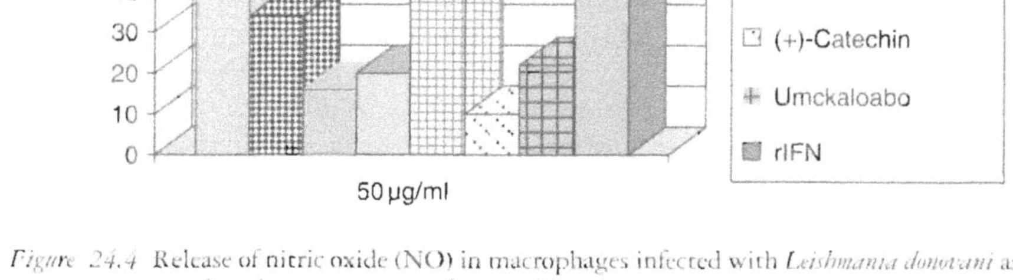

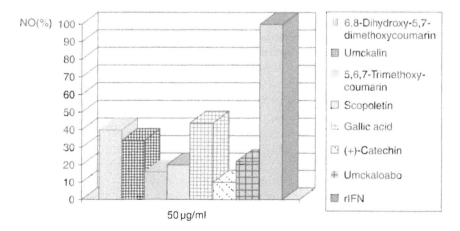

Figure 24.4 Release of nitric oxide (NO) in macrophages infected with *Leishmania donovani* as compared to the stimulus interferon (rIFN).

stimulus LPS, the NO-inducing effect of the samples accounted only for 10–45 per cent. The most potent NO-inducers of the series of samples tested were gallic acid (1) and the highly oxygenated coumarins 7-hydroxy-5,6-di-methoxycoumarin (71) and 6,8-dihydroxy-5,7-dimethoxy-coumarin (74). NO production increased in a dose-dependent manner and reached a plateau in the range of 12.5 up to 50 μg/ml. Higher concentrations had an opposite effect by slightly decreasing the intracellular NO release.

However, intracellular *Leishmania* toxicity did not correlate with NO-release. In order to further assess the role of induced NO as a cytotoxic mechanism involved in sample-induced intracellular destruction of *Leishmania*, the activity of the NO-synthase was blocked by addition of the well-known inhibitor L-NMMA to parallel incubations. Interestingly, the leishmanicidal effects of samples in experiments with and without inhibitor were found to be very similar, possibly indicating the involvement of additional leishmanicidal mechanisms in macrophages activated by umckaloabo constituents.

Induction of the release of tumour necrosis factor

Besides oxygen-dependent cytotoxic defence mechanisms (Nathan and Hibbs, 1991; Nussler and Biliar, 1993), induction of cytokines such as tumour necrosis factor (TNF) represents a further parameter of macrophage activation. Beyond its role as proinflammatory cytokine, TNF plays an important regulatory role in the cytokine network including the activation of immune cells and is required for host defence against certain pathogens.

This factor was assayed by a protocol of determining spectrophotometrically the lysis of actinomycin D pretreated TNF-sensitive L 929 fibroblasts in a classic cytotoxic TNF assay (Wagner and Jurcic, 1991). At the subtoxic concentration of 25 μg/ml, the ethyl acetate (20.2 U/ml) and *n*-butanol phase (18.9 U/ml) were found to possess a moderate TNF-inducing potential. Both phases contained significant amounts of gallic acid (1) and its methyl ester (2), which proved to exhibit the strongest TNF-inducing potential among the isolated *Pelargonium* constituents tested. Compared to the LPS stimulus, the inducing potential of gallic acid and its methyl ester accounted for 24 and 19 per cent, respectively. All coumarins exhibited only negligible effects in the concentration range up to 50 μg/ml. The considerable amounts of the simple phenols (1) and (2) in both *Pelargonium* species, but also the occurrence of coumarins, albeit very weak TNF-inducers, thus play a beneficial immune modulatory role of the underlying active principle of umckaloabo.

Interferon-like activities

In the light of the proven clinical efficacy of *Pelargonium*-containing herbal medicines in respiratory tract infections, it is noteworthy that umckaloabo extracts revealed significant interferon (IFN)-like activities in a virus protection assay (Figure 24.5).

For this, sample-treated murine encephalo-myocarditis virus (EMCV)-sensitive L929 fibroblasts were incubated with EMCV suspensions. Inhibition of the cytopathic effect (CPE) was determined spectrophotometrically using crystal violet as staining reagent for protected cells and an IFN standard (100 U/ml) as positive control. Prominent

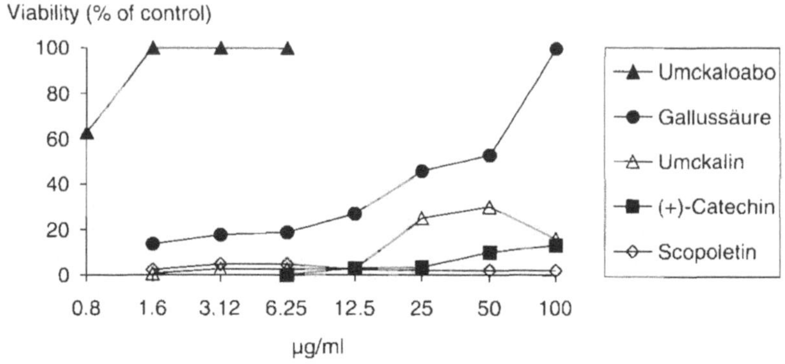

Figure 24.5 Cytoprotective effect (% of control) by Umckaloabo and its representative constituents as assessed in the fibroblast/EMCV protection assay.

cytoprotective effects were observed for umckaloabo, as evidenced not only by complete inhibition of CPE, but also significant proliferation of fibroblasts. In a modified experimental design which permitted differentiation between direct and indirect cytoprotective effects, the strong CPE-inhibitory potential of sample-treated cells was clearly located in supernatants, giving credence to IFN-like activities. Subsequent bioassay-guided fractionation led to the detection of gallic acid (1) as potentially active constituent that significantly reduced indirectly the cytopathic effect of EMCV on L929 cells in a dose-dependent manner. At 100 µg/ml, not only complete inhibition of cytopathic effects, but also proliferation of fibroblasts was noticed. In contrast, (+)-catechin (67) did not show any significant IFN-like activities. Of the coumarins, umckalin (71) and 5,6,7-trimethoxycoumarin (73) showed the relatively strongest inhibitory effects in the range of 25 up to 100 µg/ml with 12 and 30 per cent respectively, while the remaining compounds of this group of metabolites exhibited only negligible effects at all concentrations. From these findings, it can be concluded that the phenolic constituents, gallic acid (1) and its methyl ester (2), enhanced non-specific immune response in terms of macrophage activation for intracellular parasite kill and release of NO and TNF. The results also provide strong evidence for IFN-like activities of umckaloabo, but here it is reasoned that co-substances synergistically contribute to its powerful overall action. Clearly, further relevant experiments are needed to elucidate the mode of action of these phenols.

CLINICAL STUDIES

To obtain information about the efficacy and safety of the phytomedicine Umckaloabo®, elaborated from the traditional herbal medicine, a series of clinical studies were performed (Heil and Reitermann, 1994; Dome and Schuster, 1996; Haidvogl *et al.*, 1996). First, a multi-centre postmarketing surveillance study was carried out in 1991/92 in 641 patients with ENT and respiratory tract infections, e.g. tonsillitis, rhinopharyngitis, sinusitis, and bronchitis. Outcome criteria were the change in the subjective and objective symptoms during treatment and an assessment of treatment outcome by both physicians and patients

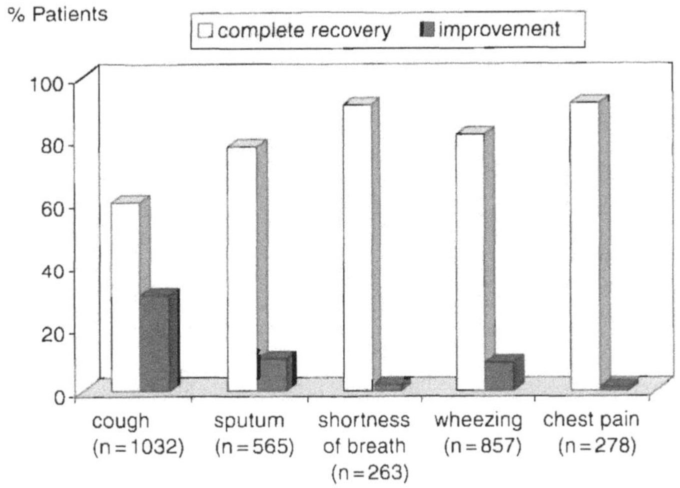

Figure 24.6 Umckaloabo multi-centre study in bronchitis: Bronchitis-specific symptoms.

on a 4-point rating scale. After 14 days of treatment, a total of 85 per cent of the patients were back to normal or showed a major improvement. No adverse drug reactions were observed.

Umckaloabo also represents a valuable therapeutic alternative in childhood infections, particularly respiratory tract infections. Its efficacy and safety were investigated in a prospective, open multi-centre study in 1042 children up to 12 years old suffering from acute bronchitis. For all individual symptoms such as cough, expectoration, difficulty in breathing, wheezing, and chest pain, the response rate (remission/improvement) was over 80 per cent (Figure 24.6). In a prospective, randomised controlled trial, the efficacy and safety of Umckaloabo was compared with Acetylcysteine in 60 children with acute bronchitis. The primary outcome criterion was the change in the total score of bronchitis-specific symptoms at day 7. After 7 days of treatment, the total score decreased similarly in both groups. The number of patients with complete recovery was 76.7 per cent in the group treated with Umckaloabo compared to 56.8 per cent in the group treated with Acetylcysteine ($p = 0.170$; Fisher's exact test, two-sided). Patients treated with Umckaloabo reported to improve faster and be more satisfied with their treatment ($p = 0.006$; Fisher's exact test, two-sided). The results suggest that Umckaloabo is at least as effective as Acetylcystein in the treatment of acute bronchitis.

This modern herbal medicine has also been found to be extremely effective for the treatment of acute tonsillitis, one of the most common conditions of children and young adults. In a prospective, randomised controlled trial involving 60 children aged between 6 and 10 years with acute tonsillitis (angina catarrhalis), the response rate after 4 days of treatment with Umckaloabo was 76 per cent compared to that of 30 per cent with symptomatic treatment ($p = 0.001$, Fisher's exact test, two-sided) (Figure 24.7). In another randomised, double-bind, placebo-controlled trial, 78 children with acute tonsillitis were treated with Umckaloabo or placebo for 6 days. The primary outcome criterion was the response rate defined as total score of the angina-specific symptoms ≤ 4 points at

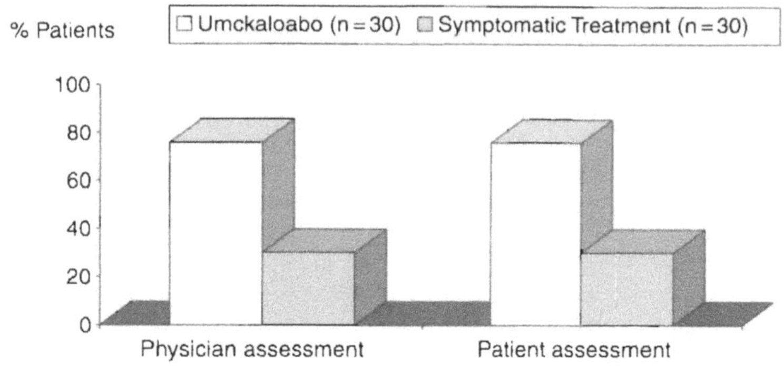

Figure 24.7 Assessment of efficacy in Angina catarrhalis (day 4; intention to treat-analysis; response rate).

day 4. After four days of treatment, the response rates were 90.0 per cent in the Umckaloabo group and 44.7 per cent in the placebo group ($p < 0.0001$; Fisher's exact test, two-sided). The higher response rates in the Umckaloabo group corresponded to a more pronounced decrease of the total score of angina-specific symptoms from 8.7 ± 1.5 points at admission to 2.0 ± 2.3 points after 4 days (last observation carried forward) and to 1.4 ± 2.6 points at the final assessment as compared to a decrease from 8.8 ± 1.6 points to 5.2 ± 3.1 points and to 4.9 ± 3.3 points in the placebo group. Based on the comparison of the individual last observation under treatment within the first 4 days with the baseline value, the mean decrease of the total score was 6.8 ± 2.8 points in the Umckaloabo group and 3.7 ± 3.3 points in the placebo group ($p < 0.0001$; Wilcoxon rank-sum test, two-sided). No assessment was available for one patient in the Umckaloabo group who terminated the study prior to the first control visit. The mean total scores during the course of treatment are depicted in Figure 24.8. Tolerability was rated as good or very good by patients and investigators in 97.5 per cent of patients treated with Umckaloabo. Treatment of acute tonsillitis with Umckaloabo is well tolerated and demonstrated to be significantly superior compared with placebo.

A viral infection spread by droplets is often present initially. The complication most feared is a bacterial superinfection by β-hemolytic streptococci. Although viral pathogens are not susceptible to antibiotics and are best treated symptomatically, in practice antibiotics are often employed as prophylactic treatment. The benefits of prophylactic antibiotic treatment is therefore questionable, but the risk of such treatment lies in the possibility of allergies to penicillin. The incidence of severe events (anaphylactic shock, localised allergic symptoms) is estimated as 25 per 100,000 administrations. Intolerance reactions caused by penicillin, e.g. disturbances of the physiological intestinal flora and similar less serious side effects, are much more common. Repeated administration of antibiotics also changes the flora in the nasopharyngeal cavity in favour of pathogenic organisms. It must, therefore, be assumed that the incidence with which tonsillitis recurs would be increased by antibiotic treatment. In view of the incidence of intolerance reactions, and of the danger of resistance development, uncritical use of antibiotics, e.g. prophylactic treatment against bacterial superinfection, is certainly questionable, particularly if one remembers

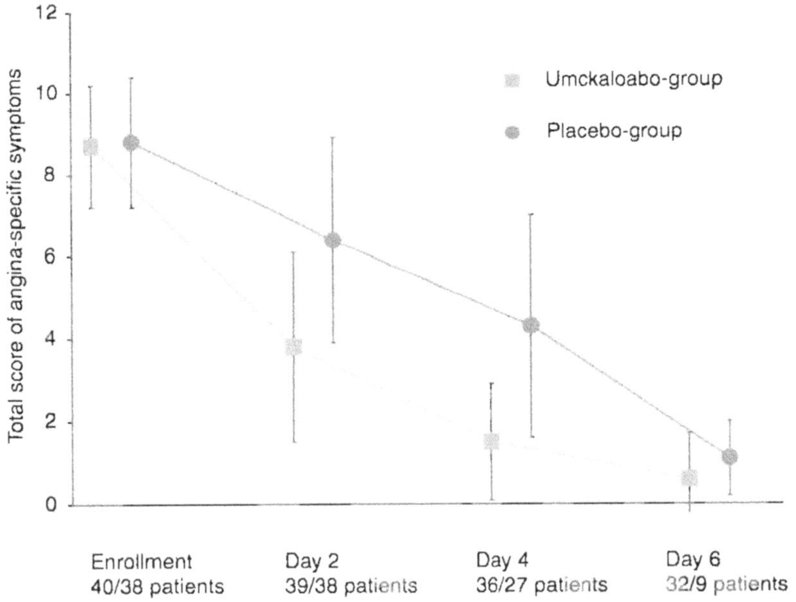

Figure 24.8 Total score of angina-specific symptoms during treatment with Umckaloabo compared to Placebo (arithmetic mean ± standard deviation).

that viral infections are not susceptible to causal treatment. Prescription of antibiotics should be weighed up very carefully, particularly in children, in whom immunological confrontation with the microbial environment is still going at full pace. It is particularly at this stage that Umckaloabo offers a valuable alternative: it strengthens the non-specific immunological response and, ideally, has bacteriostatic properties. The antibacterial effect of *Pelargonium* extracts is distinctly weaker than that of antibiotics, which is not itself surprising, but which does emphasise the therapeutic value of antibiotics in proven bacterial infections. Further positive factors in favour of Umckaloabo are the lack of toxicity and the low incidence of side effects indicating a positive risk-benefit-ratio, particularly in the treatment of children's disorders.

TOXICITY

From the long therapeutic use of umckaloabo in folk medicine and the lack of respective reports, absence of any fatal toxic effects may be anticipated. Taken into account clinical studies demonstrating just incidences of mild to moderate side effects in patients treated with Umckaloabo, an extract of the titled *Pelargonium* species may well represent a safe herbal medicinal product.

In a more direct approach to place non-toxicity beyond doubt, first evidences followed from studies on the toxicity of umckaloabo constituents in some preliminary tests. For example, this type of simple coumarins showed some antimutagenic activity against IQ-induced mutagenesis in *Salmonella typhimurium* (Edenharder *et al.*, 1995), while distinct coumarins were found to be potent inhibitors of tumorigenesis (Nair *et al.*, 1991; Noguchi *et al.*, 1993). Also, catechin and proanthocyanidins are reported to have some antimutagenic potentials (Dauer *et al.*, 1998). Using a human small cell lung carcinoma (GLC_4) and a colorectal cancer cell line (COLO 320) for evaluating the cytotoxicity, most coumarins exhibited only low cytotoxicity with IC_{50} values >100 μM (Kolodziej *et al.*, 1997). The most potent cytotoxic coumarin, 6,8-dihydroxy-5,7-dimethoxycoumarin (74) with an IC_{50} of 22 μM (GLC_4) and 9.5 μM (COLO 320) respectively, was only moderately cytotoxic, when compared with the respective IC_{50} values (1 and 2.7 μM, respectively) of the reference compound cisplatin. In a similar study, proanthocyanidins were also found to possess only low cytotoxicity. (Kolodziej *et al.*, 1995a). Finally, the brine shrimp lethality bioassay (McLaughlin, 1991) was used to get an idea of the toxicity of *Pelargonium* extracts and their phenolic constituents including benzoic and cinnamic acid derivatives, hydrolysable tannins and *C*-glycosylflavones. With LC_{50} values >1000 and >200 μg/ml for extracts and test compounds respectively (Latté, 1999), the cytotoxic potential of the samples may be negligible (Anderson *et al.*, 1992).

ACKNOWLEDGEMENTS

Sincere appreciation is expressed to Dr O. Kayser, Dr K.P. Latté (FU Berlin), Dr A.F. Kiderlen and U. Folkens (Robert Koch-Institut, Berlin) for the enthusiastic support, valuable discussions and skilled technical assistance. Special thanks are extended to Dr M. Heger, Research Center HomInt, Karlsruhe, Germany, for her great interest to this work, invaluable advice on the clinical studies section, and the assistance with graphics. Thanks are also due to TAACF for antimycobacterial tests. The plant material was kindly provided by the pharmaceutical company Dr Willmar Schwabe, Karlsruhe, Germany.

REFERENCES

Achenbach, H., Löwel, M., Waibel, R., Gupta, M. and Solis, P. (1992) New lignan glucosides from *Stemmadenia minima. Planta Med.*, 58, 270–272.

Amakura, Y., Miyake, M., Ito, H., Murakaku, S., Araki, S., Itoh, Y., Lu, C.-F., Yang, L., Yen, K., Okuda, T. and Yoshida, T. (1999) Acalphidins M_1, M_2 and D_1, ellagitannins from *Acalypha hispida. Phytochemistry*, 50, 667–675.

Anderson, J.E., Goetz, C.M., McLaughlin, J.L. and Suffness, M. (1992) A blind comparison of simple bench-top bioassays and human tumour cell cytotoxicites as antitumour prescreens. *Phytochem. Anal.*, 2, 107–111.

Bladt, S. (1974) Zur Chemie der Inhaltsstoffe der Pelargonium reniforme CURT. – Wurzel (Umckaloabo), Dissertation, Ludwig-Maximilians-Universität, München.

Burmann, J. (1738) *Catalogi duo plantarum africanum quorum prior complectitur plantas ab (Paulo Hermanno observatas, posterior vero quas (Henricus Bernadus) Olderlandus et (Johannes) Hastogius indaragunt*, Amsterdam.

Curtius, W. (1800) *The botanical magazine*, vol. 13, Couchman, London.

Dauer, A., Metzner, P. and Schimmer, O. (1998) Proanthocyanidins from the bark of *Hamamelis virginiana* exhibit antimutagenic properties against nitroaromatic compounds. *Planta Med.*, 64, 324–327.

De Candolle, A.P. (1824) *Prodromus systematis naturalis regni vegetabilis, sive enumeratio contracta ordium generum specierumque plantarum*, Paris.

Demarne, F.-E. and van der Walt, J.J.A. (1990) *Pelargonium tomentosum*: a potential source of peppermint-scented essential oil. *S. Afr. J. Plant Soil*, 7, 36–39.

Demarne, F.-E. and van der Walt, J.J.A. (1992) Composition of the essential oil of *Pelargonium vitifolium (L.) L'Hérit.* (Geraniaceae). *J. Essent. Oil Res.*, 4, 345–348.

Demarne, F.-E. and van der Walt, J.J.A. (1993a) Composition of the essential oil of *Pelargonium citronellum* (Geraniaceae) *J. Essent. Oil Res.*, 5, 233–238.

Demarne, F.-E., Viljoen, A.M. and van der Walt, J.J.A. (1993b) A study of the variation in the essential oil and morphology of *Pelargonium capitatum* (L.) L'Hérit. (Geraniaceae). Part I. The composition of the essential oil. *J. Essent. Oil Res.*, 5, 493–499.

Ding, A.H., Nathan, C.F. and Stuehr, D.J. (1988) Release of reactive nitrogen intermediates and reactive oxygen intermediates from mouse peritoneal macrophages. Comparison of activating cytokines and evidence for independent production. *J. Immunol.*, 141, 2407–2412.

Dome, L. and Schuster, R. (1996) Umckaloabo – eine phytotherapeutische Alternative bei akuter Bronchitis im Kindesalter. *Ärztezeitschr. Naturheilv.*, 37, 216–222.

Dreyer, L.L., Albers, F., van der Walt, J.J.A. and Marschewski, D.E. (1992) Subdivision of *Pelargonium* sect. *Cortusina* (Geraniaceae). *Plant Syst. Evol.*, 183, 83–97.

Dreyer, L.L., Marais, E.M. and van der Walt, J.J.A. (1995) A subspecific division of *Pelargonium reniforme* CURT. (Geraniaceae). *S. Afr. J. Bot.*, 61, 325–330.

Edenharder, R., Speth, C., Decker, M., Kolodziej, H., Kayser, O. and Platt, K.L. (1995) Inhibition of mutagenesis of 2-amino-3-methylimidazol[4,5-*f*]quinoline (IQ) by coumarins and furanocoumarins, chromanones and furanochromanones. *Mutation Res.*, 345, 57–71.

Gildemeister, E. and Hoffmann, F. (1959) *Die ätherischen Öle*, Akademie-Verlag, Berlin.

Haddock, E.A., Gupta, R.K., Al-Shafi, S.M.K., Layden, K., Haslam, E. and Magnolato, D. (1982) The metabolism of gallic acid and hexahydroxydiphenic acid in plants: biogenetic and molecular taxonomic considerations. *Phytochemistry*, 21, 1049–1062.

Haidvogl, M., Schuster, R. and Heger, M. (1996) Akute Bronchitis im Kindesalter – Multizenter-Studie zur Wirksamkeit und Verträglichkeit des Phytotherapeutikums Umckaloabo. *Z. Phytother.*, 17, 300–313.

Hänsel, R. (1991) *Phytopharmaka*, Springer, Berlin.

Harvey, W.H. and Sonder, O.W. (1859) *Flora Capensis: Systematic Description of the Plants of Cape Colony, Caffraria and Port Natal*, vol. 1, Hodges, Smith & Co, Dublin.

Haslam, E. (1988), *Plant Polyphenols – Vegetable tannins revisited*, Cambridge University Press, Cambridge.

Haslam, E. (1998) *Practical Polyphenolics*, Cambridge University Press, Cambridge.

Hegnauer, R. (1966) *Chemotaxonomie der Pflanzen*, vol. 4, Birkhäuser, Basel.

Hegnauer, R. (1989) *Chemotaxonomie der Pflanzen*, vol. 8, Birkhäuser, Basel.

Heil, Ch. and Reitermann, U. (1994) Atemwegs- und HNO-Infektionen: Therapeutische Erfahrungen mit dem Phytotherapeutikum Umckaloabo®. *Therapiew. Pädiatrie*, 7, 523–525.

Helmstädter, A. (1996) Umckaloabo – Late vindication of a secret remedy. *Pharm. Historian*, 26, 2–4.

Hör, M., Rimpler, H. and Heinrich, M. (1995) Inhibition of intestinal chloride secretion by proanthocyanidins from *Guazuma ulmifolia*. *Planta Med.*, 61, 208–212.

Hutchings, A. (1996) *Zulu Medicinal Plants*, Natal University Press, Pietermaritzburg.

Jay, M. (1994) *C*-Glycosylflavonoids. In J.B. Harborne (ed.), *The Flavonoids – Advances in Research since 1986*, Chapman Hall, London, pp. 57–93.

Kaiser, R. (1984) (5R*,9S*)- and (5R*,9R*)-2,2,9-trimethyl-1,6-dioxaspiro[4.4]non-3-ene and their dihydro derivatives as new constituents of geranium oils. *Helv. Chim. Acta*, 67, 1198–1203.

Kayser, O., Kiderlen, A.F. and Kolodziej, H. (1997) Inhibition of luminol-dependent chemiluminescence and NO release by a series of oxygenated coumarins in murine macrophages infected with *Leishmania donovani*. *Pharm. Pharmacol. Lett.*, 7, 71–74.

Kayser, O. and Kolodziej, H. (1997) Antibacterial activity of extracts and constituents of *Pelargonium sidoides* and *Pelargonium reniforme*. *Planta Med.*, 63, 508–510.

Kayser, O., Latté, K.P., Kolodziej, H. and Hammerschmidt, F.-J. (1998) Composition of the essential oils of *Pelargonium sidoides* DC. and *Pelargonium reniforme* CURT. *Flav. Fragr. J.*, 13, 209–212.

Kiderlen, A.F. and Kaye, P.M. (1990) A modified colorimetric assay of macrophage activation for intracellular cytotoxicity against *Leishmania parasites*. *J. Immunol. Methods*, 127, 11–17.

Knuth, R. (1912) *Geraniaceae*. In A. Engler (ed.), Das Pflanzenreich, vol. IV, Engelmann, Leipzig.

Kolodziej, H., Haberland, C., Woerdenbag, H.J. and Konings, A.W.T. (1995a) Moderate cytotoxicity of proanthocyanidins to human tumour cell line. *Phytotherapy Res.*, 9, 410–415.

Kolodziej, H. and Kayser, O. (1998) *Pelargonium sidoides* DC. – Neueste Erkenntnisse zum Verständnis des Phytotherapeutikums Umckaloabo. *Z. Phytother.*, 19, 141–151.

Kolodziej, H., Kayser, O., Woerdenbag, H.J., van Uden, W. and Pras, N. (1997) Structure – cytotoxicity relationships of a series of natural and semi-synthetic simple coumarins as assessed in two human tumour cell lines. *Z. Naturforsch.*, 52c, 240–244.

Kolodziej, H., Kayser, O. and Gutmann, M. (1995b) Arzneilich verwendete Pelargonien aus Südafrika. *Dtsch. Apotheker Ztg.*, 135, 853–864.

Latté, K.P. (1999) Phytochemische und pharmakologische Untersuchungen an *Pelargonium reniforme* CURT. Dissertation, Freie Universität Berlin.

Latté, K.P. and Kolodziej, H. (2000) Pelargoniins, new ellagitannins from *Pelargonium reniforme*. *Phytochemistry*, 54, 701–708.

Latté, K.P., Ferreira, D., Ventatraman, M.S. and Kolodziej, H. (2002) *O*-galloyl-*C*-glycosylflavones from *Pelargonium reniforme*. *Phytochemistry*, 59, 419–424.

Latté, K.P., Kaloga, M. and Kolodziej, H. (1998) First cyclolignan derivatives from *Pelargonium reniforme*. *Proc. 46th Ann. Congr. Med. Plant Res.*, Wien, September, 1998.

Latté, K.P. and Kolodziej, H. (1999) Diversity of structure in *O*-galloylated metabolites from *Pelargonium reniforme*. *Proc. 47th Ann. Congr. Med. Plant Res.*, Amsterdam, July, 1999, Abstract p. 292.

L'Heritier de Brutelle, Ch-L. (1788) *Geraniologie, seu Erodii, Pelargonii, Geranii, Monsoniae et Grieli. Historia iconibus illustrata*, Paris.

Lin, J.-H., Nonaka, G. and Nishioka, I. (1990) Isolation and characterization of seven new hydrolyzable tannins from the leaves of *Macaranga tanarius* (L.) MUELL. Et ARG. *Chem. Pharm. Bull.*, 38, 1218–1223.

Lis-Balchin, M., Houghton, P.J. and Woldemariam, T.Z. (1996) Elaeocarpidine alkaloids from *Pelargonium* species (Geraniaceae). *J. Essent. Oil. Res.*, 3, 99–105.

Liu, K.C.S., Lin, M., Lee, S., Chiou, J.-F., Ren, S. and Lien, E. (1999) Antiviral tannins from two *Phyllanthus* species. *Planta Med.*, 65, 43–46.

Marcucci, F., Klein, B., Kirchner, M. and Zawatzky, R. (1992) Production of high titers of interferon gamma by prestimulated spleen cells. *Eur. J. Immunol.*, 12, 787–790.

McLaughlin, J.L. (1991) Crown gall tumours on potatoe discs and brine shrimp lethality: two simple bioassays for higher plant screening and fractionation. In P.M. Dey and J.B. Harborne (eds), *Methods in Plant Biochemistry*, vol. 6, Academic Press, London, pp. 1–32.

Nair, R.V., Fisher, E.P., Safe, S.H., Cortez, C., Harvey, R.G. and DiGiovanni, J. (1991) Novel coumarins as potential anticarcinogenic agents. *Carcinogenesis*, 12, 65–69.

Nathan, C.F. and Hibbs, J.B. (1991) Role of nitric oxide synthesis in macrophage antimicrobial activity. *Curr. Opin. Immunol.*, 3, 65–70.

Noguchi, M., Kitagawa, H., Miyazaki, I. and Mizukami, Y. (1993) Influence of esculin on incidence, proliferation, and cell kinetics of mammary carcinoms induced by 7,12-dimethylbenz[a]anthracene in rats on high- and low-fat diets. *Jpn. J. Cancer Res.*, 84, 1010–1014.

Nonaka, G. and Nishioka, I. (1983) Rhubarb (2): isolation and structures of a glycerol gallate, gallic acid glucoside gallates, galloylglucoses and isolindleyin. *Chem. Pharm. Bull.*, 31, 1652–1658.

Nonaka, G., Nishimura, H. and Nishioka, I. (1982) Seven new phenol glucoside gallates from *Quercus stenophylla* MAKINO (1). *Chem. Pharm. Bull.*, 30, 2061–2067.

Nonaka, G., Nakayama, S. and Nishioka, I. (1989) Isolation and structures of hydrolyzable tannins, phillyraeoidins A–E from *Quercus phillyraeoides*. *Chem. Pharm. Bull.*, 37, 2030–2036.

Nussler, A.K. and Biliar, T.R. (1993) Inflammation, immunoregulation, and inducible nitric oxide synthase. *J. Leuk. Biol.*, 54, 171–178.

Okuda, T., Yohida, T. and Hatano, T. (1995) Hydrolyzable tannins and related polyphenols. *Progr. Nat. Prod.*, 66, 1–117.

Rodriguez-Saona, C.R., Maynard, D.F., Phillips, S. and Trumble, J.T. (1999) Alkylfurans: effects of alkyl side-chain length on insecticidal activity. *J. Nat. Prod.*, 62, 191–193.

Saijo, R., Nonaka, G. and Nishioka, I. (1989a) Isolation and characterization of five new hydrolyzable tannins from the leaves of *Mallotus japonicus*. *Chem. Pharm. Bull.*, 37, 2063–2070.

Saijo, R., Nonaka, G. and Nishioka, I. (1990) Gallic acid esters of bergenin and norbergenin from *Mallotus japonicus*. *Phytochemistry*, 29, 267–270.

Saijo, R., Nonaka, G., Nishioka, I., Chen, I. and Hwang, T. (1989b) Isolation and characterization of hydrolyzable tannins from *Mallotus japonicus* (THUNB.) MUELLER-ARG. and *M. philippinensis* (LAM.) MUELLER-ARG. *Chem. Pharm. Bull.*, 37, 2940–2947.

Singh, A.K., Pathak, V. and Agrawal, P.K. (1997) Annphenone, a phenolic acetophenone from *Artemisia annua*. *Phytochemistry*, 44, 555–557.

Sitte, P., Ziegler, H., Ehrendorfer, F. and Bresinsky, A. (1998) *Strasburger – Lehrbuch der Botanik*, Fischer, Stuttgart.

Southwell, I.A. and Stiff, J.A. (1995) Chemical composition of an Australian Geranium oil. *J. Essent. Oil Res.*, 7, 545–547.

Thieme, H. and Benecke, R. (1969) Isolierung und Konstitutionsaufklärung eines Cyclolignanglucosids aus den Blättern von *Populus nigra* L. *Pharmazie*, 24, 567–572.

Thorsell, W., Mikiver, A., Malander, I. and Tunón, H. (1998) Efficacy of plant extracts and oils as mosquito repellents. *Phytomed.*, 5, 1540–1541.

Thunberg, C.P. (1794) *Prodromus plantarum carpensium, quas, in promontorio bonae spei Africes, annis 1772–1775, collegit Carolus Pet{rus} Tunberg*, Edman, Upsala.

van der Walt, J.J.A. and Demarne, F. (1988) *Pelargonium graveolens* and *P. radens*: a comparison of their morphology and essential oils. *S. Afr. J. Bot.*, 54, 617–622.

van der Walt, J.J.A. and Vorster, P.J. (1981) *Pelargoniums of Southern Africa*, vol. 2, Juta, Cape Town.

van der Walt, J.J.A. and Vorster, P.J. (1983) Phytogeography of *Pelargonium*. *Bothalia*, 14, 517–523.

van der Walt, J.J.A. and Vorster, P.J. (1988) *Pelargoniums of Southern Africa*, vol. 3, National Botanic Gardens, Kirstenbosch.

Van Wyk, E., Van Oudtshoorn, B. and Gericke, N. (1997) *Medicinal Plants of South Africa*, Briza Publications, Pretoria.

Wagner, H. and Bladt, S. (1975) Cumarine aus südafrikanischen *Pelargonium*-Arten. *Phytochemistry*, 14, 2061–2064.

Wagner, H. and Jurcic, K. (1991) Assays for immunmodulation and effects on mediators of inflammation. In P.M. Dey and J.B. Harborne (eds), *Methods in Plant Biochemistry*, vol. 6, Academic Press, London, pp. 195–217.

Walters, D.S., Craig, R. and Mumma, R.O. (1990) Fatty acid incorporation in the biosynthesis of anacardic acids of *Geraniums*. *Phytochemistry*, 29, 1815–1822.

Walters, D.S., Grossman, H., Craig, R. and Mumma, R.O. (1989) Geranium defensive agents. Iv. Chemical and morphological bases of resistance. *J. Chem. Ecol.*, 15, 357–372.

Watt, C. and Breyer-Brandwyk, M.G. (1962) *Medicinal and Poisonous Plants of Southern and Eastern Africa*. Livingstone, Edinburgh.

Wettstein, R. (1935) *Handbuch der systematischen Botanik*, Deiticke, Leipzig.

Yoshida, T., Haba, K., Nakata, F., Okano, Y., Shingu, T. and Okuda, T. (1992a) Nobotanins, G, H, and I, dimeric hydrolyzable tannins from *Heterocentron roseum*. *Chem. Pharm. Bull.*, 40, 66–71.

Yoshida, T., Itoh, H., Matsunaga, S., Tanaka, R. and Okuda, T. (1992b) Hydrolyzable tannins with $^{1}C_{4}$ glucose core from *Phyllanthus flexuosus* MUELL. ARG. *Chem. Pharm. Bull.*, 40, 53–60.

Yoshida, T., Namba, O., Chen, L. and Okuda, T. (1990) Euphorbin C, an equlibrated dimeric dehydroellagitannin having a new tetrameric galloyl group. *Chem. Pharm. Bull.*, 38, 86–93.

25 Interactions between arthropod pests and pelargoniums

Monique S.J. Simmonds

INTRODUCTION

Species of *Pelargonium* vary in their susceptibility to predation by insects and mites. This variation depends in part in their chemistry and their morphology. Pelargoniums (colloquially known also as geranium), like most species have evolved an array of defence mechanisms that will deter a high proportion of phytophagous insects and mites. However, some insects will usually overcome these defence systems and then cause damage and become pests. The success of these pests often depends on whether the defence mechanisms that the pest has overcome will also protect it from predation. Thus the pest is able to exploit a new plant host until a predator or parasitoid is able to exploit the pest host on the plant. The defence systems can be seen as two types: direct and indirect. The former will consist of morphological features such as the presence of trichomes and the latter with the presence of plant secondary metabolites that deter or are toxic to insects.

Understanding the mechanisms by which geranium plants modulate insects will assist in the design of pest control strategies. In natural conditions wild geraniums are not often reported to suffer greatly from herbivory, the main problems occur when plants are cultivated in high densities under glass. Under these conditions geraniums vary in their susceptibility, especially to pests such as whitefly and mites. Mites and whitefly appear to have a preference for Regals, Angels and some scented varieties. This chapter outlines our knowledge about some of the pests of geraniums and it also provides an overview of pest resistant mechanisms in geraniums and how geraniums could be used in the control of insect pests.

RESISTANCE TO ARTHROPOD PESTS

Anacortes acids are important in determining the susceptibility of zonal geranium plants to attack by insects and mites. These compounds are secreted from the glandular trichomes on leaves and inflorescences. They can prevent insects and mites from moving about on the plant by acting as sticky traps (Walters *et al.*, 1989). They can kill pests via their toxic properties (Gerhold *et al.*, 1984) and they can decrease the build up of pest populations by reducing the fecundity of pests (Grazzini *et al.*, 1991). These compounds also occur in representatives of other families, for example cashews (Anacardiaceae) and ginkgo (Ginkgoaceae) (Walters *et al.*, 1990).

Up to 1990 there were thought to be about six of these 6-alkyl salicyclic acid derivatives called anacardic acids, in which the composition and the length of the side chain varied (Walters *et al.*, 1988). The glandular trichomes on resistant plants (*Pelargonium* × *hortorum* red-flowered plant line 71-17-7) contain higher levels of the C_{22} and C_{24} unsaturated anacardic acids with olefinic groups at the ω-5 position in the side chain, whereas trichomes in susceptible plants contain higher concentration of C_{22} and C_{24} saturated anacardic acids (Figure 25.1). These plants also contain C_{23} and C_{25} anteiso anacardic acids, but in low concentrations (Hesk *et al.*, 1992). The anacardic acids were thought to be synthesised by the addition of acetate units to fatty acid precursors that occur in high concentrations in these cultivars. The synthesis of these compounds was thought to be similar to that used to produce fatty acids, but in the production of anacardic acids the keto groups formed during chain elongation are not all reduced and dehydrated (Geissman, 1963, Gerhold *et al.*, 1984). The evidence for this scheme was provided by Walters *et al.* (1990). They were able to show, using ^{14}C amino acids, that anacardic acids were biosynthesised from saturated, unsaturated C_{16} and C_{18} fatty acids by the addition of three acetate groups and the production of the aromatic ring via reduction, dehydration and intra-molecular aldol condensation, a process very similar to that proposed for the biosynthesis of salicylic acids (Walters *et al.*, 1990).

Initially researchers investigating arthropod–geranium interactions concentrated their research on the profile of anacardic acids in resistant plants but Hesk *et al.* (1992) used high performance liquid chromatography (HPLC) to compare the profiles of anacardic acids in pest-resistance and susceptible plants. Hesk *et al.* (1992) characterised 19 anacardic acids in the geranium plants. In this study, they also used a wide range of ^{14}C-labelled fatty acids to study the biosynthesis of anacardic acids in geranium plants resistant to mites and aphids (plant line 71-17-7) and mite and aphid susceptible plants (plant line 85-26-8). Hesk *et al.* (1992) showed that both lines of plants produced anacardic acids from fatty acids and amino acid precursors in a similar way. However, the resistant plants can oxidise and re-incorporate the acetate groups from whatever fatty acid or amino acid precursor more readily than the susceptible geraniums. Hesk *et al.* (1992) suggest that the resistant plants use this indirect method to produce ω-5 unsaturated anacardic acids, which are not found in susceptible plants. In an earlier study, Winner (1975) showed the insect susceptible and resistant plants differed by one or two genes. It could be that the susceptible plants lack the gene associated with the production of the enzyme involved in the oxidation of the saturated anacardic acids.

Researchers at Penn State University, USA have concentrated on investigating the mechanisms associated with the inheritance of the resistant factors. A selection of 16 South African species of *Pelargonium* were screened for the presence of anacardic acids and glandular trichomes. The ω-5 anacardic acids and presence of glandular trichomes were restricted in their distribution to *P. frutetorum* and *P. inquinans* (Grazzini *et al.*, 1995a). In another study, 13 diploid and 25 tetraploid cultivars of

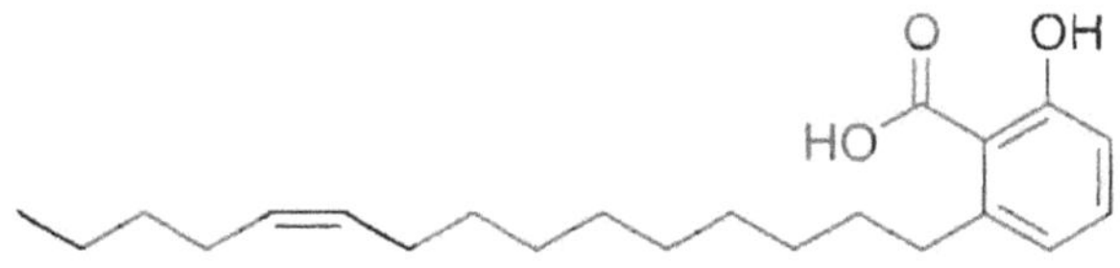

Figure 25.1 Structure of anacardic acid.

Pelargonium × *hortorum* were screened for anacardic acids (Grazzini *et al.*, 1995b). All 38 cultivars contained anacardic acids but no diploid cultivars produced both ω-5 and ω-9 anacardic acids, although three of the tetraploid cultivars did produce both forms of anacardic acids. Overall, there was no relationship between the production of anacardic acids and ploidy levels. However, there were significant differences in the production of specific anacardic acids in different cultivars. This suggested that traditional breeding methods could be used to increase the resistance of *Pelargonium* cultivars to insect pests. Further horticultural experiments have been undertaken with crosses between diploid zonal geraniums that differ in their susceptibility to two-spotted mites (Grazzini *et al.*, 1997). In these experiments the level of anacardic acids and density of the long glandular trichomes that secrete anacardic acids in F-1, F-2 and backcross generations were measured. A selection of the F-2 plants were also tested for resistance to mites. They found that the mite-resistant plants had low densities of trichomes, high levels of ω-5 unsaturated anacardic acids and this condition was controlled by a single dominant allele.

Recently, Grazzini *et al.* (1999) studied the distribution of ω-5 fatty acids in tissue from mite-resistant plants, as these fatty acids could be precursors of the ω-5 anacardic acids. In resistant plants the ω-5 fatty acids were found in the glandular trichomes and pedicel trichomes but not in trichome-free tissue or seeds. Grazzini *et al.* (1999) demonstrate that these fatty acids are available for the biosynthesis of anacardic acids. Thus the current data shows that anacardic acids play a key role in the resistance of geraniums to mites, aphids and whitefly. The levels of these compounds vary among cultivars and also within a resistant cultivar depending on the part of the plant being attacked by the pest and the environmental conditions. For example, a resistant plant could become susceptible if rain washed off the acids from the leaf surface or the temperature increases to over 25.5 °C resulting in a reduction in the viscosity of the exudates secreted by the glandular trichomes and thus reducing the ability of the exudates to trap insects (Walters *et al.*, 1991). Overall, anacardic acids play an important role in the ecology of geranium–arthropod interactions (Scultz *et al.*, 2000). Any resistance mechanisms can alter with time. Harmen *et al.* (1996) developed a short-term bioassay to evaluate the factors in geraniums that influenced their susceptibility to mites. These authors were able to show that leaves can loose their resistance to mites but it can be regained after 14 days. This resistance can be induced. The fact that the fatty acids precursors of anacardic acids occur in the trichomes on the leaves of resistant geraniums ensures their replacement and thus the leaves' potential resistance to pests.

Anacardic acids are not the only compounds in geraniums that modulate insects. Geraniums also contain a range of phenolics that could contribute to their susceptibility to attack by herbivores. For example, some of the phenolics associated with the medicinal properties of geraniums could modulate insect–geranium interactions, such as the hydrolysable tannins (ellagitannins), geraniin (Okuda *et al.*, 1982), corilagin (Tanaka *et al.*, 1985) and elaeocarpusin (Nonaka *et al.*, 1986). Although, as yet we have no direct experimental data to indicate they influence the susceptibility of geraniums to insects.

Geraniums also contain indole alkaloids, a group of compounds with known insecticidal activity. However, as yet there is very little information about the role of these compounds in geranium–insect interactions. Observations in the glasshouses at Kew showed that when the glasshouse whitefly lay their eggs on the leaves of zonal varieties they often lay their eggs on the margin of the distinct dark horse shoe-shaped region on the leaves. The factors that influence this behaviour have not been fully studied.

Lis-Balchin *et al.* (1996) identified two indole alkaloids elaeocarpidine 1 and epielaeocarpidine 2 in zonal cultivars of *Pelargonium* but not in non-zonal varieties. In a preliminary study (Simmonds *et al.*, unpublished), these compounds were applied to the centre of leaves from non-zonal varieties of *Pelargonium* and the number of whitefly eggs laid on the treated plants decreased. A laboratory experiment also showed that elaeocarpidine 1 but not epielaeocarpidine 2 was toxic to whitefly adults at 1000 ppm (Figure 25.2). Whether the levels of these compounds influence the susceptability of geraniums to herbivory by insects still needs to be evaluated.

IMPORTANT PESTS OF GERANIUMS

Whitefly

The glasshouse whitefly *Trialeurodes vaporariorum* (Westwood) can be an important pest of geraniums. Adult and nymphs feed on the plant sap and when present in high numbers they can cause a decrease in plant vigour as well as spread viruses. The insects also produce a sticky secretion called honeydew that acts as a substrate for fungal growth.

Adult whitefly are usually found on the underside of leaves were they lay their eggs in a horse-shoe configuration. These eggs turn from a light yellow to black in about 2–4 days. Pale green nymphs hatch from the eggs, they are mobile and move over the surface of the plant to find a suitable feeding place. They then settle and form what are often called 'scales'. The immature nymphal stages feed on plant sap via stylets that they insert into the plant tissue. The duration of the lifecycle depends on the temperature and the suitability of the plant. For example, at 21 °C the life cycle of the whitefly from egg to adult takes about 27 days.

Whitefly can be controlled by insecticides or via beneficial organisms such as the parasitoid *Encarsia formosa* or fungal pathogens. The chalcid wasp, *E. formosa*, is a very effective parasitoid and is used commercially to control whitefly in many glasshouses. The female wasp lays eggs into the second or third nymphal stage of the whitefly. The parasitoid then develops within the scale and emerges from the whitefly pupae after the whitefly has pupated. The presence of the parasitoid in the whitefly pupae

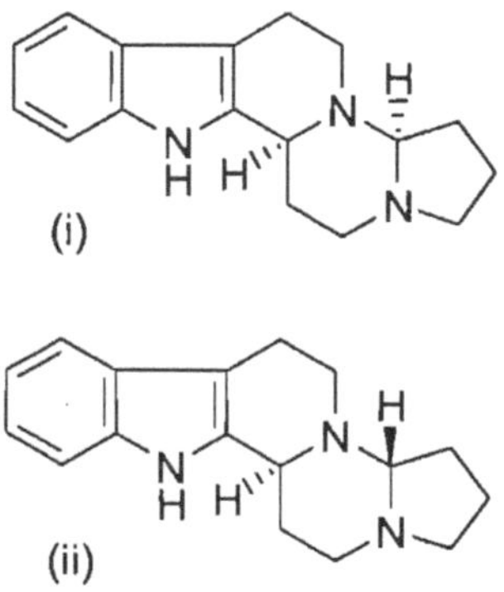

Figure 25.2 Structure of indole alkaloids: (i) elaeocarpidine and (ii) epielaeocarpidine.

causes the pupae to turn black. This colour change in the whitefly pupae can be used to monitor the effectiveness of *E. formosa* in controlling whitefly on different plants.

E. formosa can parasitise whitefly on geraniums. However, we do not know what influence the exudates produced by glandular trichomes on geranium leaves have on the ovipositing behaviour of these parasitoids. Recent laboratory based studies have shown that *E. formosa* and fungi can be used in an Integrated Pest Management (IPM) strategy to control whitefly on geraniums (Avery, unpublished data). Further research is needed into how whitefly-resistant lines of geraniums might influence the interactions among whitefly, their parasitoids and fungal pathogens.

Mites

Two-spotted mite, *Tetranychus urticae* (Kock), formerly known as the red-spider mite is an important pest of geraniums, especially those grown in glasshouses. The susceptibility of geranium plants to mites appears to be similar to their susceptibility to whitefly. Many of the zonal geraniums (*Pelargonium* × *hortorum*) that have resistance to mites, are associated with the density of tall glandular trichomes, which secrete anacardic acids (Gerhold *et al.*, 1984; Harman *et al.*, 1996).

Thrips

The pest status of thrips has increased in the last decade, especially the damage caused by the western flower thrips, *Frankliniella occidentalis*. These insects can directly damage plants by feeding on the flowers and indirectly by spreading viruses, such as *Pelargonium* flower break carmovirus (Krczal *et al.*, 1995). As yet there are no validated IPM programmes available for the control of thrips on geraniums, as the effect of mite- and whitefly-resistant cultivars on predators of thrips has to be still evaluated.

The level of thrips infestations can be monitored by the use of blue sticky traps, which should be placed just above the height of the plants. Currently there is a range of biological control agents that are available for the control of thrips. These include predatory mites, bugs such as species of *Orius* as well as nematodes and fungi. Traditionally predatory mites such as *Iphiseius degenerans* (Berlese) have been used for thrips control. These mites can be effective against the immature nymphal stages but do not attack eggs or adults. Predatory mites are usually supplied in culture packs which can be placed on plants or the contents sprinkled over plants when flowers first appear.

Orius are more effective foragers than mites and are able to attack thrips' nymphs and adults. However, it is difficult to establish a populations of *Orius* in a glasshouse as they breed slowly and lay eggs in parts of plants that are frequently removed during routine pruning (Jacobson, 1993).

Vine weevil

Larvae of the vine weevil *Otiorhynchus sulcatus* (Fabricius) can cause damage to roots of geraniums. The adult weevils are nocturnal and feed on leaves. They lay their eggs in the soil. After an incubation period of 8–24 days the larvae emerge from the egg and search for suitable roots to feed on. The larvae are white, legless with a brown head. When fully grown the larvae construct earthen cells in the soil and pupate in these cells.

The flightless female adults emerge from these pupae. As yet, no male beetles have been discovered. It is the larvae that do the most damage to the plants, although adults can cause damage to leaves by biting notches along the edge of the leaves. Larvae are usually targets for control programmes: they can be killed by beneficial nematodes such as *Heterorhabditis*. However, if the damage is extensive due to high densities of larvae then insecticides will have to be used and these will need to be applied to the soil to kill the larvae.

Other pests

The glasshouse and potato aphid *Aulacorthum solani* (Kaltenbach) is a common pest of geraniums in glasshouses. The adult and nymphal stages are greenish-yellow with brown markings. *Acyrthosiphon malvae* (Mosley), the *Pelargonium* aphid, is another common pest of geraniums, however, unless infestations are high or the aphids are transmitting viral infections, these insects do not normally cause economic loss.

The cyclamen mite, *Phytonemus pallidus* (Banks), can be a another pest of geraniums in glasshouses. The mites prefer young leaves and flower buds and their life cycle can be between 2 and 3 weeks at 20–25 °C. Infestations can result in the margin of the leaves curling inwards, leaves become brittle and the plants become stunted with deformed leaves and buds.

There are a few Lepidopteran pests that can damage geraniums. For example, the larvae of the geranium bronze butterfly can damage geraniums by boring into their stems. The adults lay their eggs on the flower buds and the larvae when they emerge move from the flowers into the stems. Here they feed on the stem causing wilting and if the infestation is high, the death of the plant. This pest was reported in Europe in the late 1980s. It is a native of South Africa, where the natural predators and parasitoids help to keep caterpillar numbers down. However, in Europe there are no natural predators and the pest is through to be on the increase. Other lepidoptera pests include the silver moth, *Autographa gamma* (Linnaeus), which can attack plants in glasshouses in Europe. It can cause damage to plants growing outside in the southern parts of Europe but it will not survive winters in northern Europe.

The flax tortrix moth, *Cnephasia asseclana* (Denis and Schiffermuller), is a less important pest and can, like other tortrix moths, be identified by the fact that larvae produce silk to spin a web between leaves or flowers. The web provides protection for the feeding larvae and the pupae.

The glasshouse leafhopper *Hauptidia maroccana* (Melichar) can cause speckling, silvering or mottling to the foliage. However, unless the infestations are very high this pest does not usually result in the death of the plant. The common green capsid, *Lygocoris pabulinus* (Linnaeus), can cause damage to new foliage and flowers. The new shoots become red to brown and the flowers are often distorted and the buds can be aborted.

USE OF GERANIUM EXTRACTS IN PEST CONTROL

The control of insect pests continues to challenge scientists. As the concern about the overuse of broad spectrum synthetic insecticides increases, there is a need to find alternative strategies. These alternatives could include plant-derived compounds which are usually more host specific in their activities and are biodegradable. For example, plants have already provided a range of insecticides such as nicotine, rotenone, pyrethrum and azadirachtin. However, there is an increased interest in aromatic plants and many essential oils (EOs) have been screened for insecticidal activity (Regnault-Roger, 1997).

The EOs obtained by steam distillation of the aerial parts of *Pelargonium* species, especially. *P. graveolens* (L.), *P. capitatum* Ait., *P. odoratissimum* and *P. radula* (Cav.) (syn. *P. radens* and *P. roseum* Willd.) used in perfumes can be used to deter insects. The oil can contain over 120 different mono- and ses-quiterpenoids (Vernin *et al.*, 1983). The composition of the oils differ depending in part on the source of the plants. The main compounds in EOs are citronellol, geraniol, linalol, nerol, myrene, camphene, limonene, carvone and farnesene. Many of these compounds have known insecticidal or deterrent activity against insects. For example, citronellol has known insecticidal activity and citronellol extracted from cloves has been used as a mothprooofing agents (Riedel *et al.*, 1989).

Extracts from wild geraniums are repellant to the cockroach *Blatella germanica* and extracts from the rose geranium are repellant to flies such as *Musca domestica* and *Culex fatigans* (Jacobson, 1990). However, to date very little research has been undertaken on the use of geranium extracts in pest control. Although from our knowledge about the anti-insect activity of the compounds in the plants the extracts from geraniums should have some potential as insect deterrents. A problem encountered in comparing some of the ethnobotanical reports about the use of Geranium oil relates to inadequate information about the species or variety of geranium used. As yet there has been very little research into the effect of geranium-derived compounds on non-geranium feeding insects. However, the information we have suggests they could be sources of active molecules. For example, geraniin the bound form of ellagic acid found in geraniums inhibits the development of *Heliothis virescens* (Klocke *et al.*, 1986).

ACKNOWLEDGEMENTS

I thank Dr Nigel Veitch for drawing the chemical structures and to Dr Maria Lis-Balchin for introducing me to the complexity of *Pelargonium*–insect interactions.

REFERENCES

Geissman, T.A. (1963) The biosynthesis of phenolic plant products. In P. Bernfield (ed.), *Biogenesis of Natural Compounds*. Macmillan, New York, pp. 563–616.

Gerhold, D.L., Craig, R. and Mumma, R.O. (1984) Analysis of trichome exudates from mite-resistant geraniums. *J. Chem. Ecol.*, 10, 713–722.

Grazzini, R., Hesk, D., Heininger, E., Hildenbrandt, G., Reddy, C.C., CoxFoster, D., Medford, J., Craig, R. and Mumma, R.O. (1991) Inhibition of lipoxygenase and prostaglandin endoperoxide synthase by anacardic acids. *Biochem. Biophys. Res. Commun.*, 176, 775–780.

Grazzini, R., Hesk, D., Yerger, E., CoxFoster, D., Medford, J., Craig, R. and Mumma, R.O. (1995a) Species distribution of biochemical and morphological characters associated with small pest resistance in *Pelargonium* × *hortorum*. *J. Am. Soc. Hort. Sci.*, 120, 336–342.

Grazzini, R., Hesk, D., Yerger, E., CoxFoster, D., Medford, J., Craig, R. and Mumma, R.O. (1995b) Distribution of anacardic acids associated with small pest resistance among cultivars of *Pelargonium* × *hortorum*. *J. Am. Soc. Hort. Sci.*, 120, 343–346.

Grazzini, R., Walters, D., Harmon, J., Hesk, D.J., CoxFoster, D., Medford, J., Craig, R. and Mumma, R.O. (1997) Inheritance of biochemical and morphological characters associated with two-spotted spider mite resistance in *Pelargonium* × *hortorum*. *J. Am. Soc. Hort. Sci.*, 122, 373–379.

Grazzini, R.A., Paul, P.R., Hage, T., CoxFoster, D.L., Medford, J.I., Craig, R. and Mumma, R.O. (1999) Tissue-specific fatty acid composition of glandular trichomes of mite-resistant and susceptible *Pelargonium* × *hortorum*. *J. Chem. Ecol.*, 25, 955–968.

Harman, J., Paul, P., Craig, R., CoxFoster, D., Medford, J. and Mumma, R.O. (1996) Development of a mite bioassay to evaluate plant resistance and its use in determining regeneration of spider mite resistance. *Entomol. Exper. Applic.*, 81, 301–305.

Hesk, D., Craig, R. and Mumma, R.O. (1992) Comparison of anacardin acid biosynthetic capability between insect-resistant and -susceptible geraniums. *J. Chem. Ecol.*, 18, 1349–1364.

Jacobson, M. (1990) Glossary of Plant-derived insect deterrents. CRC Press, Florida.

Jacobson, R.J. (1993) Egg laying sites of *Orius majusculus*, a thrips predator, on cucmber. In: B.L., Parker, M. Skinner and T. Lewis (eds), *Thrips Biology and Management*. Plenum Press, New York, pp. 241–244.

Klocke, J.A., Van Wagenen, B. and Balandrin, M.P. (1986) The ellagitannin geraniin and its hydrolysis products isolated as inset inhibitors from semi-arid land plants. *Phytochem.*, 25, 85.

Krczal, N.K.K., Albouy, J., Damy, I., Kusiak, C., Deogratias, J.M., Moreau, J.P., Berkelmann, B. and Wohanka, W. (1995) Transmission of pelargonium flower break virus (PFBV) in irrigation systems and by thrips. *Plant Dis.*, 79, 163–166.

Lis-Balchin, M., Houghton, P. and Woldermarian, T. (1996) Elaeocarpidine alkaloids from *Pelargonium* species (Geraniaceae). *Natl. Prod. Lett.*, 8, 105–112.

Nonaka, G., Morimoto, S. and Nishioka, I. (1986) Elaeocarpusin, a proto-type of geraniin from *Geranium thunbergii*. *Chem. Pharmaceut. Bull.*, 34, 941–943.

Okuda, T., Yoshida, T. and Hatano, T. (1982) Constituents of *Geranium thunbergii* Seb. Et Zucc. Part 12. Hydrated sterostructure and equilibration of geraniin. *J. Chem. Soc. Perkins Trans.*, 1, 9–14.

Regnault-Roger, C. (1997) The potential of botanical essential oils for insect pest control. *Integr. Pest Manag. Rev.*, 2, 25–34.

Riedel, G., Heller, G. and Voigt, M. (1989) Detia Freyberg GmbH, Germany (Federal Republic). Citronellol and eugenol as mothproofing agents. Patent DE 98-39013 i1 890118.

Scultz, D.J., Medford, J.I., Cox-Foster, D., Grazzini, R.A., Craig, R. and Mumma, R.O. (2000) Anacardic acids in trichomes of *Pelargonium*: Biosynthesis, molecular biology and ecological effects. *Adv. Bot. Res. (incorp.) Adv. Plant Path.*, 31, 175–192.

Tanaka, T., Nonaka, G. and Nishioka, I. (1985) Punicafolin, an ellagitannin from the leaves of *Punica granatum*. *Phytochem.*, 24, 2075–2078.

Vernin, G., Metzger, J., Fraisse, D. and Scharf, C. (1983) Etude des huiles essentielles par GC-SM-banque specma: essences de geranium. *Parfum. Cosmet. Arom.*, 52, 51–61.

Walters, D.S., Minard, R., Craig, R. and Mumma, R.O. (1988) Geranium defensive agents III. Structural determination and biosynthetic considerations of anacardic acids in geranium. *J. Chem. Ecol.*, 14, 743–751.

Walters, D.S., Grossman, H., Craig, R. and Mumma, R.O. (1989) Geranium defensive agents IV. Chemical and morphological bases of resistance. *J. Chem. Ecol.*, 15, 357–372.

Walters, D.S., Craig, R. and Mumma, R.O. (1990) Fatty acid incorporation in the biosynthesis of anacardic acids in Geraniums. *Phytochem.*, 29, 1815–1822.

Walters, D.S., Harman, J., Craig, R. and Mumma, R.O. (1991) Effect of temperature on glandular trichome exudates composition and pest resistance in *Geraniums*. *Entomol. Exper. Applic.*, 60, 61–69.

Winner, B.L. (1975) Inheritance of resistance to the two-spotted spider mite, *Tetranychus uriticae* (Kock) in the geranium *Pelargonium* × *hortorum* (Bailey). MS Thesis. Pennsylvania State University, University Park, Pennsylvania.

26 Correlation of the chemical profiles of the essential oil of *Pelargonium* (Geranium oil) and others separately and in mixes, with their relaxant or stimulant properties in man and smooth muscle preparations *in vitro*

Maria Lis-Balchin and Stephen Hart

INTRODUCTION

Geranium oil, like other plant essential oils (EOs) has been used for many years in the manufacture of perfumes and these have been designed to either stimulate or relax the wearer or possibly even to attract the opposite sex. Perfumes are made up of a large number of EOs and nowadays also numerous synthetic chemical components, but the basis of a good perfume is to have a specific blend of 'top notes', 'middle notes' and 'base notes'. Perfumes were originally designed for either men or for women individually, but nowadays there is a tendency for unisex perfumes, which suggests that the emphasis lies more on the stimulant or relaxant nature of the perfume rather than sexual attraction.

Essential oils have also been individually categorised by aromatherapists as being either relaxant (sedative) or stimulant. It is not clear whether this refers to the action on the brain or to some or all of the muscles, as many EOs are also classified as anti-spasmodic. Most EOs are used by aromatherapists in mixes of 3 or more, diluted by 95 per cent with almond or other carrier oil and then massaged into the skin. The mixture almost inevitably has the three 'notes' as its theme, in order to give the correct 'balance'.

The overall effect of this 'perfume' together with the massage is supposed to relieve stress and in so doing may also 'cure' or at least alleviate many of the stress-related conditions like eczema, stomach ache, back ache, headache. Whether or not the aroma of the perfume mix has any effect is unclear as clinical studies have not provided any statistically significant results (see Chapter 21).

There are many studies on the effects of aroma on psychology as well as physiology of the 'recipient', however, there is so far no published evidence for the role of the chemistry of the EO, component and mixture of these in a perfume or 'aromatherapy mix'. These parameters were therefore studied using smooth muscle.

Stress: biochemical and physiological implications

Changes in the body which occur outside of the brain, as a result of stress, are not under conscious control but are mediated by the sympathetic branch of the autonomic nervous system. The activity of most organs of the body is controlled by the autonomic nervous system and as a general rule the sympathetic system may be considered to be activated in times of flight or fight which will include stress. Stress-related changes in the body will also be mediated by hormones, such as those released from the adrenal gland. Stimulation of the sympathetic system, and adrenaline released from the adrenal gland, will increase heart rate and stroke volume and by dilating and contracting different blood vessels will cause blood to be distributed to those organs such as skeletal muscle, heart and lungs which are involved in exercise.

Smooth muscle will also be either contracted or relaxed such that the body is prepared for exercise, thus bronchial muscle relaxes and sphincters of the gastro-intestinal system contract. If one considers the fight response in animals, smooth muscle contracts to give dilated pupils and make hair stand on end. In both man and other animals, stimulation of the sympathetic system will cause metabolic changes which favour activity, such as an increase in blood glucose. The nerves of the sympathetic system which innervate smooth, cardiac and vascular smooth muscle all release noradrenaline as their neurotransmitter and the differential response, either contraction or relaxation, is brought about by the presence of different adrenoceptors on the innervated tissue.

In general, alpha adrenoceptors mediate contraction and beta-adrenoceptors relaxation, but of course there are exceptions to this rule. Further differentiation and control of the system is obtained by the presence of sub-types of alpha and beta adrenoceptors. Occupation of a receptor by an appropriate agonist results in a change in cell activity (such as contraction or relaxation) which is mediated via a secondary messenger within the cell. Alpha-2 adrenoceptors mediate their actions via a fall in cyclic AMP (cAMP), whilst beta-adrenoceptor activation is associated with a rise in cAMP. Alpha-adrenoceptors are linked to the phosphoinositide pathway. In general, contraction is associated with an increase in the concentration of calcium ions within the muscle fibre whilst relaxation involves either a removal of calcium, the blocking of calcium channels or the opening of potassium channels.

Many tissue of the body receive a dual innervation from the two branches of the autonomic nervous system (sympathetic associated with activity and parasympathetic with feeding and the restoration of energy). In the gastro-intestinal tract, we have this dual innervation plus an additional plexus of nerves in the wall of the intestine, often called the enteric nervous system, which involves several other neutotransmitters.

It is on account of this rich innervation of the intestine that we have studied the action of EOs on the smooth muscle of the guinea-pig ileum *in vitro*.

Smooth muscle preparation

The preparation of the smooth muscle of the guinea-pig ileum will remain viable for several hours after removal from the animal and will respond to electrical field stimulation with reproducible contractions which are due to the stimulation of the parasympathetic nerve with the release of acetylcholine.

Essential oils which stimulate smooth muscle contraction can be recognised immediately whilst the site of action of those which reduce the size of the electrically-induced contraction can be determined. Possible sites of action include inhibition of the release of acetylcholine, or relaxation of the tissue via stimulation of adrenoceptors, action on secondary messengers or on calcium or potassium channels. This preparation thus allows us to recognise spasmogenic and spasmolytic activity, to determine whether or not the activity is dose-related, to measure duration of action and also attempt to determine the mechanism of action.

The question arises whether the knowledge of the activity of EOs on smooth muscle gives us any clues about the likely actions of these compounds if and when they enter the central nervous system (CNS).

A famous English pharmacologist suggested that the intestine could be considered a paradigm of the CNS, but it still remained almost impossible to infer action in the CNS from activity on isolated smooth muscle.

The reason for this is simply the complexity of the CNS, with the interaction between excitatory and inhibitory fibres being such that reduced activity in one neurone can lead either to sedation or excitation. Thus alcohol appears to stimulate some behaviour although it is a CNS depressant, the explanation being that the inhibition of inhibitory pathways removes a normal break and behaviour therefore changes.

Another aspect of the complexity of the CNS is that any one particular behaviour is controlled by several neurotransmitters, each of which is likely to be able to bind to different sub-groups of receptors.

If one for example considers pain, this involves neurotransmitters in the afferent pathway such as Substance P, glutamate and nitric oxide (NO) and this afferent pathway can be modulated by neuronal pathways releasing a range of neurotransmitters including opioid peptides, acetyl choline, histamine, 5-hydroxytryptamine and cholecystokinin.

With so many neurotransmitters involved in the pain pathway it is not surprising that the experience of pain can be influenced by many different compounds. For example, monoterpenes like menthone and α-terpineol (administered by the subcutaneous route) showed activity similar to that of accepted analgesics e.g. indomethacin and naproxen in reducing the behavioural activity of the mouse to a noxious stimulus (Hart *et al.*, 1994).

In experiments studying the motor activity of mice after exposure to the aroma of various EOs (Buchbauer, 1991, 1992; Jager *et al.*, 1992) rosemary, jasmine and Ylang ylang increased activity whilst lavender, neroli, lime-blossom, passiflora, citronellol and linalool (both the latter found in Geranium oil in high concentrations) decrease motor activity. The presence of components in the blood when applied by inhalation has also been demonstrated (Jager *et al.*, 1992). The effect on the motor activity has been shown to be similar to that when the EO was injected. It has been assumed that changes in motor activity are a central effect but the possible action on neuromuscular transmission has not been investigated.

However, recent experiments on the motor-nerve skeletal muscle preparation (rat phrenic-nerve diaphragm) by the authors has shown that geranium, lavender and tea tree oils cause a reduction in the size of the twitch of the skeletal muscle in response to electrical stimulation of the motor nerve.

Linalool, which was shown to reduce motor activity in the mouse has also been shown to have an action within the brain itself: using membranes from rat cerebral cortex, linalool exhibited a dose-related inhibition of the binding of glutamate, a main excitatory neurotransmitter of the CNS (Elisabetsky *et al.*, 1995).

The effect of EOs in man has been studied in several different ways including measuring the alertness and reaction times (Manley, 1993) and human brain activity (Torii *et al.*, 1988; Kubota *et al.*, 1992) using contingent negative variation (CNV). The latter is the brain potential which occurs between a warning stimulus and an imperative stimulus i.e. when the subject is expecting something to happen. The CNV amplitude is increased by caffeine, jasmine and peppermint and decreased by chlorpromazine, lavender and marjoram. There is some discrepancy between results from different groups regarding many oils.

The present experiments were designed to see whether the effect of stress on man could be mimicked using isolated segments of small intestine and by monitoring their spasmogenic or spasmolytic effects, it could be possible to assess their relaxant or stimulant nature.

MATERIALS AND METHODS

Essential oils were obtained from various commercial sources and each oil was analysed by GC using a Shimadzu GC 8A with a 50 m × 0.32 mm OV101 column; the temperature program was set at 4 °C min^{-1} from 100 to 230 °C.

The percentage of all the components was calculated in each selected retention time (RT) interval of under 10 min, 11–15 min, 16–20 min, 21–30 min and 30 + min. The main components present in each RT interval was also determined. The EOs were diluted in methanol (usually × 1000) and 0.1 to 0.2 ml was applied to the tissue preparations in the organ bath giving a final dilution of ×200,000 to ×400,000 (a concentration of 2.5×10^{-6} to 5×10^{-6}).

Pharmacological studies, carried out on guinea-pig ileum were contrasted against many practising Aromatherapists' predictions of the effect of EOs on the patient (alone or as mixtures).

RESULTS OF THE STUDIES

Monoterpenes versus contractions of smooth muscle

Previous comparisons of the pharmacological activity of many components and EOs suggested that monoterpenes were responsible for contractions in the guinea-pig ileum *in vitro* (Lis-Balchin *et al.*, 1996a,b).

This was best illustrated by work on two New Zealand EOs Manuka and Kanuka. The former was largely composed of sesquiterpenes and produced a relaxation in the gut, whilst the latter was composed largely of monoterpenes and produced a contraction (Lis-Balchin *et al.*, 1996a). Further work on over 70 EOs suggested that there was a considerable correlation of contraction of the small intestine with a high percentage of monoterpenes and not sesquiterpenes (Lis-Balchin *et al.*, 1996b, 1998).

These results therefore suggested that it was simply the actual percentage composition of the monoterpenes, but not sesquiterpenes, which determined whether the effect on the smooth intestinal muscle would be contractile or relaxant. This was therefore investigated further.

Table 26.1 The predicted effect on guinea-pig ileum based on the percentage of components at different retention time (RT) intervals

RT	>10	11–15	16–20	21–30	30+	*Effect*
a	19	66	–	5	–	R
b	13	64	12	7	–	R
c	–	3	–	17	45	R
d	82	1	–	5	–	S
e	97	3	–	–	–	S
f	57	–	–	32	–	S/R
g	69	25	2	2	–	S/R
h	2	–	85	2	–	R
i	82	–	–	4	–	S
j	6	41	45	5	–	R
k	46	22	30	–	–	S/R
l	8	30	5	39	10	R
m	–	–	–	59	48	R
n	–	–	–	78	34	R
o	7	29	58	–	–	R
p	–	92	2	–	–	R
q	–	52	37	6	–	R
r	99	–	–	–	–	S
s	1	30	51	5	–	R
t	14	–	1	69	–	S/R
u	16	–	76	–	–	S/R
v	67	7	2	9	–	S
w	1	–	34	56	–	R
x	–	–	–	89	4	R
y	58	32	–	–	–	S/R
z	4	–	1	76	1	R

Retention times versus percentage of components

This hypothesis was put to the test, using EOs alone or in mixes, by calculating the total per cent of components in different RT intervals (Table 26.1) and predicting what the effect on the smooth muscle would be.

It was noted that monoterpenes were in the RT > 10 interval, with the exception of 1,8-cineole which was also found here, whilst alcohols, ketones and aldehydes occurred in the 11–15 min interval, esters and phenols in the 15–20 min interval and sesquiterpenes thereafter.

Chemical predictions versus actual effect on ileum

Whenever there was a considerable percentage of components in the > 10 min interval, this would be associated with a small to large contraction of the ileal muscle (depending on the actual percentage). Predictions of pharmacological activity could therefore be easily made based on the chemical composition, with the exception of EOs containing 1,8-cineole e.g. *Eucalyptus globulus* (Table 26.2). Geranium oil showed a very high proportion of the non-stimulating components and therefore was stated to be relaxant according to the chemical prediction.

Table 26.2 Comparison of the actual effect of essential oils on Guinea-pig ileum and the predicted effects using chemical composition and Aromatherapists' predictions

	Actual effect on tissue	*Chem prediction*	*Aromather. prediction*
a. Tea tree	R & S/R	S/R	S/R
b. Neroli	R	S/R	S/R
c. Camomile German	R	R	R
d. Frankincense	S	S	R
e. Camphor	S	S	S
f. Black Pepper	S/R	S/R	S/R
g. Rosemary	S/R	S/R	S
h. Lemongrass	R	R	S/R
i. Juniper	R	S	S
j. Lavender	R & S/R	R	R
k Bergamot	S/R	S/R	S/R
l. Ylang Ylang	R	R	R
m. Sandalwood	R	R	R
n. Vetivert	R	R	R
o. Petitgrain	S/R	R	S/R
p. Rosewood	R	R	R
q. Geranium Bourbon	R	R	R
r. *Eucalyptus globulus*	R	S	S
s. Clary Sage	S & S/R	R	R
t. Ginger	R	S/R	S/R
u. Dillweed	S/R	S/R	R
v. Nutmeg	S	S	R
w. Manuka	R	R	R
x. Spikenard	R	R	R
y. Camomile Roman	S/R & S/R	S/R	R
z. Valerian	R	R	R

Chemical predictions versus Aromatherapists' predictions of effect on patient

The chemical predictions were largely similar to both the actual observed effect on the smooth muscle and also similar to the Aromatherapists' prediction on the 'patient'. The latter effect was either a relaxant effect or a stimulant effect on the 'patient'; the stimulant effect could be directly related to a contraction on the isolated muscle. The chemical prediction for Geranium oil was confirmed by the actual effect and that of the Aromatherapists' predictions.

Effect of mixtures of EOs

Mixtures of two or more EOs also showed the same trend, some of which are shown in Tables 26.3 and 26.4. This proves that contractions of smooth muscle are largely as a result of a high monoterpene concentration, regardless of the actual monoterpene component. Geranium oil added to other components caused a swing towards chemical readjustment towards a greater concentration of relaxant components and therefore it was predicted to be a more relaxant mixture than the components like frankincense (mix 6), frankincense and bergamot (mix 7) would have been alone, due to their high concentration of monoterpenes.

Table 26.3 The predicted effect on guinea-pig ileum based on the percentage of components at different retention time (RT) intervals

RT	>10	11–15	16–20	21–30	30+	*Effect*
1	–	85	2	–	7	R
2	41	5	49	–	–	S/R
3	85	1	–	6	–	S
4	78	22	1	14	–	S/R
5	86	2	2	1	–	S
6	44	45	–	5	–	S/R
7	37	47	8	1	–	S/R
8	4	90	–	2	–	R
9	9	56	12	15	–	S/R
10	–	–	–	58	37	R
11	98	–	–	–	–	S
12	99	–	–	–	–	S
13	–	–	–	71	19	R
14	22	69	–	–	–	S/R
15	10	26	4	41	6	S/R
16	10	39	36	9	–	S/R
17	78	9	3	5	–	S

Table 26.4 Comparison of the predicted effect of essential oil blends on 'clients/patients' by Aromatherapists and by the chemical composition with their actual effect on guinea-pig ileum

Blend	*Effects predicted*		
	Actual effect	*Aromatherapist*	*Chemical prediction*
1. Orange 2: Nutmeg 1: Dill 1	S	S	S/R
2. Lemongrass 1: Juniper 1: Rosemary 2	S	S	R
3. Frankincense 1: Rose Abs. 1: Clary Sage 2	R/S	R/S	S/R
4. *Eucalyptus glob.* 1: Black pepper 1: Ginger 1	S	S	S/R
5. Ginger 1: Tea Tree 1: Rosemary 2	S	S/R	S/R
6. Frankincense 2: Ylang Ylang 1: **Geranium** 1	R/S	S/R	R
7. Frankincense 1: **Geranium** 1: Bergamot 2	S/R	S/R	S/R
9. Camomile Roman 1: Lavender 1: **Geranium** 1	R/S	S/R	S/R
10. Frankincense 1: Mandarin 2: Scotch Pine 1	S/R	S	S/R
11. Camomile Roman 1: Valerian 1: Rose Abs. 1	R/S	S/R	R
12. Ylang Ylang 1: Marjoram 1: Thyme Red 1	R	S/R	R
13. Petitgrain 1: Melissa 1: Sage Dalmatian 1	S/R	R	R
14. Kanuka 1: Lavender 1: Frankincense 1	R	S/R	R
15. Manuka 1: Lavender 1: Frankincense 1	S	S/R	R
16. Fennel 1: Orange 1: Bergamot 1	S/R	S/R	S/R
17. Basil 1: Bergamot 1: Clary Sage 1: Jasmine 1	S	S/R	S/R

As before, the correlation broke down if 1,8-cineole was involved e.g. in mixtures with rosemary or *Eucalyptus globulus*. There is no easy explanation for this discrepancy. It is also of interest that if *Eucalyptus globulus*, containing 95 per cent of 1,8-cineole is presented to the smooth muscle preparation it will cause a relaxation, whereas if 1,8-cineole alone is presented it causes a contraction.

Studies on well known commercial perfumes were only effected on their chemical composition. There seemed to be a very positive correlation between the chemical distribution and the product's intention, as determined by the publicity information. Thus, the lavender-containing Eau de colognes, were refreshing and stimulating only because there was a predominance of monoterpenes (largely limonene) from the citrus EOs used in greater concentration than lavender in the formulation. Paris by Kenzo, showed a similar over-preponderance of monoterpenes again due to the limonene of its citrus components. These are obviously stimulating oils and their predicted effect would be contractile on smooth muscle. 'Geranium oil' – containing perfumes, with a more rose-like odour (with either the synthetic components or real Geranium oil added) would have a preponderance of relaxant components and therefore be predictably relaxant on smooth muscle. The effect on the wearer or those smelling the wearer would also be predictably of a relaxant nature i.e. its holistic effect would also be of a relaxant nature, unless the wearer or person sniffing that perfume had a great dislike for that particular odour…but that is an unpredictable idiosyncratic, psychological effect, which could completely contradict the chemical findings.

CONCLUSION

The present results indicate that there is a very close correlation between the pharmacological activity of EOs on the isolated smooth ileal muscle of the guinea pig and the predicted effect on the human psyche by aromatherapists, as with the use of Geranium oil.

This 'holistic' effect would probably have originated from the direct effect of EOs on the CNS with the concomitant effect of the massage (and of course counselling and possible placebo effect). The actual effect on the isolated smooth muscle is less complex and probably involves various adrenoceptors, but there could also be a simple direct action of components on the membrane with all monoterpenes initiating a rise in calcium levels which cause contraction of the muscle.

It is suggested that perfumes containing large proportions of relaxant components, i.e. other than monoterpenes, would be of a relaxant nature to the wearer and those around the wearer.

REFERENCES

Buchbauer, G., Jirovetz L., Jager, W., Dietrich, H., Plank. C. and Karamat, E. (1991). Aromatherapy: Evidence for sedative effects of the essential oils of lavender after inhalation. *Z. Naturforsch.*, 46, 1067–1072.

Buchbauer, G., Jirovetz, L. and Jager, W. (1992) Passiflora and Limeblossom: Motility effects after inhalation of the essential oils and some of the main constituents in animal experiments. *Arch. Pharm. (Weinheim)*, 325, 247–248.

Elisabetsky, E., Marschner, J. and Souza, D.O. (1995) Effects of linalool on glutamatergic system in the rat cerebral cortex. *Neurochem. Res.*, 20, 461–465.

Hart, S.L., Gaffen, Z., Hider, R.C. and Smith, T.W. (1994) Antinociceptive activity of monoterpenes in the mouse. *Can. J. Physiol. Pharmacol.*, 72, Sup.1, 344.

Jager, W., Buchbauer, G., Jirovetz, L. and Fritzer, M. (1992) Percutaneous absorption of lavender oil from a massage oil. *J. Soc. Cosmet. Chem.*, 43, 49–54.

Kubota, M., Ikemoto, T., Komaki, R. and Inui, M. (1992) Odor and emotion-effects of essential oils on contingent negative variation. *Proc. 12th Int. Congress on Flavours, Fragrances and Essential Oils*, Vienna, Austria, Oct. 4–8, pp. 456–461.

Lis-Balchin, M., Deans, S.G. and Hart, S.L. (1996a) Bioactivity of New Zealand medicinal plant essential oils. *Acta Hort.*, 426, 13–30.

Lis-Balchin, M., Hart, S.L., Deans, S.G. and Eaglesham, E. (1996b) Comparison of the pharmacological and antimicrobial action of commercial plant essential oils. *J. Herbs, Spices Med. Plants*, 4, 69–86.

Lis-Balchin, M., Deans, S.G. and Eaglesham, E. (1998) Relationship between the bioactivity and Chemical composition of commercial plant essential oils. *Flav. Fragr. J.*, 13, 98–104.

Manley, C.H. (1993) Psychophysiolical effect of odor. *Crit. Rev. Food. Sci. Nutr.*, 33, 57–62.

Torii, S., Fukuda, H., Kanemoto, H., Miyanchio, R., Hamazu, Y. and Kawasaki, M. (1988). Contingent negative variation and the psychological effects of odor. In: S. Toller and G.H. Dodd (eds), *Perfumery: The Psychology and Biology of Fragrance*. New York: Chapman and Hall.

Index